An international team of four authors, led by distinguished philosopher of science, Nancy Cartwright and leading scholar of the Vienna Circle, Thomas E. Uebel, has produced this lucid and elegant study of a much-neglected figure. The book, which depicts Neurath's science in the political, economic and intellectual milieu in which it was practised, is divided into three sections: Neurath's biographical background and the socio-political context of his economic ideas; the development of his theory of science; and his legacy as illustrated by his contemporaneous involvement in academic and political debates. Coinciding with the renewal of interest in logical positivism, this is a timely publication which will redress a current imbalance in the history and philosophy of science, as well as making a major contribution to our understanding of the intellectual life of Austro-Germany in the inter-war years.

IDEAS IN CONTEXT

# OTTO NEURATH: PHILOSOPHY BETWEEN SCIENCE AND POLITICS

IDEAS IN CONTEXT

*Edited by* QUENTIN SKINNER *General Editor*
LORRAINE DASTON, WOLF LEPENIES, RICHARD RORTY and J. B. SCHNEEWIND

The books in this series will discuss the emergence of intellectual traditions and of related new disciplines. The procedures, aims and vocabularies that were generated will be set in the context of the alternatives available within the contemporary frameworks of ideas and institutions. Through detailed studies of the evolution of such traditions, and their modification by different audiences, it is hoped that a new picture will form of the development of ideas in their concrete contexts. By this means, artificial distinctions between the history of philosophy, of the various sciences, of society and politics, and of literature may be seen to dissolve.

The series is published with the support of the Exxon Foundation.

*A list of books in the series will be found at the end of the volume.*

# OTTO NEURATH: PHILOSOPHY BETWEEN SCIENCE AND POLITICS

NANCY CARTWRIGHT, JORDI CAT, LOLA FLECK
and
THOMAS E. UEBEL

Published by the Press Syndicate of the University of Cambridge
The Pitt Building, Trumpington Street, Cambridge CB2 1RP
40 West 20th Street, New York, NY 10011-4211, USA
10 Stamford Road, Oakleigh, Melbourne 3166, Australia

© Cambridge University Press 1996

First published 1996

*A catalogue record for this book is available from the British Library*

*Library of Congress cataloguing in publication data*

Otto Neurath: philosophy between science and politics / Nancy
Cartwright ... [et al.].
    p.    cm. (Ideas in context: 38)
ISBN 0 521 45174 4
1. Neurath, Otto, 1882–1945.  I. Cartwright, Nancy.  II. Series.
B3309.N394O88  1996
193–dc20  95-8505  CIP

ISBN 0 521 45174 4 hardback

Transferred to digital printing 2003

# Contents

| | | |
|---|---|---|
| *List of figures* | | *page* ix |
| *Preface* | | x |
| *Acknowledgements* | | xii |
| Introduction | | 1 |
| Part 1 | A life between science and politics | 7 |
| 1.1 | Before Munich | 8 |
| | 1.1.1 Early years | 8 |
| | 1.1.2 War economics | 14 |
| | 1.1.3 During the First World War | 19 |
| 1.2 | The socialisation debate | 22 |
| | 1.2.1 Setting the problem | 22 |
| | 1.2.2 Bauer and Korsch | 24 |
| | 1.2.3 The standard of living | 29 |
| | 1.2.4 Neurath on the structure of the socialist economy | 32 |
| | 1.2.5 The road to socialisation | 37 |
| | 1.2.6 Neurath's position in the debate | 41 |
| 1.3 | In the Bavarian revolution | 43 |
| | 1.3.1 The appointment | 43 |
| | 1.3.2 In office | 49 |
| | 1.3.3 On trial | 53 |
| 1.4 | In Red Vienna | 56 |
| | 1.4.1 People's education | 56 |
| | 1.4.2 The housing movement | 60 |
| | 1.4.3 The Museum of Economy and Society | 63 |
| | 1.4.4 The Vienna Circle | 72 |
| 1.5 | Exile in The Hague and Oxford | 82 |
| Part 2 | On Neurath's Boat | 89 |
| 2.1 | The Boat: Neurath's image of knowledge | 89 |
| 2.2 | In the first Vienna Circle | 95 |
| | 2.2.1 Three hypotheses | 95 |

viii *Contents*

|  |  |  |  |
|---|---|---|---|
|  |  | 2.2.2 Mach's legacy | 102 |
|  |  | 2.2.3 The 1910 programme | 107 |
|  | 2.3 | From the Duhem thesis to the Neurath principle | 111 |
|  |  | 2.3.1 Normative anti-foundationalism | 111 |
|  |  | 2.3.2 Radical descriptive anti-foundationalism | 115 |
|  |  | 2.3.3 Metatheoretical anti-foundationalism | 125 |
|  | 2.4 | Rationality without foundations | 131 |
|  |  | 2.4.1 The primacy of practical reason | 131 |
|  |  | 2.4.2 Determining the conventions of science | 133 |
|  |  | 2.4.3 The second Boat: one world | 136 |
|  | 2.5 | A theory of scientific discourse | 142 |
|  |  | 2.5.1 Anti-philosophy, Marxism and radical physicalism | 143 |
|  |  | 2.5.2 The forward defence of naturalism | 148 |
|  |  | 2.5.3 Science as discourse: the theory of protocols | 158 |
|  | 2.6 | Towards a theory of practice | 163 |
| Part 3 | Unity on the earthly plane | | 167 |
|  | 3.1 | Two stories with a common theme | 167 |
|  | 3.2 | Science: the stock of instruments | 171 |
|  |  | 3.2.1 From representation to action | 171 |
|  |  | 3.2.2 Unity without the pyramid | 179 |
|  | 3.3 | The attack on method | 188 |
|  |  | 3.3.1 Boats and *Ballungen* | 188 |
|  |  | 3.3.2 Protocols, precision and atomicity | 193 |
|  |  | 3.3.3 The two Neurath Principles | 202 |
|  | 3.4 | Where *Ballungen* come from | 208 |
|  |  | 3.4.1 Duhem's symbols | 208 |
|  |  | 3.4.2 The congestion of events | 213 |
|  |  | 3.4.3 The density of concepts | 224 |
|  |  | 3.4.4 The separability of planning and politics | 229 |
|  |  | 3.4.5 How Marxists think of history | 235 |
|  | 3.5 | Negotiation, not regulation | 244 |
| Conclusion | | | 253 |
| *List of references* | | | 257 |
| *Index* | | | 283 |

# Figures

| | | |
|---|---|---|
| 1.1 | Otto Neurath, c. 1910 | page 8 |
| 1.2 | Schematic model of economic organisation (after Neurath 1920c, p. 18) | 33 |
| 1.3 | Handbill announcing the establishment of the Central Economic Administration, Munich 1919 | 48 |
| 1.4 | Marriages in Germany per year. (Neurath 1936b [1991], p. 382) | 64 |
| 1.5 | Marriages in Germany per year. (Neurath 1936b [1991], p. 383) | 64 |
| 1.6 | Births and deaths in Vienna. (Neurath 1933e [1991], p. 320) | 66 |
| 1.7 | Infant mortality and social conditions in Vienna. (Neurath 1933e [1991], p. 311) | 66 |
| 1.8 | Unemployment in Berlin. (Neurath 1933b [1991], p. 233) | 68 |
| 1.9 | Economic crises and industrial furnaces in March 1932. (Neurath 1933b [1991], p. 238) | 69 |
| 1.10 | Pigiron production and employment in the US automobile industry. (Neurath 1933b [1991], p. 234) | 70 |
| 1.11 | (a) Steel production in the USA. (b) Strikes in the USA. (Neurath 1939 [1991], p. 531) | 71 |
| 1.12 | Silhouettes of war economy. (Neurath 1939 [1991], p. 519) | 73 |
| 1.13 | Silhouettes of war economy. (Neurath 1939 [1991], p. 520) | 74 |
| 1.14 | Otto Neurath, 21 December 1945 | 88 |
| 3.1 | Pyramid | 170 |
| 3.2 | Balloons | 170 |
| 3.3 | Birds (pre-1931) | 189 |
| 3.4 | People (pre-1931) | 189 |
| 3.5 | Birds (post-1931) | 192 |
| 3.6 | People (post-1931) | 193 |

*Preface*

This book is made up of three parts. Part 1, Neurath's intellectual biography, is primarily from Lola Fleck's Graz dissertation, with additions by the other authors to fill in missing links with the account of parts 2 and 3. The translation from the German of Fleck's original work is by Martin Anduschuss with revisions by Nancy Cartwright and Thomas Uebel. Part 2 of the book was written by Thomas Uebel and part 3 by Nancy Cartwright and Jordi Cat. Although the book has four different authors, listed alphabetically on the title page, there is a common point of view among them. The work on the book has been a close collaborative effort among Cartwright, Cat and Uebel, and these three authors would like especially to thank Lola Fleck for her generosity in allowing additions and revisions to integrate her dissertation more fully with the remainder of the text. Existing translations have been used where available; translations of previously untranslated materials are by the present authors. Timothy Childers has served as editorial assistant throughout. Figure 1.1 was recreated by George Zouros. Original drawings are by Rachel Hacking. The index was contributed by Mauricio Suárez.

Neurath's own distinctive idea of *Ballungen* – congested concepts with fuzzy edges – plays a special role in our philosophic discussions. It enters the work of parts 2 and 3 by independent routes. In 1990 Nancy Cartwright went to talk to C.G. Hempel about Neurath. Hempel reported that there were two themes that he felt were really dear to Neurath and central to his thought: the moneyless economy and *Ballungen*. So Cartwright was delighted to find just after that what appears to be Neurath's first public airing of his ideas on *Ballungen*, first among Rudolf Haller's personal copies of Neurath's papers in Graz and later officially in the Pittsburgh archives. Neurath's discussion is recorded in the minutes of a meeting attended by many members of the Vienna Circle on 4 March 1931 (not one of the Circle's regular Thursday evening meetings). The document is called *Besprechung über Physikalismus*.

After that Cartwright and Cat set about trying to understand the source of *Ballungen* in the nexus of Neurath's political, scientific and philosophic work at that time.

In 1991 Thomas Uebel went to Pittsburgh as a Mellon Fellow, having completed a dissertation on Neurath in the Vienna Circle and edited a collection of contemporary Austrian writings on the topic. In the Carnap papers there he also discovered the *Besprechung*. He too was delighted. He knew not only how crucial *Ballungen* were to Neurath's increasingly radical attack on the myth of scientific certainty and scientific method, but he also found there the first public use of Neurath's private language argument which he knew to be of central importance to the protocol sentence debate. Uebel and Cartwright and Cat had already been discussing putting their ideas together in a single book with Fleck's 1979 dissertation. The excitement over the *Besprechung* settled the matter.

# Acknowledgements

Nancy Cartwright and Jordi Cat wish to thank Hasok Chang and Peter Galison for helpful discussions, and Rachel Hacking for her illustrations.

Lola Fleck would like to thank all who helped her research with information about Otto Neurath, in particular his widow, Marie Neurath, and his niece, Gertrud Neurath. She also wishes to thank Rudolf Haller for his support and critical attention when the work here translated in its greater part was completed as a dissertation.

Thomas Uebel gratefully acknowledges the support of a Walther Rathenau Fellowship at the Verbund für Wissenschaftsgeschichte, Berlin, during the tenure of which part 2 was drafted. His personal thanks go to colleagues and friends in Berlin, Vienna, Graz and London for numerous discussions, especially to Axel Fuhrmann, Camilla R. Nielsen, Max Urchs and Susan Watt.

The authors are grateful to Gerald Heverley of the Archive of Scientific Philosophy, Special Collections Department, Hillman Library, University of Pittsburgh, and Prof. A. J. Kox, the curator of the Vienna Circle Archive, formerly in Amsterdam, now at the Rijksarchief in Noord-Holland, Haarlem, for permission to quote from their holdings (for which they reserve the rights). The authors are also grateful to Dr Reinhard Fabian of the Forschungsstelle und Dokumentationszentrum für österreichische Philosophie, Graz, Dr Heinz Jankowsky of the Österreichisches Gesellschafts- und Wirtschaftsmuseum, Vienna, Herrn Erich Pehm of Verlag Hölder-Pichler-Tempsky, Vienna, and to Frau Stöckinger of the Stadtbibliothek München, Monacensia-Bibliothek, for help with the illustrations in part 1.

Finally, all the authors would like to thank Timothy Childers for his great efforts in producing the book and Mauricio Suárez for the index.

# Introduction

> Science and life can be connected, above all by the setting which encompasses both.[1]

Among economists, Otto Neurath is most well known for his views about moneyless economies; among historians, for his work on full socialisation during the short-lived Bavarian revolution of 1919; among educators, for his work on social museums. Among philosophers, he is surely most widely known for his metaphor of sailors rebuilding their ships at sea, which he used to attack foundational accounts of knowledge. This book is about Neurath's philosophy and it is about his political life. And it is about the connection between the two, which left neither of them unchanged.

As a young man near the beginning of his career, fifteen years before the official founding of the Vienna Circle, Otto Neurath wrote his first extended attack on the hunt for certainty in science and in life. There he urged: 'The thinking of a man during his whole life forms a psychological unity, and only in a very limited sense can one speak of trains of thought *per se*.'[2] This description perfectly fits Neurath himself. Otto Neurath was a philosopher, a publicist, an activist, a bureaucrat, a scholar, a social scientist and a Marxist. His philosophy will be our central topic. But philosophy for Neurath was not a discipline. His philosophical thought did not evolve within a closed system, new philosophical views emerging from older ones adjusted by new philosophical insights and arguments. Indeed, as Neurath saw it, the disciplines themselves crystallised into separate self-contained systems as a result of useful but false abstraction. According to Neurath's 1913 vision, 'The phenomena that we encounter are so much interconnected that they

---

[1] Neurath to Reichenbach, 22 July 1929, HR 014-12-04 Archives of Scientific Philosophy, Pittsburgh (hereafter referred to as ASP).   [2] Neurath 1913b [1983], p. 3.

cannot be described by a one-dimensional chain of statements.'[3] That is the thesis of this book with respect to Neurath's own life and work, and especially with respect to the connections of his science with his politics and with his philosophy.

In 1906 Neurath wrote a scholarly dissertation on the history of ancient economies, particularly that of Egypt; in 1912 and 1913 he worked in the Balkans studying how economies operate under the stress of war. Between 1913 and 1916 he also wrote two philosophical pieces still of interest today, one against foundationalism in epistemology and the other on the structure of science. During the First World War he was deputed by the Austro-Hungarian war office to study the organisation of the war economy. As the war ended he co-authored a scheme for the full conversion to socialism of Saxony, and then Bavaria. During the socialist revolution in Bavaria Neurath headed the Central Planning Agency with the hope of implementing his plans for full socialisation. Awaiting deportation to Austria in 1919 in place of serving an eighteen-month sentence for assisting high treason in Bavaria, he wrote a long and rich attack, partly sociological, partly philosophical, on Spengler's *Decline of the West*.

In Red Vienna after the war Neurath helped in city planning, especially for the co-operative housing movement. He thought hard about how to make statistics intelligible to ordinary people and set up a workers' museum to make statistical information more readily available. And he joined the Vienna Circle, where he argued against Schlick that science could not be based on personal experience but required the public and objective knowledge of a large community of scientific workers and against Carnap that an *Aufbau* of knowledge was a false ideal. He founded the Unity of Science movement. He wanted unity without reduction, unity for action not for representation; and from his lifelong recognition of the necessity for co-operation and joint action he worked not just for the unity of science but for the unification of scientists to oppose the rising Nazi threat.[4] His philosophy changed. It evolved and grew richer. But, as Neurath said about phenomena in general, the change 'cannot be described by a one-dimensional chain of statements'.[5] Instead, Neurath's philosophy finds expression in a body of thought and work, both practical and theoretical, that evolves as one whole in an exacting political setting. The same point of view was expounded in one format in *Erkenntnis*, in another in *Der Kampf*. This is the story that we will tell here.

[3] Ibid.   [4] Cf. Galison 1990.   [5] Neurath 1913b [1983], p. 3.

Knowledge has no foundations. The things we believe can only be checked against other beliefs; nothing is certain; and all is historically conditioned. These are lessons to be learned from Neurath and they are lessons that are commonly taught by philosophers in Britain and America at this moment. But Neurath's attack on the prevailing pictures of scientific knowledge went beyond this. For Neurath not only has science no foundations; it has no cast-iron method either. Falsificationism, confirmation, the hypothetico-deductive method are all pseudo-rational. Like the story of God in his heaven, they set a fairy-tale ideal; and the harm is not just an epistemological one, that the purity of our cognition will be threatened by false beliefs, but a problem of how we make our life in this world. Under the banner of certainty we deduce when we should negotiate; we reason when we should flip a coin; we call for more information when we should act. We use cautious inductions, generalising from what we see to what must be. This kind of empiricism spawns conservatism, as Neurath argued in his 1911 monograph on political economy: 'Those who stay exclusively with the present will very soon only be able to understand the past.' Instead, just as 'mechanics gives information to the machine builder about machines which have never yet been built', so too 'science can give to the social planner information about social orders that have never yet existed'.[6] For Neurath the false ideal of certainty, like all empty metaphysical claims, does not merely lead to farcical philosophical constructions, like the Kingdom of Ends or Plato's heaven; it robs us of the power to change our lives.

These kinds of remarks are familiar in contemporary post-modern discourse. Yet Neurath had one striking difference from most post-moderns. Although he was a pluralist about knowledge-systems and took seriously their historical and cultural roots, he trusted firmly in the power of science. Not science on its own as an abstract system of thought, but science in the hands of the social technician, who can 'orchestrate'[7] the different systems of knowledge to build new social orders. The Unity of Science movement was a tool for social change; the possibility of it being so depended on Neurath's special views about what unity means. The canonical Positivist account pictures science as one giant pyramid set on the sure foundations of physics. For Neurath the sciences were more like a heap that could be rummaged through to solve problems and construct change. His aim was to replace the heap by a tool kit, to arrange the tools and to ready them for action.

---

[6] This aspect of Neurath's views has been studied by Nemeth 1981. The Neurath remarks are quoted on pp. 61 and 226.   [7] Neurath 1946a.

Neurath had not always seen unity as a matter of assembling a tool-kit and 'orchestrating' the sciences. Before the Bavarian revolution he too hoped that science would provide a single unified picture so that 'the world would stand before us as a whole again'.[8] This attitude changed in tandem with his belief in full socialisation. The account of Neurath's views on unity of science and how they shifted during the Bavarian revolution is one of our three central philosophical topics. The second is his rejection of scientific method, which took shape during his simultaneous participation in two separate debates: a narrowly philosophical one in the Vienna Circle on the proper form for protocol, or data-reporting, sentences; the other a Marxist debate in the pages of *Der Kampf* on the materialist conception of history. The third is his anti-foundationalism and its interweaving with the delicate issues of the protocol sentence debate. The issues are philosophical. But as we have already suggested, Neurath's grounds for them will not always seem so. We take that to be a special virtue of Neurath's work.

Neurath read voraciously and was keenly interested in a vast variety of subjects. Of the more than 400 articles, pamphlets and monographs that he wrote, hardly a dozen and a half could be counted as works of professional philosophy. He never had a proper job as an academic and his career as a philosopher, if one wants to ascribe one to him, was confined to his interaction with the Vienna Circle and the Unity of Science movement. Neurath wrote and spoke easily, with vigour and conviction. Indeed during the Bavarian revolution he was accused by the opposition of being 'equipped with all the art of a demagogue. In great gatherings he was able to win the restless masses to his full socialisation plans, even to fanaticism.'[9] Once when he was invited to lecture to the Philosophy Department at Yale University, he startled his hosts by arriving the day before and asking, 'What do you think I should talk about?'[10] His writing is discursive, repetitive and loosely structured. In short Otto Neurath was far from the paradigm of the analytic philosopher.

Given this, those who admire his philosophical work are often at pains to expose a structure of subtle and convincing argument beneath his rambling discourse. We do that too. That is one of the chief purposes of part 2, where we will be especially concerned with the development of his 'philosophy' of science up to his disagreements with Carnap in the

---

[8] Neurath 1910a [1981], p. 46.  [9] Müller-Meiningen 1923.
[10] From John Watkins; conversation Dec. 1992.

protocol sentence debate. But we want to urge that in large part Neurath's philosophical work is valuable precisely because it is not standard philosophy. Neurath's philosophical positions are not derived by decisive arguments from first principles, a procedure which is altogether suspect from his own philosophical standpoint. Instead they are formed out of his immersion in day-to-day life and his attempts to understand and change the social structures around him. They ring of soundness and good sense, not of subtlety and depth. Neurath may be accused of being utopian, but he could never be accused of being absurd, as philosophers propelled by chains of argument often are.

One case we shall discuss which illustrates the double-sidedness of Neurath's philosophic approach is his insistence that the data-base for science consists in publicly observable, intersubjectively available facts, not in items of personal experience as Schlick and Carnap would have it. On the one hand Neurath defended his view by a series of detailed criticisms of Carnap's *Aufbau*, to which he added his own pre-Wittgenstein version of the private language argument. On the other hand he pointed to the obvious fact that intersubjectivity is required by the nature and tasks of modern science. Neurath was above all a social scientist, and he, like the other members of the Verein für Sozialpolitik, saw the great power of statistics. The seven volumes of their house journal between 1908 and 1912 are devoted to detailed studies of the economic and social conditions of all the regions of Germany plus a number of different countries around the world: Austria, Italy, Switzerland, Belgium, Australia, France, England, Hungary. How many plants, cars, double-decker buses are to be found in Manchester? What kind of gas works does the city have? Water supply? Sewage? Market gardening? Understanding the social order depends on a mammoth community effort, not on the certainty of an individual's inner experience.

But how can that be a philosophical argument? The conclusion, against subjective foundations for knowledge, may be a philosophical conclusion; but the contents of the leading empirically oriented social science journal hardly seems a stock philosophical premise. That is just the point. Neurath did not believe in philosophy. He believed only in science, or in positive knowledge. There is a well-known story that in the discussions of the Vienna Circle Neurath was always on guard against metaphysical claims. When he heard another member make a scientifically empty claim, he called out 'Metaphysics!' During the discussions of Wittgenstein's *Tractatus* Neurath interrupted continuously. Finally it was suggested that he should hum 'M-m-m-m' instead. Neurath answered that it would be more efficient if he were to simply say 'not-M', whenever they were not misled into talking metaphysics. Epistemology is equally bad: 'within a consistent physicalism there can be no "theory of knowledge" ... Some problems of the theory of knowledge will perhaps be transformed into empirical questions so that they can find a place within unified science.'[11] For Neurath there is only science; there is no metaphysics, no epistemology and no philosophy. Hans Reichenbach wanted to join his Berlin group with the Vienna Circle under the title 'the Scientific Philosophy'. Neurath was appalled. 'Scientific philosophy?' he said. 'We might as well talk of scientific theology.'[12]

Our book has three parts. Part 1 is an intellectual and political biography of Neurath. Part 2 tells the history of Neurath's five different uses of the boat metaphor. It is about Neurath's anti-foundationalism and the development of his conception of science up to the seminal debates in the Vienna Circle. Part 3 examines the unity of science and the attack on method. The Bavarian revolution, the question of how Marxists should conceive of history and the debates about the differences between the natural and the social sciences all play a role.

[11] Neurath 1932a [1983], p. 65.  [12] Neurath to Carnap, 23 Sept. 1932, RC 029-12-32 ASP.

PART I

# A life between science and politics

> A tall, handsome man, bald, with a large forehead, beautifully shaped head, red full beard, and small brown eyes, a slightly Semitic (rather Egyptian) type, with fluent, intelligent, captivating speech.[1]

This is how Ernst Niekisch described Otto Neurath in his diary in March 1919. But it was not only Neurath's outer appearance, his gigantic figure, his behaviour – considered rude by some – that impressed his contemporaries. Over and over again he was characterised as charming, friendly, witty; often as energetic, ambitious and full of purpose, a man of action. William M. Johnston wrote:

> Otto Neurath (1882–1945) is one of the most neglected geniuses of the twentieth century. He made innovations in so many fields that even his admirers lost count of his accomplishments.[2]

Over the course of his life Neurath was a business-school teacher, a military officer, a junior university professor, a commissioner for socialisation, a secretary of a housing movement, a museum director – to name but a few of his professions. Neurath's manifold activities correspond to the many stereotypes applied to him. One Austrian minister of education considered him a Communist; an old Social Democratic librarian reported that Neurath was considered an eccentric in the Social Democratic movement of the first Austrian republic. Another contemporary viewed Neurath as a Marxist who had hardly read Marx. Older citizens of Vienna still remember his permanent exhibitions in the town hall. By philosophers he is classed as a radical empiricist, while amongst economists he counts as a daring innovator, if not a misguided visionary. Part 1 of our book will give an overall picture of Neurath's life and work, paying special attention to those aspects which we think have been

---

[1] Niekisch 1958 [1974], p. 56.   [2] Johnston 1972, p. 192.

Fig. 1.1 Otto Neurath, c. 1910

largely neglected in the literature, such as his war economics and his socialisation models.

## 1.1 BEFORE MUNICH

### 1.1.1 Early Years

Otto Neurath was born into a family of the educated bourgeoisie. His father, Wilhelm Neurath (1840–1901), came from near Bratislava (Slovakia) but his native language was German.[3] Wilhelm Neurath grew up under the influence of his parents' strict Mosaic beliefs. He attended

---

[3] Wilhelm Neurath 1880c [1973], pp. 2–4.

elementary school for only four years. It was poverty and his father's religiously motivated fear of mundane knowledge that forced him to become an autodidact. He struggled along as a teaching assistant and copyist, starved, was often sick and spent the little money he had on books. Nevertheless he graduated with great success, presumably as an external student.

Even before he began his studies in Vienna in 1866 Wilhelm Neurath was intensely interested in religion, physics, astronomy, mathematics and philosophy, studies 'which attracted [him] in a dreamlike way, with irresistible power'.[4] In August 1871 he received his Ph.D. from the School of Philosophy at the University of Vienna; later he received a second doctorate in political science from the University of Tübingen. He became a university lecturer in 1880.[5] He gained a reputation in industrial circles from his economic and social-political writings. In 1889 he became associate professor, and in 1893 full professor of economics at the Agricultural Academy (Hochschule für Bodenkultur) in Vienna.

In his writings Wilhelm Neurath repeatedly addressed the problems of why poverty and want cannot be eliminated despite increases in production, and why crises of overproduction lead unavoidably to inflation and unemployment. He was not a Marxist. He sought a solution for these economic and social problems which did not involve changes in the prevailing social order and the institution of private property. In his curriculum vitae Wilhelm Neurath pointed out that in his youth he was an atheist and a materialist and even believed that Communism was the only social solution possible. But he also stated that he had been 'liberated' from these ideas through his study of Kant.[6] He said in his lecture on the 'true causes of overproduction and unemployment' in 1892:

> It is not the institution of private ownership of land and capital, but rather the need to make a profit and the fragmentation of the entrepreneurial sector that leads to the exploitation of the working classes, both of those who work with ideas and of those who work with their hands, and to the paralysis of production and the deepest deformation of society, including that of the capitalist class.[7]

Wilhelm Neurath saw the profit motive as the primary cause of economic crises. A monetary economy and the credit system force entrepreneurs to strive for profit, since firms that are not profitable lose their credit. Thus he argued against the contemporary form of credit as the

---

[4] Ibid., p. 2.   [5] Von Schullern zu Schrattenhofen 1902.
[6] Wilhelm Neurath 1880c [1973], p. 3.   [7] Wilhelm Neurath 1892, p. 30.

common means of financing enterprises. Economic crises, wage reduction, unemployment and related problems would be avoided if the important branches of mass production were to be organised into trusts and these in turn united under one umbrella organisation: 'Organising the entrepreneurial sector into a system of producers' associations in healthy solidarity would liberate the production process from its dependence on the creation of surplus value and eliminate the shocks of economic crises.'[8] Because of these views Wilhelm Neurath has been designated a 'pan-cartelist'. According to him wealth and prosperity in human society depend on 'the richness, the harmony and the degree of the development of man's sensual, intellectual and moral life.'[9] For Wilhelm Neurath an ideal world lay hidden behind appearances; all of history represented the 'development from the material to the spiritual'.[10]

Otto Karl Wilhelm Neurath was born on 10 December 1882. He was the first child in the marriage between Wilhelm Neurath and Gertrud Kaempfert (1847–1914), the daughter of a Viennese lawyer and notary public. He spent his childhood browsing at will amongst his father's books, which filled the walls of their apartment. From an early age he had long and serious discussions with his father. Perhaps the roots of the moral aversion to the dominant economic system and the radical disapproval of the monetary and credit systems that are to be found in Otto Neurath's writings were planted in these early discussions.

For a while the family enjoyed financial stability and spent several months every year in the country, a habit common in bourgeois circles. In 1902, a year after his father's death, Neurath graduated from secondary school (Staatsgymnasium) and began studying mathematics and physics at the University of Vienna. Presumably this was what his father had wanted, since Neurath did not consider himself mathematically gifted. He soon switched to economics, history and philosophy.[11]

During his studies Neurath came to know Philipp Frank, Hans Hahn, Olga Hahn and Anna Schapire.[12] Together with Anna Schapire, Neurath followed the recommendation of Ferdinand Tönnies, whom he had met at a summer academy in Salzburg, that he should go to Berlin,

---

[8] Ibid., p. 30.  [9] Ibid., p. 31.
[10] Wilhelm Neurath 1891, p. 10; for more on the intellectual father–son relation see Uebel 1993b and 1995b.
[11] See Neurath's letter to Josef Frank. He writes: 'He himself [Wilhelm Neurath] tried hard to interest me in mathematics and physics, which he had loved very much and had studied quite seriously.'  [12] Marie Neurath 1973b, p. 7; Paul Neurath 1982, p. 230.

where he continued his studies under Eduard Meyer and Gustav Schmoller.[13] He later recalled in a letter to Josef Frank:

> I do not remember for sure but I believe it was only after I had already distinguished myself with difficult seminar papers that it was discovered I was only in my third semester. Schmoller was dumbfounded, but he thought one could not undo everything at this point and I thus had the advantage of attending the finalists' seminar throughout my time as a student.[14]

Since the death of his father Otto Neurath had been forced to earn his own living. His mother received only a meagre pension on which she had to raise his younger brother Wilhelm. Neurath was nearly destitute and even spent days in bed because he was too weak to get up.[15] Some friends financed a vacation for him in Switzerland, where he cured his malnutrition 'with eggs and jam'.[16]

In 1904, while he was still at university, Neurath published an essay on monetary interest in antiquity.[17] He also edited a new edition of Marlow's *Faust* while on his convalescent holiday in Bern in 1906, having discussed this book three years earlier in the *Goethe-Jahrbuch*.[18] The name 'Marlow' served as a pseudonym for Ludwig Hermann Wolfram (1807–53), who had used it in 1839 for his *Faust. Ein dramatisches Gedicht in drei Abschnitten (Faustus: A Dramatic Poem in Three Parts)*. In the preface Neurath says about his monumental introduction which runs to more than 500 pages: 'This introduction provides for the first time detailed information about the life and work of Ludwig Hermann Wolfram.'[19] In addition to a detailed list of Wolfram's ancestors the introduction contains an elaborate internalist interpretation of his works. Neurath later described his introduction as 'the nonsensical, all-comprehensive Wolfram biography'.[20]

On 29 September 1906, after four years of study, Neurath received his doctorate at the School of Philosophy of the Friedrich-Wilhelms-Universität in Berlin.[21] For unknown reasons Neurath submitted two dissertations: one on the economic history of antiquity and the other on ancient conceptions of commerce, trade and agriculture. Eduard Meyer decided to accept the latter dissertation; Gustav Schmoller acted as second reader. In the accepted dissertation Neurath translated and interpreted a passage of Cicero (*De Officiis* I C.45). In order to illustrate the

---

[13] Ibid.; see also the letter to Josef Frank.    [14] Neurath, letter to Josef Frank.
[15] Conversation with Heinrich Neider.    [16] See letter to Josef Frank.    [17] Neurath 1904a.
[18] Neurath 1903c, 1906c.    [19] Neurath 1906c, p. V3.    [20] Letter to Josef Frank.
[21] Neurath 1906a.

changing conceptions in question he compared the interpretations of this passage offered by Cicero's contemporaries with those advanced by later authors. He also investigated the influence of different conceptions of history and of different perspectives on forms of income on the interpretations of the passage in Cicero. He restricted himself to the enumeration and comparison of these various interpretations and did not delve into the development or purpose of the underlying theories of history. Neurath said of his work that it was intended to 'increase . . . the understanding of a document'.[22] With the approval of the school only the first chapter was submitted in printed form for the doctorate.[23] All three chapters appeared in the respected political economy and statistical journal *Jahrbücher für Nationalökonomie und Statistik* in two parts in 1906 and 1907. The dissertation was graded *summa cum laude*. Neurath later commented: 'This was a lot at that time in Berlin, especially for a foreigner who always had disagreements with Schmoller.'[24]

Neurath returned to Vienna, where he joined the army as 'a one-year volunteer for active service at his own expense'.[25] Following a two-month military training, Neurath attended a military school in Vienna from December 1906 until September 1907, and passed the exams for rations and provisions officials with great success. During the summers of 1910 and 1911 he served three weeks each year at the military rationing and provisions office in Vienna.[26]

It is most likely that Neurath wanted to follow in the footsteps of his father: to qualify as a university lecturer and become a scientist. Instead he became an assistant teacher at the New Business School in Vienna (Neue Wiener Handelsakademie), and remained there until the war. Twenty-five years later he commented: 'I simply accepted this and did not expect anything substantial from my future except quiet scientific work.'[27] Neurath taught economics and history; the teaching load left enough time for him to work as a scientific author as well.

On 22 November 1907 Neurath married his long-term fellow student, Anna Schapire.[28] She was born in Galicia, but spent long periods abroad in her youth, becoming proficient in Polish, Russian, French and English. Together with Neurath she studied philosophy, literature and economics in Vienna and Berlin. She also accompanied him to Switzerland, where in 1906 she completed her doctorate on the positions

---

[22] Neurath 1906a, p. 33.   [23] Neurath 1906/1907, p. 205.   [24] Letter to Josef Frank.
[25] His rank was that of Unterkanonier (gunner) (Letter from the War Archive, Vienna).
[26] Letter from the War Archive, Vienna.   [27] Letter to Josef Frank.
[28] See Letter from the War Archive, Vienna. On Anna Schapire-Neurath see also Fleck 1982b.

## Before Munich

of German political parties *vis-à-vis* the maintenance of industrial health and safety standards. Previously she had published both poems and novellas, but after two works on Hebbel she turned to social themes and women's rights. She also published on the history of the women's movement and herself became active in it.

In 1910 Neurath and Schapire published a two-volume reader on economics, containing passages from Plato to Marx. Presumably it evolved in connection with Neurath's occupation as a teacher, because Neurath published a textbook of economics at the same time.[29] The textbook was officially approved as a teaching text for higher business schools in January 1911, as was, for final-year students, their reader.

On 12 September 1911 Paul Neurath was born. He is Otto Neurath's only child. Shortly afterwards Anna Schapire died, probably from complications during birth. Neurath was deeply struck by her death and even contemplated suicide.[30] Paul had to be taken to a home where he stayed until he was ten.[31]

In 1912 Neurath married Olga Hahn, the daughter of state councillor Ludwig Hahn, who headed the telegraph and communications office in Vienna and also edited the journal called *Politische Korrespondenz*.[32] Olga's brother Hans later became well known as a mathematician and was one of the founders of the Vienna Circle. Olga Hahn was an old friend of Anna and Otto. Olga had become blind in 1904, at the age of twenty-two, perhaps due to typhus.[33] When Anna and Otto had returned from Berlin to Vienna they found their friend deeply depressed because she could not pursue any of the activities she was interested in. Together with other friends, Otto and Anna organised a reading service for her. Since Olga was highly intelligent and had an excellent memory, this allowed her to finish her studies of mathematics. In 1909 and 1910 she published essays on Schröder's algebra of logic, both on her own and together with Otto.[34]

Otto and his wife Olga lived in modest conditions. They had no electricity and no running water.[35] Both were very sociable and enjoyed hosting stimulating discussions. Later, during the time of the Vienna Circle, discussions of the Thursday group were often continued at their place in the Schloßgasse, and Carnap, Feigl, Neider and Hahn were frequent guests.

---

[29] Neurath 1910d, Neurath and Schapire-Neurath 1910a.  [30] Schumann 1973, p. 15.
[31] Neurath 1973, p. 15.
[32] Notification of the Police Headquarters in Vienna to the Standgericht (court martial) in Munich of 17 June 1919, State Archives Munich.  [33] Schumann 1973, p. 15.
[34] Neurath and O. Hahn 1909a, 1909b, 1910.  [35] Neurath 1973, p. 31.

While Neurath worked as a teacher at the New Business School he published mainly in economics. He was a member of the Verein für Sozialpolitik (Social Policy Association),[36] but at that time there is no evidence of a specific party-political commitment. In 1912/1913 Neurath received a one-year stipend from the Carnegie Endowment for International Peace in order to investigate the economic and social conditions in the Balkan states and the changes caused by the war which was then going on. He made two journeys, which not only resulted in a series of newspaper reports, but also allowed him to test his first tentative conclusions about war economies.[37]

### 1.1.2 War economics

Already in 1909, in the published version of his economic history of antiquity, Neurath compared different economic systems in order to bring out the reciprocal effects between social and economic developments.[38] For him, the point of economics lay in the investigation of the relation between the production and the distribution of goods, and prosperity. As a starting point he considered how economic systems influence standards of living, that is, the levels of housing, nutrition, education, entertainment, labour and health in the population.

By 1913 Neurath faulted the dominant economic theories of his day for 'construing one specific economic order as *the* economic order and regarding empirically encountered variations as unimportant'.[39] Even economic changes as significant as those caused by war are characterised only as 'deviations from the normal state'.[40] 'But a war may occasionally even improve a population's living standard.'[41] For Neurath this was reason enough to investigate the effects that war has on an economy.

Different nations have over time had differing attitudes towards war. On the basis of his historical investigations Neurath claimed that it was only subsequent to British mercantilism that wars were considered as disturbances of economic life. The Greeks and the Romans, by contrast, considered war as one among many ways of securing an income. Neurath himself took the economy of war to be a special type of economy. He perceived in it the motor for a general economic change

---

[36] See the list of members in *Verhandlungen des Vereins für Sozialpolitik in Wien 1909, Schriften des Vereins für Sozialpolitik* 132, Leipzig 1910.
[37] See the annual report of the Neue Wiener Handelsakademie, Vienna 1913, p. 97, fn. 1.
[38] Neurath 1909.   [39] Neurath 1913e [1919a], p. 29.   [40] Ibid.
[41] Neurath 1914 [1919a], p. 148.

allowing a full development of production capacities. Neurath thus proposed to introduce war economics as a special sub-discipline which would investigate the changes that a war brings about in production, distribution and the structure of the market, as well as its actual effects on the social conditions of a population.[42]

Neurath named two aspects that differentiate peace-time and war economies: in the case of war (1) 'questions of profitability must take second place to questions of productivity';[43] (2) monetary exchanges are replaced by barter and thus an economy in kind (*Naturalwirtschaft*) slowly supersedes a monetary economy.

Consider first the replacement of profit by productivity maximisation. In the capitalist economy (which Neurath often called 'the free market economy') overproduction occurs periodically. When productivity rises, profitability declines; the net profit does not increase and might even decline if production were to rise any higher. Through deliberate cuts in production or the formation of cartels entrepreneurs try to counteract overproduction. Profits remain constant since less is produced than is demanded or than could be consumed. For Neurath restrictions on production are a normal feature of a peace-time economy. 'If disturbances of a certain kind occur, e.g. war, the restrictions can be lifted and the productive forces are liberated.'[44]

According to Neurath's pre-war views the use of productive forces up to capacity (the principle of maximum productivity) is a central feature of war economies. In this way it is possible to raise the national income despite the war: 'One can thereby achieve a prosperity that far exceeds that of the preceeding peace.'[45] War is a means of preventing or delaying the crises that arise from overproduction. All available capacities are concentrated on the production of war material, which is subsequently destroyed. A clear increase in a population's standard of living obtains when the use up to capacity not only satisfies the demands of the war but leads to a general increase in consumption. (The USA may serve as an example from recent times.)

Neurath's views on this subject are not uncontroversial. In general

---

[42] On Neurath's war economics see the collection of essays Neurath 1919a.
[43] Neurath 1910c [1919], p. 16.  [44] Ibid., p. 13.
[45] Ibid. On the following page Neurath quotes Henry George's *Social Problems* from the German translation (Berlin 1885): 'Perhaps nothing illustrates more strikingly the ongoing waste of productive forces than the fact that the most prosperous time that ever occurred in this country in every branch of business was the time of the Civil War, when we had to subsidise huge naval units and armies and when millions of our industrialised population had sufficient work in supplying them with goods for unproductive consumption, or careless destruction.'

goods traded on the market in peace time have a use value that serves either for the reproduction of manpower, or for investments, or for the increase in the stock of capital. In contrast war material is destroyed and with it machinery, natural resources, labour, etc. The intensified production eventually leads to a decrease of the stock of capital because the economy does not produce enough long-lasting investment goods to be able to secure or to renew the stock of capital. A decrease in reproductive capacity is a long-term characteristic of a war economy: the amount of goods produced either does not suffice to feed all citizens or does not maintain the current stock of tools, or both.[46]

A second controversial point in Neurath's view concerns the connection he draws between the increase of productivity during war, and the rise in the standard of living for many parts of the population. Neurath says, for example:

For the workers remaining at home, chances are not too bad and in some cases even better than during peace. This is because the number of competitors decreases. During a war employees currently working in minor positions are often promoted. For those in the field, the war is a momentary distraction of their worklife, since the state supports their relatives, at least to some extent.[47]

The effects of the First World War seem to have contradicted him. During the first months of the war mass dismissals caused the number of unemployed to increase tenfold, and during the first year the independent unions spent 21.5 million marks on unemployment benefits.[48] The government support for the family of an average soldier (i.e., with a wife and two children) amounted to one-tenth of the average monthly income of a metal worker in Berlin. While the number of women employed in war industries rose by 541 per cent in comparison with pre-war times, women were paid a quarter or at most half of a man's wage.

Let us now turn to the second aspect of war economy that Neurath names, the increase of barter. In 1914 he wrote: 'There are some signs that a world war would radically change the current monetary and credit system, namely in the direction of a large-scale economy in kind controlled by the state.'[49] Past wars were almost always connected from their beginning with deep economic disturbances resulting in bankruptcies and disruptions of the credit system. Termination of credit by foreign countries, withdrawal of money by small savers and similar effects

---

[46] Compare here the views of Mandel 1969, chapter 10.    [47] Neurath 1913e [1919a], p. 21.
[48] Compare R. Müller 1925 [1974], pp. 40–5.    [49] Neurath 1914 [1919a], p. 103.

rendered a warring state unable to satisfy its higher demand for goods within the constraints of the monetary and credit economy. According to Neurath, in order to counteract such problems a state is likely to attempt to obtain sufficient funds at the start of a war either by taking out loans or raising taxes, or by drawing on its stock of gold. If these measures do not suffice, the state could also, by 'administrative measures', commandeer the required commodities. 'The comandeering of commodities mainly concerns [the allocation of] manpower.'[50] In the case of other consumer goods, Neurath considered compulsory loans or taxes in the form of commodities. For example, during the Seven Years War requisitions formed the economic basis of the army. This is why commanders tried to operate in enemy territory. According to Neurath 'the Napoleonic wars first helped this method to achieve renewed recognition'.[51] Neurath did not approve of allowing troops to loot. He argued for the 'introduction of a requisition system comparable to a tax system in those regions that are subordinate to an organised administration, be it foreign or home territory'.[52]

One should not consider Neurath's ideas in isolation from the situation that existed immediately before the First World War. He wrote: 'The bellicose atmosphere of recent years has led all states to perform more or less comprehensive calculations concerning the question of how it could produce or obtain the necessary goods in case of war.'[53] This historical setting makes intelligible his demand that 'one ought to work out a general mobilisation plan covering not only the army but in addition the whole social structure, including all allied states'.[54] In this pre-war situation Neurath asked the state to learn from previous wars, to draw consequences from the failure of the monetary economy and to start preparing a large-scale economy in kind. He believed that to be the necessary economic condition for winning a future world war.

The advantage of a large-scale economy in kind is that it allows more accurate control over the movement of commodities. 'There are indeed extremely primitive systems of economy in kind, but there are also types of that kind of organisation that rank higher than any type of monetary organisation.'[55] An economy in kind requires an economic plan – and this is a measure alien to free market economies. An economic plan implies that decisions are made by a central institution

---

[50] Neurath 1910c [1919a], p. 21.   [51] Neurath 1914 [1919a], p. 96.   [52] Ibid., p. 96.
[53] Ibid., p. 103.   [54] Ibid., p. 104.   [55] Ibid., p. 101.

overseeing the whole economy. Neurath calls the transition period between a free market economy and an economy in kind an 'administrative economy'. In this phase it is not individuals and their considerations concerning purchases that influence the economy, but rather a central institution run by a cartel or the government. 'The administrative economy may perfectly well pursue class interests' but it does not have to result in police terror.[56] Neurath cited the tax system as an example of a measure taken in the spirit of an administrative economy.

A necessary condition for an economic plan is 'calculation in kind', that is, one has to ascertain the quantities of goods produced, consumed, imported, kept in stock, etc. By contrast, a monetary economy is exclusively geared to prices and profits, and thus does not require calculations in kind. Neurath urged that 'statistics in kind' be substituted for present monetary statistics so as to make possible the transition to calculations in kind.

According to Neurath economies in kind solve the problems of free market economies, for they obtain and distribute products directly. With his socialisation plans Neurath showed how one could arrive at an ideal society based on an economy in kind. We will describe his conception of an economy in kind in greater detail below in connection with his socialisation models. Yet even before 1914 Neurath conceived of a world war as a transition period towards such an economy. He had a concrete vision of the complete organisation of a state at war, starting with the centralisation of bread production via an expansion of the storehouse system up to the standardisation of railway wagons.[57] His past involvement in the rations and provisions service in the army reveals itself here.

The experiences of the First World War cured Neurath of the illusion that war might increase the prosperity of a population. But he continued to believe that, as a matter of fact, war could be the midwife for a future economy in kind. In 1918 he said in a lecture:

Wartime economy was a period of time when numerous events that could never have been performed experimentally occurred by chance. War economics enables us to overcome the confinement that surrounds us [applause], and perhaps we will the face the future better equipped [lively applause] than we did at the beginning of the war.[58]

---

[56] Neurath, 1917b [1919a], p. 149.   [57] Neurath 1914 [1919a], pp. 106 ff.
[58] Neurath 1918b, pp. 19 f.

### 1.1.3 During the First World War

The Balkan war began in 1912 and ended the following year with Bulgaria, Serbia, Greece and Montenegro triumphing over Turkey. Disputes over the distribution of the former Turkish territory prompted a second war between the former allies in which Bulgaria lost Macedonia to Serbia and Greece. During this period Neurath journeyed twice to the Balkans for the Carnegie Endowment for International Peace, finding ample opportunity for empirical studies, especially comparative ones of the economic structures of the warring states. For the first time Neurath was able to examine his ideas about war economies in practice. Neurath published reports on the economic and the social conditions in these countries in numerous articles for different newspapers and journals.

The Balkan wars were only a prelude to the First World War, which began in the summer of 1914. Neurath was drafted in August 1914 as part of the general mobilisation and positioned with the Reserve Provisions Unit 2, responsible for technical projects at the communications headquarters. In March 1915 he was promoted and shortly afterwards was commended for 'excellent performance in battle' and decorated in July.[59] Neurath served as a rations and provisions officer at the Second Army Service Corps, from December 1915 to January 1916 in Podkomiec and after that until June 1916 in Radzwillow.[60] Neurath described his war experience as follows:

> My service in the army corps was not hard, but boring. Then I had an idea and I put it before the general. 'Your Excellency' I said, 'we are in the middle of the war. We do very little here and should do so much.'
>
> 'What do you mean?'
>
> 'We should found a department for war economy and start to work seriously. We are learning a lot from successes, even more from failures; but nobody has time to keep a record of our experiences for the benefit of the future. All we do will be lost and wasted through such neglect.'
>
> The general pondered and said, 'There is something in that, but what would you suggest?'
>
> 'Probably it would be best for a special department of the War Ministry to concentrate on this task. It should have an intelligent open-minded and experienced head; there could, I think, be no better person than your Excellency.'
>
> 'Hm, this is worth considering. Please write a detailed memorandum for me.'[61]

---

[59] Letter from the War Archive, Vienna. Neurath was promoted to *Militär-Verpflegeoffizial in der Reserve*, his decoration was the *Goldene Verdienstkreuz mit der Krone am Band der Tapferkeitsmedaille*.
[60] On Neurath's activities in Radzwillow see Schumann 1973, pp. 16 ff.   [61] Neumann 1973, p. 10.

In 25 July 1916 Neurath was transferred to the Scientific Committee for War Economics at Department 10 of the War Ministry and promoted to Head of the General War and Army Economics section.[62] At the same time he became director of the Museum of War Economy (*Kriegswirtschaftliches Museum*) in Leipzig.[63] This museum was founded by representative bodies of German businesses, the German Trade Association, the German Agricultural Association and the German Trade and Business Association.[64] It was supposed to 'commemorate all economic achievements of the First World War' and create a 'centre-point for the dissemination of knowledge and research in war economy'.[65] Neurath's repeated calls for a systematic science of war economics had caught the founders' attention. With the approval of his superiors at the Austrian War Ministry, Neurath worked for fourteen days in Vienna and for fourteen days in Leipzig.[66]

In Leipzig Neurath began to develop the conception of museum design that was to reach fruition in his Viennese museum ten years later. For Neurath a museum should not simply collect curiosities but rather educate by offering visual information. In his Leipzig museum Neurath sought 'to make as clear as possible to everyone how a peace economy gradually changed into a war economy, how the latter changed in turn, and was replaced by a new peace economy that was partly shaped by its predecessor'.[67] In Leipzig Neurath tried to provide visualisations of the processes by which every product on display changed on its way from production to use. He wanted to display the whole 'mechanism of economy', to illustrate the way of life of a population at a certain time. To do so he also used statistical tables and models, techniques of representation that he would later develop into his Vienna method of picture statistics.

Neurath appointed Wolfgang Schumann as the general secretary of the Leipzig museum. Schumann was the editor of the cultural journal *Kunstwart und Kulturwart*, in which Neurath had previously published. He was also a socialist and an enthusiastic follower of Neurath's war economics. In his eyes it constituted a possible theoretical foundation for the future economy of a socialist society.

---

[62] Letter from the War Archive, Vienna. Neurath's idea to create a department for war economics does not seem to have been unique. A little-known engineer, Hermann Beck, proposed to the German War Ministry that an Economy-Organisation Office be founded (Beck 1919, p. 21). Beck's proposals for an economic plan and an economic administration are reminiscent of Neurath's socialisation theory.   [63] Schumann 1973, p. 16.   [64] Neurath 1918b, p. 3.
[65] Ibid., pp. 5, 8.   [66] Letter to Josef Frank.   [67] Neurath 1918b, p. 16.

When his army service ended in August 1918 Neurath received the 'highest recognition for an excellent performance during the war'.[68] This commendation refers mainly to his scientific work: 'With his pioneering studies of war economics, he performed an extraordinary service for the army command and for the sciences, both before and during the war.'[69] From the end of Austria's participation in the war, Neurath lived in Dresden, where the Schumann family also lived, and commuted from there to Leipzig.[70]

Despite his diverse activities Neurath qualified as a university lecturer at the University of Heidelberg on 21 July 1917.[71] His *Habilitation* was granted for previous work and the usual requirement of a second dissertation was waived. Neurath received the warm support of Eberhart Gothein and was to lecture on the war economies of the Balkans.[72] But he did not start lecturing, because of his commitments in Leipzig. From the winter of 1917/18 onwards the faculty record at Heidelberg lists Neurath as on leave.[73]

The German revolution began with strikes, demonstrations and armed rebellions in early November 1918 in the ports of Northern Germany and soon spread south. Emperor Wilhelm II resigned, as did King Ludwig III of Bavaria. The Bavarian Republic was declared on 8 November, and the German empire soon followed this example.

The war museum in Leipzig was dissolved in the course of the general breakdown.[74] Fierce discussions and fights between the various political groups were now on the general political agenda. Schumann thought the time had come for the overthrow of the old social system. In long discussions he convinced Neurath to join the Social Democratic Party and work on a plan for a comprehensive socialisation of the economy. This decision was a difficult one for Neurath to make. A lectureship was waiting for him in Heidelberg and his wife Olga Hahn-Neurath was vehemently opposed to his entry into politics.[75] Several months later he wrote:

The hesitations and vacillations of those called upon to act, the advice of my friends and sundry accidental circumstances finally moved me, after much reflection, to conclude my life of contemplation and to begin one of action to help to introduce an administrative economy that will bring happiness.[76]

---

[68] Letter from the War Archive, Vienna.   [69] Ibid.   [70] Conversations with Marie Neurath.
[71] Letter from the University Archives, Heidelberg.
[72] Letter from Otto Neurath to Ferdinand Tönnies, 26 June 1917, Tönnies Nachlaß.
[73] According to the letter from the University Archives in Heidelberg the documents of the Institut für Philosophie on Neurath's habilitation were lost in 1945.   [74] Schumann 1973, p. 16.
[75] Conversations with Marie Neurath.   [76] Neurath 1919a [1973], p. 124.

## 1.2 THE SOCIALISATION DEBATE

### *1.2.1 Setting the problem*

It is characteristic of the period following the First World War that someone like Neurath, who had not been very close to the Social Democratic movement, felt pressed to develop a socialisation plan. The Social Democrats on the other hand had long been opposed to utopianism and consequently failed to generate any concrete plans for the future. Karl Kautsky, a leader of the Social Democratic movement, expressed this attitude as follows:

> Whoever grasps our point of view clearly sees that it is impossible to forecast the shape of the socialist industrial economic co-operative. It will not be ready-made the day after the revolution; instead it will be the product of a steady continuous development that will produce ever new questions and problems. We may leave this worry to our children and grandchildren.[77]

This opinion places Kautsky in the orthodox Marxist tradition. Marx took himself to have provided an exact analysis of the driving forces of capitalism with the conclusion that the dynamics of the capitalist society would lead to its downfall. As he wrote in the *Communist Manifesto*: 'Not only has the bourgeoisie forged the weapons that bring death to itself; it has also called into existence the men who are to wield those weapons – the modern working class – the proletarians.'[78] Marx and his followers tended to think of socialism as the product of the historical process.

Yet historical determinism can also be joined by activist components. Class war can speed up the process. Marxists and Social Democrats constructed a number of different political conceptions and strategies, but their emphasis was on strategies for the seizure of power. They argued with anarchists in the First International, discussed the failure of the Paris Commune, debated the concepts of the mass strike and of councils or soviets (*Räte*). Few thought about programmes for constructing the socialist economy.

In the wake of the collapse of Tsarist Russia, the Habsburg empire and Wilhelmian Germany, the chances for the workers' movement to seize power increased considerably. Previously neglected questions concerning the shaping of society, the distribution of goods and the management and steering of the economic system became urgent. The

---

[77] Karl Kautsky and Bruno Schoenlank, *Grundsätze und Forderungen der Sozialdemokratie*, Berlin n.d., quoted in Wernitz 1966, p. 69. [78] Marx and Engels 1848.

## The socialisation debate

workers' movement split into Social Democrats and Communists and still further. The answers and solutions that were proposed diverged correspondingly.

The term 'socialisation' refers to a variety of theoretical considerations concerned with the problem of how to transform a private economy to a collective economy. The term is used in a number of different ways corresponding to different tendencies. Consider the following definitions given by leading theorists of the left. For Karl Korsch 'socialisation means a new organisation of production, guided by the goal of replacing the capitalist private economy by a socialist communal economy'.[79] Otto Bauer equated socialisation with the nationalisation of selected means of production, also placing less emphasis on workers' control than Korsch.[80] Otto Neurath's position was different:

Socialisation means the reshaping of a way of life. Nationalisation means changing the balance of power. A company not yet subject to social influences [i.e., collective decisions] may be nationalised, for instance, by taking it away from the entrepreneur and handing it over to a corporation representing the society.[81]

All conceptions of socialisation were influenced both by an analysis of the current social order and by a picture of the new social order to be established. The different conceptions can be classified according to two criteria: (1) the attitude towards property ownership of the means of production and (2) questions concerning the management of production. These criteria help to distinguish two types of socialisation plans.

The first aimed for a rapid transformation from capitalism to socialism by radical alterations in the basis of society. All privately owned means of production were to be collectivised. Production was to be organised according to the interests of the collective and goods distributed according to socialist maxims. This is the revolutionary conception of socialisation which differs from the reformist conception. The adherents of the latter viewed socialisation as a continuous process in which individual entrepreneurs would only be gradually replaced by socialist institutions. This typology is meant to characterise two extremes in the socialisation debate; it does not cover all possible views.

In order to understand Neurath's position we shall first consider Karl Korsch, an exponent of the revolutionary conception, and Otto Bauer,

---

[79] Karl Korsch 1919 [1969], p. 15. In this context 'production' not only covers the material aspects of the production of goods but also the relations of production, and 'that means the social conditions of production'. [80] O. Bauer 1919 [1976], p. 96. [81] Neurath 1922a, p. 54.

an adherent of the reformist approach. (We may note that Bauer and Korsch were also close to what Neurath later called 'the scientific world conception' which he defended in the Vienna Circle. In the early 1930s Korsch gave talks to the associated Berlin Society for Scientific (later: Empirical) Philosophy, led by Hans Reichenbach and others, and participated in the Unity of Science congress in Cambridge, Massachusetts, in 1939. During his exile he also intended to publish about the relationship between Marxism and logical empiricism but never realised this plan because of a serious disease. Otto Bauer was a personal friend of Neurath's and testified on Neurath's behalf at his trial in Munich; he also was an editor of the Austrian socialist journal *Der Kampf* in which Neurath published regularly after the war and lectured to the Ernst Mach Society in Vienna around 1930.)

### *1.2.2 Bauer and Korsch*

Bauer served as the first Foreign Secretary of the first Austrian Republic and he also headed the committee for socialisation set up by the parliament from 15 March until 17 October 1919.[82] At the same time he sought to present his ideas to the wider public in a series of articles in the *Arbeiter-Zeitung* (*Workers' Times*), later collected and published as *Der Weg zum Sozialismus* (*The Road to Socialism*). The writings constitute the most compact presentation of his ideas on socialisation which, it should be noted, he developed only after he became president of the committee for socialisation.

The collapse of the Habsburg empire had paved the way for democracy. 'Nevertheless the political revolution is but half the revolution. It removes the political suppression but leaves economic exploitation intact.'[83] According to Bauer a political revolution may be 'a day's work', the result of a putsch; completing the social revolution, however, would be a long and difficult process. All the same, it was the political revolution that set the conditions, since it determined how much power the proletariat commanded. The socialist society was to be constructed step by step in a planned and organised fashion; socialism would be achieved 'solely by creative legislative and administrative work'.[84]

According to Bauer the first step on the way to socialism was the nationalisation of heavy industries. The previous owners were to be compensated from property taxes levied on all capitalists and land-

---

[82] Fischer 1968, p. 23.  [83] O. Bauer 1919 [1976], p. 91.  [84] Ibid., p. 95.

owners. After that an administrative committee would appoint managers, determine prices, design employment contracts and decide upon investments. The members of the committee would be drawn equally from three groups: blue and white-collar workers employed in the factories, consumers and delegates of the state (representing the people). Nationalisation would not only improve the situation of those who worked in these companies, it would also create new sources of income for the state, for Bauer proposed dividing the net profits between, on the one hand, the state and the nationalised businesses, and on the other the employees of these businesses. This measure was also meant to increase the worker's enthusiasm and productivity.

In addition Bauer discussed the socialisation of other branches of the economy and concluded that 'a lot of industries are not yet ready for socialisation'.[85] A business was 'not yet ready' if expulsion of the entrepreneurs threatened loss of vital managerial skills which would badly disturb production. No measure towards socialism should 'destroy the capitalist organisation unless it also builds a socialist organisation that is at least as capable of managing the production process'.[86] Businesses 'not yet ready' were to be prepared for socialisation in the future by gradual integration into industrial associations run by administrative committees. These associations were to promote technical development, support rationalisation, cut production costs, etc., in the expected transition to labour-saving mass production. They were also to determine prices and organise the sale of products. This socialisation of the industries was to be followed by the socialisation of commercial trade, especially wholesale and export.

The next step on the road to socialism was to be the socialisation of the large estates. Bauer held that compensation should be paid for these as well. The running of the estates would be handed over to administrative committees on the model of the socialised industries. In parallel Bauer advocated supporting small-scale agriculture and sought to improve the social security of small farmers and agricultural workers by mandatory insurance schemes. Another step was the socialisation of housing. Since everybody had a right to housing, local authorities had to provide it and thus were to be given legislative means to socialise building land, again with compensation. Tenant committees were to staff the administration, and communal facilities were to ease the burden of private households. The final step of Bauer's socialisation plan was the

[85] Ibid., p. 100.   [86] Ibid., p. 95.

socialisation of the banks, whose power was to be transferred to representatives of the people. Summing up, Bauer wrote about his scheme: 'Socialism aims to return to the people what has been appropriated by capitalists and land owners . . . Expropriating those who expropriated the people thus constitutes the primary condition for a socialist society.'[87]

Bauer's road to socialism respected the framework of formal democracy. But the question remained whether the framework of formal democracy would suffice to safeguard the possibility of defeating capitalism. Bauer did not believe that the Austrian workers had enough authority to follow the Russian example and declare the 'dictatorship of the proletariat' or a soviet republic. He therefore recommended his scheme as a peaceful alternative realisable within formal democracy. Bauer's scheme became the Austrian socialisation programme; yet only a few measures were actually implemented. The bourgeoisie quickly recovered its nerve and reclaimed the political leadership of Austria.[88]

Karl Korsch was an important theoretician of the German council or soviet (*Räte*) movement. Immediately following the First World War he became a member of a soldiers' council. In 1919 he acted as a scientific advisor to the Independent Social Democrats (USPD), who had split in 1917 from the majority Social Democrats (SPD) because of the latter's support for Germany's war aims. In 1920 Korsch joined the newly founded Communist Party (KPD) but was expelled in 1926.

Production figures as the most important factor in Korsch's model of socialisation because it 'epitomises the social conditions'.[89] In capitalist society the 'social process of production is basically viewed as a private matter'.[90] In contrast, socialisation was to produce an economic system in which production becomes a 'public matter of the producing and the consuming community'. Ownership of the means of production not only implies control of the production process, but also power to appropriate profit. According to Korsch, 'to socialise production is to abolish the contrast between capital and wage labour that dominates the current

[87] Ibid., p. 124.
[88] During the first year of socialisation in Austria, in 1919, Otto Bauer's socialisation committee advanced four bills. Three were accepted by the national assembly with some minor changes (a bill on the formation of factory councils, a bill on co-operative institutions and a bill on the expropriation of industrial businesses). But the bourgeoisie increasingly opposed the socialisation measures and all attempts to socialise ended with the dissolution of the coalition in 1920. On the effects of socialisation in Austria compare Bauer 1920 with an opposing voice: 'The debates and legislative measures did not do much to promote the workers' demands to socialise businesses, they rather blocked and hindered these demands' (Rothschild 1961, p. 72). For a comprehensive treatment of the Austrian socialisation movement see Weissel 1976.
[89] Korsch 1919 [1969], p. 15.   [90] Ibid., p. 16.

## The socialisation debate

capitalist economies. It also means to abolish the class division that results from this contrast and to end the class struggle.'[91] Korsch attacked those who wanted to socialise without expropriation, for example, Eduard Bernstein, who did not envisage expropriation but rather increased control by society. By contrast, Korsch was sure of one thing: 'No socialisation of the means of production without the complete elimination of entrepreneurs from the production process!'[92] For him, only the revolutionary conception of socialisation was viable. The expropriation of the means of production was to constitute its first phase of socialisation.

The socialist economy thus required the relations of production to be reorganised. The difficulty remained that even in socialism the interests of the workers who produce and those who consume the goods may still conflict. For Korsch, this meant that a socialisation plan to create common property had to consider the interests of both groups equally. The elimination of private ownership of the means of production has different meanings for the workers and the consumers. For the workers it means: '(a) the right to claim part of the profit resulting from the labour, (b) participation in the control of the production process, corresponding to the importance of their contribution to the production'.[93] For the consumers it means, by contrast: '(a) division of the profit from the overall production among the totality of consumers, (b) a transition of authority from private owners to the institutions of the community'.[94]

Korsch was especially opposed to nationalisation because he felt it would lead to putting consumer interests first. He therefore did not recognise nationalisation as a form of socialisation unless it was to be joined by other measures. He also rejected the formation of production co-operatives because he believed that these favoured the producers. To socialise and arrive at common property one ought not transmit authority from entrepreneurs to either workers participating in production, or to officials representing consumer interests, but rather each measure should be accompanied by the public control of production in the interest of society as a whole. For Korsch socialisation consists 'in a transfer of the means of production to social functionaries and the legislative curtailment of the powers of the new managers in the interest of society as a whole'.[95] Korsch labelled the form of socialisation that is thereby created 'industrial autonomy'. Businesses would be incorporated into the property of the community but retain some independence from the central

[91] Ibid., p. 20.   [92] Ibid., p. 22.   [93] Ibid., p. 26.   [94] Ibid.   [95] Ibid., p. 33.

government or administration. The overall profit of a business would be divided into three parts: the workers receive one part, the consumers another, with the rest reserved for investments.

This first phase of socialisation, the socialisation of the means of production, would be followed by a second phase: the socialisation of labour. During the first phase formerly unfree and exploited labour would be liberated. But wages still depended on performance: different performances would be paid differently. In the course of time a community spirit would evolve so that everybody's labour would become communal property as well. All members of society would contribute to production according to their abilities and receive from the overall output according to their needs.

Korsch's conception of socialisation remained a mere programme. Otto Bauer could at least point to three legislative reforms that resulted from his socialisation model, but Korsch achieved no comparable political results. He traced this failure to the lack of a clear strategy and an underdeveloped will for revolutionary activities on the part of the political leadership of the proletariat.

A comparison of Bauer's and Korsch's plans reveals the following differences. (1) Bauer conceived of socialisation as a long-term process. Korsch thought that the first phase of his model could be realised right away. Only the second phase required a long series of cultural and political measures, especially ones directed to transforming the education system. (2) According to Bauer, once the working class had gained power through a political revolution, it should use this power in a peaceful way; that is, it should implement reforms by legislative means. For Korsch, by contrast, force is also a way to achieve control of the production process. (3) Bauer argued for compensation for property that was expropriated, whereas Korsch did not even consider it. (4) Only 'ripe' businesses were to be socialised according to Bauer; 'unripe' businesses had to undergo a lengthy process of preparation. By contrast Korsch would socialise even 'unripe' businesses since some may 'perhaps never achieve ripeness'.[96] (5) Bauer argued for nationalisation. In his system, workers employed in a business participate in its administrative committee with one-third of the votes, while representatives of state and consumers have a two-thirds majority. Korsch opposed mere nationalisation for he saw in it the danger of 'consumer capitalism' or 'state capitalism' (consumers might appropriate control of the production process). His model of 'industrial

---

[96] Ibid., p. 35.

autonomy' was to avoid the dangers of bureaucratisation and stagnation which nationalisation encouraged.

### 1.2.3. The standard of living

Neurath described socialisation as 'the conscious creation of a "new way of life."'[97] He conceived of a 'way of life' as 'the totality of measures, institutions and customs of a person or group of people'.[98] The current way of life is dominated by capitalism or 'the free market economy'. Neurath's critique of capitalist economies starts from two points. First, a capitalist economy does not strive for high productivity but rather for an increase in net profits. 'It is not the demands of the majority of the people that are decisive but a minority's desire to gain profits.'[99] Second, the free market economy treats natural resources as irresponsibly as it treats people. It exploits technical and scientific inventions only if they can be transformed into profit. A system like this necessarily produces crises, economic depressions and unemployment. It also displays systematic 'underuse'. Neurath's examples are the multiplicity of similar products which result from competition rather than demand, and what is now called 'planned obsolescence' which reflects only commercial interest.[100]

A socialist society would have fundamentally different goals. Economic standards would be directed not towards profit but happiness. 'Maximum profit is the purpose of the individual business in the capitalist economy, . . . a maximum of happiness, of the enjoyment of life in a community and of utility is the purpose of a socialist economy.'[101] The new way of life would become the basis for the spiritual and intellectual development of the people. 'The fight for freedom from capitalist dependency [is] a fight for a better material and intellectual existence and a free personality.'[102]

For Neurath the chief drawback to traditional economics is its failure to treat questions of happiness, except perhaps in 'the opening and closing fanfares' of public lectures. 'What is more natural than to examine certain institutions, such as monetary economy and economy in kind, free competition or mercantilism, to see whether under given circumstances they can realise the maximum of happiness or not?'[103] Neurath aims at maximising happiness. In contrast with other economic theories, his was to be guided not by 'objectified' categories like the

---

[97] Neurath 1920a, p. 45. [98] Neurath 1925a, pp. 30 f. [99] Neurath and Schumann 1919, p. 15.
[100] Neurath 1920c, pp. 9 f. [101] Neurath 1925b, pp. 394 f. [102] Neurath 1925a, p. 9.
[103] Neurath 1912 [1973], p. 113.

growth of the national product and the prosperity of nations, but by the concrete circumstances of people's lives. That is why he developed a sociological theory of the standard of living.

The concept of a standard of living (*Lebenslage*) and of distribution of the standard of living played a central role in Neurath's assessment of economic systems. A number of factors are relevant indicators: 'housing, nutrition, clothing, education, amusement, working hours and the trials and tribulations of life'.[104] The standard of living in turn determines the life mood (*Lebensstimmung*). 'The joyfulness or lack of joy of living shall be called the "life mood" – or "mood" in general. We are using a word that includes both happiness and sorrow and prosperity and poverty.'[105]

In section 1.1.2 we saw that Neurath was interested in the question: 'How do different ways of life and different measures influence the living standard of people and thereby their happiness and sorrow?'[106] Two schools of thought treat the question of happiness from an economic point of view, utilitarianism and social Epicureanism. Neurath declared himself an adherent of social Epicureanism, since its principal concern was not with the individual's desire for happiness, but rather with the creation of social conditions that would provide the possibility for the pursuit of happiness. It is 'the structure of the community' that mattered to Neurath; the individual appeared as a member of the community whose life mood depends on the community's standard of living. Neurath wanted to distance himself from utilitarianism. According to Neurath utilitarianism tends to leave outcomes to the free interplay of different powers. All individuals strive for happiness; the only laws envisaged are ones that prevent one person's efforts towards happiness from constraining another's freedom. By contrast social Epicureanism assumes a collective interest in the maximisation of happiness. It is the task of social construction to generate the necessary social and economic framework.

Utilitarians talk 'in a somewhat indeterminate manner'[107] of the greatest happiness of the greatest number of people. They thus confront the problem of how to measure and compare the amount of happiness of various individuals or groups. 'If pleasure were measurable in meters or kilograms, nothing would prevent an unlimited use of this principle.'[108] In 1912 Neurath discussed this question in a lecture on the

---

[104] E.g. Neurath 1920c, p. 11.   [105] Neurath 1925a, p. 29.   [106] Ibid., p. 28.   [107] Ibid., p. 26.
[108] Neurath 1912 [1973], p. 119.

## The socialisation debate

problem of calculating the pleasure maximum (further discussed in 2.3.1). He arrived at the following conclusion: 'We have not yet succeeded in calculating pleasure sums of groups of persons under all circumstances, nor even in stating how such a calculation is to proceed; the principle of maximal happiness . . . can never be the basis of a moral or legal system, or of a whole order of life.'[109] In this passage an idea emerges that will later be important in Neurath's defence of physicalism: science must not employ fictions and ungrounded idealisations but should stick to what actually occurs. We want to learn about new possibilities, but these must be grounded in scientific knowledge and not be figments of the imagination.

Neurath conceived of his theory of living standard as a extension of Marxist thought. As he saw it, the misery of the proletariat and its poor living standard functioned both as a starting point for Marx's critique and as a motivating force for the drive to overcome capitalism. He maintained that Friedrich Engels had already formulated an approach towards a theory of living standard in his empirical study on the condition of the working class in England.[110] Comparing the living standard of pre-capitalist weavers with that of the British proletariat, Engels took the crucial variables to be nutrition, clothing, housing, disease, death and criminality.

Neurath went beyond this. He did not rest content with the enumeration of factors but tried to relate them systematically. An 'Inventory of the Standard of Living'[111] should make it possible to gain an overall picture of the distribution of welfare. This would not only alert us to unjust distributions of standards of living in free market economies. It would also be important for the construction of the new order of life since the socialism that Neurath aimed for would distribute the standard of living consciously. 'Here, the distribution of living standard is not only an effect, it is also the goal of human measures.'[112] According to Neurath we need a comparative study that will look at 'the standard of living that results for a population from the realisation of various economic measures concerning production and distribution'.[113] In socialism the measures would be chosen according to their effects on the general standard of living.

---

[109] Ibid., p. 119, translation altered.  [110] Neurath 1925a, p. 20.
[111] This is the title of his 1937b.  [112] Neurath 1925a, p. 25.
[113] Neurath and Schumann 1919, pp. 76 ff.

### 1.2.4 Neurath on the structure of the socialist economy

For Neurath the socialisation of an economic order consists in 'a transformation from an economy of domination to a social economy, from a market economy to an administrative economy, from an unplanned economy to a planned economy'.[114] It also consists in a development from a monetary economy to an economy in kind and from an incomplete use to the complete use of resources.

The hopes Neurath put into the reforming power of war economy had faded by the end of the First World War. In 1919 he declared soberly:

> World war broke out, prompted by frenzy and illusion. Power was gained by army commanders and their party comrades, by the producers of cannons and shoes and tinned food. Those who knew tricks and deceptions that bring ample intermediate profit became powerful. But the people bled and starved – and then everything collapsed.[115]

But something remained – a lesson about how 'to achieve systematic control of a national economy, down to the smallest bits'.[116]

Neurath aimed to put these experiences to the service of socialisation. It was not enough for leaders of the working class to seize authority after the First World War and to take steps towards 'freedom from old bondage'.[117] Of the German revolution of 1918 Neurath wrote: 'It was no good to praise and glorify the November triumph alone. The working class felt that the political achievements remained illusory unless they were accompanied by economic success.'[118] It might be possible to seize political power relatively quickly by taking control of decisive positions. But it takes more time to transform the economy. Fig. 1.2 shows how Neurath conceived of (part of) the new economy after the transformation of capitalist institutions. (This model of the economic organisation is embedded in a still larger political structure comprising voters, their representatives and the government and its agencies.)[119]

The socialist economy is an administrative economy, that is, an economy based on an economic plan. This plan would have to be designed by a Central Economic Administration (Zentralwirtschaftsamt) and concern 'all economic positions and departments ... that are distributed among the various ministries and other authorities and organisations'.[120] In order

---

[114] Neurath 1921c, p. 7 ff.   [115] Neurath 1920c, p. 5.   [116] Neurath 1919c, p. 10.
[117] Neurath 1920c, p. 5.   [118] Ibid., p. 5.
[119] See table 2 of Neurath 1921c. Thus the government commissions the Central Economic Administration and receives input from the Centre for Organisation for its various agencies.
[120] Neurath 1920c, p. 17.

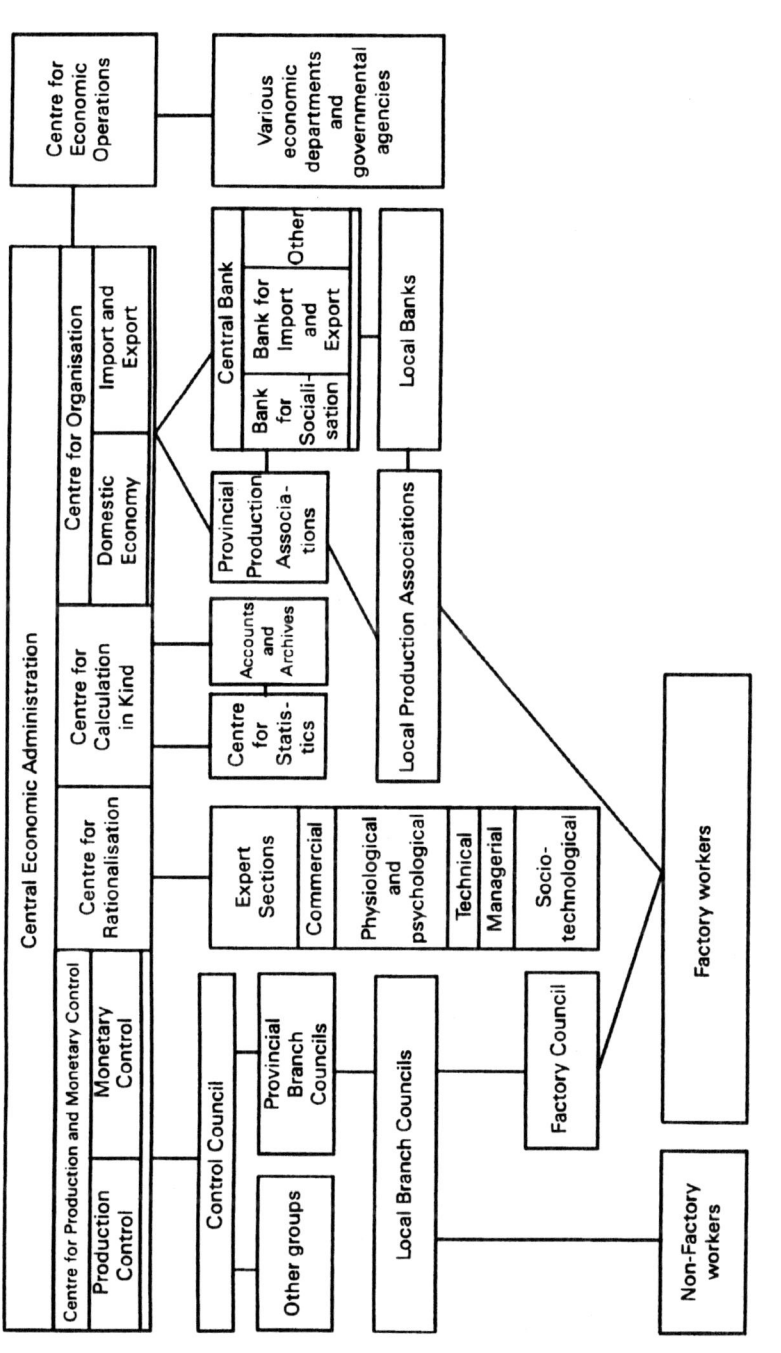

Fig. 1.2 Schematic model of economic organisation (after Neurath 1920c, p. 18)

to carry out a single ground plan for the whole economy, tight collaboration would be required. 'The Centre for Calculation in Kind (Naturalrechnungszentrale) will have to develop a universal statistic that encompasses the foundations of the economy in their interrelations and displays its relevance for the standard of living.'[121] This universal statistic was to summarise more specific statistics and compare them with one another.

The economic plan should display the distribution of all the different types of things that make up the standard of living:

> The calculation of the average amount of meat, bread and housing available per person is entirely unsatisfactory. Even the statistical distribution of housing and meat consumption remains in itself almost useless, since it is essential to determine whether bad housing conditions are compensated for by good nutrition or worsened by bad nutrition. The Centre for Calculation in Kind will delineate types of living conditions and it will indicate how many people fall into each type depending on the various economic plans one could realise.[122]

Neurath had a variety of suggestions for improving statistical information. He warned first against the uncritical use of the statistical mean: it provides a very imprecise picture of the actual distribution. And, as in the quotation above, he also cautioned against the combination of averages deriving from different dimensions, since additions of this kind result in false depictions of the 'average person'.

Continuing one of the interests he had pursued in his Leipzig museum, Neurath explored different ways of visually displaying statistical information. Production statistics serve as an example: 'Which natural resources we use to produce a certain amount of goods are displayed with "rootstocks"; "hereditary trees" are used to display what kinds of things are made out of which natural resources.'[123] With these considerations about the role of statistics in socialisation plans, Neurath was moving towards the picture statistics he was to develop during the 1920s in Vienna.

In addition to the Centre for Calculation in Kind, the Centre for Rationalisation (Rationalisierungszentrale) was to work in close collaboration with the Central Economic Administration. Its job would be to increase 'economic effectiveness, worker's performance and their health and well being'.[124] The Centre for Organisation (Organisationszentrale) of the Central Economic Administration would unite the various Provincial Production Associations (Landesverbände) and the Central

---

[121] Ibid., p. 19.   [122] Ibid., p. 21.   [123] Ibid., p. 20.   [124] Ibid., p. 22.

Bank (Bankkonzern) with all smaller organisations. It should be mainly concerned with the development of a general wage system that 'determines . . . wages and salaries depending on danger, risk, comfort and strain, as well as the location, the working method and the age of the worker, and so forth'.[125] Neurath disapproved of employee profit sharing because it constituted a concession to a net-profit economy and also connected 'the worker with the economical success of the traditional system'.[126] This proved a relatively radical position that was strongly opposed after the First World War when the handy slogan 'creation of wealth in workers' hands' was used to replace socialisation plans that were thought too risky. The Centre for Organisation would also create a general price system for displaying the value of monetary wages in terms of wages-in-kind. For Neurath both systems would be transitory steps towards an economy in kind, that is, towards the direct supply of people with the products they need. Regular publications by the Centre for Production and Monetary Control (Kontrollzentrale) should allow everyone to check whether the 'use of money, natural resource and energy accords with the overall principles'.[127]

In order to promote a better standard of living Neurath planned to unite all production. The structure of the Provincial Production Associations was to reflect the differences between the different branches of the economy. For example, the Provincial Production Association for Housing would unite 'the production of building material, the administration of land, the construction of houses and the adminstration of apartments'.[128] In addition to the Provincial Production Associations for Housing, Neurath planned Associations for Nutrition, Clothing, Education and Amusement, Health, Production Equipment and Semi-manufactured Articles, Mining, Forestry and Agriculture, Distribution and so forth.

Individual banking institutions, he thought, should be left intact to serve as a tool for overall socialisation. The Central Bank would have the task of introducing moneyless foreign trade and arranging compensation transactions and moneyless transactions. 'The more we replace monetary management with direct management of materials, the more the banks will cultivate the distribution of production equipment instead of the distribution of credit.'[129]

In order to control and ensure socialisation Neurath wanted to create

---

[125] Ibid., see also Neurath 1921b.   [126] Neurath 1921c, p. 20.   [127] Neurath 1920c, p. 23.
[128] Ibid., p. 24.   [129] Ibid., p. 26.

a novel system of economic councils. This system would be rooted in Factory or Works Councils (Betriebsräte) elected by the workers and it would culminate in the Centre for Production and Monetary Control. Neurath disapproved of the alternative *political* council movement, to the extent that it aimed to hand over the management of companies to their workers. 'It is not the power of the worker in the factory, but in the national economy that is decisive.'[130] Even after the completion of socialisation, Factory Councils would not take over the management of businesses, since 'a committee consisting of several people is usually unsuited to rule a huge industrial business . . . even more so if the composition of the committee changes constantly. Also the principles of tight organisation do not allow that the leadership be constantly subject to removal by those who are led.'[131] The Factory Councils should rather control the management, the working conditions and the wages. Since Neurath believed that all workers should be represented in Factory Councils, he insisted that housewives should send representatives too. In addition he demanded a salary for housewives. 'Housewives bring up children and thereby perform a work that is acknowledged as necessary by the community; it is thus in accordance with socialist principles to pay them like full-time workers and to have them be represented in the committees.'[132]

Together with the representatives of housewives and non-factory workers, the Factory Councils would send representatives to the Local Branch Councils (örtliche Fachräte), which were to unite delegates from within a single branch of the economy. Local Branch Councils would elect Provincial Branch Councils (Landesfachräte) who would send delegates to the Provincial Production Associations but mainly work in the (Workers' and Farmers') Control Council (Arbeiter-und-Bauern-Kontrollrat).[133] In addition, in order to counteract the spirit of rigidification threatening to affect even the organisation of Workers' Councils, other groups were to send representatives to the Control Council. Neurath here had in mind not only housewives, but also workers on leave from production tasks, representatives of 'associations of a non-capitalist character' as well as representatives of the administrative agencies.[134] He obviously was aware that complex organisations tend to become inflexible, but he did not discuss the problems of bureaucratisation any further.

---

[130] Ibid., p. 26.   [131] Neurath 1921c, p. 19.   [132] Ibid., p. 28.
[133] The former delegative relation is not represented in the model of fig. 1.1 but figures in table 2 of Neurath 1921c. The complex control function of economic councils is detailed in tables 2, 6, 7 of Neurath 1920c.   [134] Neurath 1920c, p. 28.

Experts and scientists of all kinds play a major role in Neurath's plans for socialism. 'Since the rationalisation of the economy advances without considering net profits, there will be an unforeseen demand for psychologists, physiologists, technicians, industrial engineers and social engineers, who have to be distributed purposefully within the society.'[135] The experts would provide specialist advice to the controlling councils. They were to be united in Expert Sections (Sachverständigengruppen) and represented at the Central Economic Administration via the Centre for Rationalisation.

Neurath's economic model for a future society is oriented towards consumption, aiming to secure a better satisfaction of demand than can be achieved in a free market economy. Neurath's plan organises production from the top down and it provides for the central distribution of products. 'Moneyless economy' is a very special idea of Neurath.[136] In the ideal socialist economy, money would not serve as the connecting link between production and distribution. That is, money would not be used as a unit of calculation, nor as a standard, nor as a circulation device, nor for any of the other purposes money usually fulfils on standard economic theories. 'Socialism means economy in kind', says Neurath in a lecture on the socialisation of Saxony in 1919.[137] He puts it even more clearly in an essay of 1923:

> The theory of the socialist economy acknowledges only one manager or producer – the society – who organises the production and shapes the standard of living on the basis of an economic plan, without calculations of losses or profits and without taking the circulation of money as a basis, be it in the form of coins or labour. This is an opinion that corresponds with the basic ideas of Marx and Engels.[138]

In Neurath's socialism there is nothing to buy and nothing to sell; there is no market any more. Everything relevant for the standard of living is to be distributed on command. It is remarkable that Neurath said 'we may even consider the distribution of those goods that do not cost any labour but that are available in small amounts only'.[139]

### 1.2.5 The road to socialisation

In his writings Neurath rarely commented on the principles determining the distribution of goods. He indicated the role of Workers' Councils

---

[135] Ibid., p. 29.  [136] Cf. Martinez-Alier 1987, ch. 14.  [137] Neurath 1919c, p. 75.
[138] Neurath 1923a, pp. 153 ff.  [139] Ibid., p. 146.

and others at the Control Council, and so at the Central Economic Administration, but he never indicated how much say the people were to have in the distribution or whether they would have to take what they were given. 'It is in part a question of power how far the controlling, how far the directing force of the councils will reach.'[140] Perhaps the following passage discloses how broadly he conceived of his model functioning:

> The socialist economy does not have any of the usual ways of calculating success [profit] for the individual businesses. It can only perform the following calculations for the economy as a whole: If we command a certain amount of forests, meadows, swamps, streets, rocks, animals, people and machines at a certain time, then we may act according to economic plan I, or II, or III. Plan I will lead to the following distribution of living standard among the people, the following machines will be produced and the following stocks will be available. In contrast, plan II will lead to different conditions, and the same holds good for the other cases investigated. It is the task of the society, through the appropriate central organs, to determine the planned management of production and distribution that is best for the community, or, to speak with Marx, that exhibits the greatest 'utility' or 'use value' for the community in a certain period and for the immediate future. This decision process can call on no unitary measure. One can no more add means of production than one can add 'utility' or 'use value.' Also one cannot say that the differences between economic plans are determinable by some unitary measure. All one can do is to try to capture one process with statistical means as well as possible and contrast it with another one which is also captured in statistical terms.[141]

Beyond his remarks about the Workers' and Farmers' Control Committee, he intentionally left it unclear how the task of society was to be set. Since Marx talked in several passages about the abolition of capitalist goods, and money is only one good among others, Neurath saw himself in agreement with Marx in rejecting monetary economies. But we must note that the division of production and distribution is not to be found in Marx. 'By taking the individual working hours as a standard for the [worker's] share of the product, Marx also determined the basis for the relation between producer and product.'[142] Even though Marx allowed neither market nor money in socialism, working hours remain as a connecting link between producer and product.[143] Neurath evidently did not accept Marx's labour theory of value.[144]

Let us return to Neurath's model. The Central Economic Administration is to determine production and distribution. In accor-

---

[140] Neurath 1920c, p. 28.   [141] Neurath 1923a, p. 153.   [142] Mergner 1971, p. 29.
[143] Ibid., p. 32.   [144] Neurath 1920a, p. 71.

dance with the economic plan thought best for the community and irrespective of different occupations people have, the central institution would distribute all the material goods individuals need. It is hard to conceive that a central distribution could be geared to each and every one of the various different individual needs. So it seems that if the Centre for Organisation is to succeed with the difficult task of distribution, it has to assume that everyone has roughly the same needs, or at least that such a state is worth implementing. The former can easily lead to totalitarianism, while the latter is based on a false assumption and may well lead to the creation of a black market economy to satisfy individual needs. Neurath does not answer this worry, although he certainly would have been deeply opposed to the first outcome: 'Socialisation can only last if it respects men in their variety and does not impose a new subjugation.'[145] Perhaps the key to Neurath's silence lies in his conception of what can be established *scientifically*:

It is not difficult to join this new [socialisation] movement to the traditional Marxist trains of thought and to continue the party tradition which is so important in practice. To work out the necessary interpretations and explanations, qualifications and generalisations is no longer the task of strict science.[146]

It may thus be noted that the Distribution section at the Centre for Organisation would have had to answer to the Workers' and Farmers' Control Council at the central level, and to the Branch Workers' Councils at the local levels.[147] To what degree the 'equal treatment of all' means 'socialist tyranny' is up for negotiation. Similarly it must not be overlooked that the schema above portrays only the *economic* organisation of society (which Neurath wanted to keep relatively independent of its political organisation). The Central Economic Administration was to be accountable also to a government which in turn was answerable to representatives of all the voters.[148]

What steps then are necessary to implement the new social order and who is to undertake them? Neurath believed in joint action by all those who have 'suffered' under capitalism: 'the leaders of the factory workers' who will pursue the fight, together with 'manual workers, farmers, teachers, civil servants and technicians'.[149] The fight is not directed against those who own the means of production, and their henchmen, but against a bad economic system. Nor does Neurath's fight

---

[145] Neurath 1920c, p. 41. This excerpt also in Neurath 1973, p. 156.   [146] Neurath 1920a, p. 72.
[147] Neurath 1921c, p. 22.   [148] Ibid., table 2.   [149] Neurath 1920c, p. 2.

against capitalism proceed via a destruction of its institutions. Instead 'the road to socialism proceeds via capitalist institutions'.[150] Consider, for example, the first phase of socialisation. In this stage it is even possible that 'entrepreneurs will survive via some form of inheritable management,'[151] because in his opinion it would not be easy to find qualified replacements. Nevertheless the expropriation of capital is necessary during the later stage of complete socialisation. Unlike ownership and management, distribution can be changed right away: with the help of an economic plan, vital goods should be distributed among the people already during the phase of initial socialisation.

In Neurath's scheme there is no compensation for expropriations. Monetary compensations would be pointless because money is to be slowly transformed into useless paper as socialisation proceeds. Neurath considered 'additional pay'. 'For example, a former director would maintain the management of the business and also receive compensation in the form of less work and perhaps additional pay and pensions.'[152] Entrepreneurs would thereby be enabled to maintain their standard of living. It goes without saying that they would have to work as well. Neurath's model provides free wages only for old and sick people. 'An entrepreneur who does not want to work after socialisation would have to be expropriated without compensation, or perhaps even be punished for economic desertion.'[153] Although he realised that 'the best socialisation plan remains . . . ineffective unless there is power to implement it',[154] Neurath rarely ever considered questions relating to the seizure of power. His writings neglect strategic considerations, focusing entirely on presenting a consistent alternative model for the organisation of the economy. His great point of optimism was his belief that not only did science at last provide a powerful enough tool to construct socialism, but that the political situation after the First World War was ripe for socialisation.

Neurath also did not take a stand on what political form a socialist society should have:

One may be of different opinions about the way to obtain power: by elections, general strikes or bloody street fighting, with the help of the parliamentary democracy or by a council dictatorship. One may nevertheless agree on a socialisation plan which is basically independent of ways to seize power. Contemporary socialists are more deeply divided by their opinions about the

---

[150] Neurath 1923a, p. 153.  [151] Neurath 1920c, p. 41.
[152] Neurath and Schumann 1919, pp. 63, f.  [153] Neurath 1921c, p. 12.  [154] Ibid., p. 12.

forms for obtaining and maintaining power than by their opinions about the structure of the economic order to be strived for. This suggests constructing the socialist economic order independently from the political order, in order to protect it from everyday political disruptions.[155]

### 1.2.6 Neurath's position in the debate

Even though Neurath approved of Marx and even though he conceived of his socialisation model as a development of basic ideas of Marx and Engels, there are several points on which he differs from Marxist orthodoxy.

In section 1.2.1 we distinguished between reformist and revolutionary conceptions of socialisation. This distinction relates to the attitudes one takes towards the questions of ownership of the means of production and of the management of production. From its beginning in the First International the Marxist part of the workers' movement made much of the difference between reform and revolution, and opposing groups defined themselves accordingly, even though some occasionally questioned the tenability of this distinction.[156] Other socialist movements not directly influenced by Marxism, such as the Syndicalists and the Fabians, evaded such a distinction. Neurath differed significantly from orthodox Marxism in his views about the relations between the economic and social order.

Orthodox Marxists stress the unity between economy and politics because they believe in a materialistic conception of history according to which the political structure derives from the economic basis. The motor of social change thus consists in changes in the productive forces. Those who command the means of production also command the power in society. The space a political system leaves for variation is restricted by economic factors. Marxist theories apply this point of view to political strategies. The productive forces have to exhibit a certain stage of development if a political revolution is to be successful. This idea underlies, for example, Bauer's conception of 'businesses not yet ready for socialisation' and Korsch's model of autonomy.

By contrast Neurath had a more flexible conception of the relation between politics and economy. This may be connected with his heterodox version of the materialist conception of history, which resembled the views of Plekhanov (see section 3.4.2). He believed his model of

---

[155] Ibid.  [156] Luxemburg 1898.

socialisation to be compatible with various political systems. Erwin Weissel accused Neurath of political naivety:

> Neurath's weakness lay in the political sector and it contrasted in a curious way with his strict logic in the economic sector. In politics he exhibited a naivety that puts him in a camp with the old Utopians. For instance, he told the representatives of the Bavarian industry in all seriousness that the proletarians had obtained the power to enforce their will and that 'the capitalists therefore ought to be reasonable enough to hand over its power'. One would not want to 'enforce the socialisation of Bavarian industry from the office desk, but rather collaboration with the industry itself'.[157]

This is in keeping with Neurath's deep commitment to co-operation and negotiation (see section 3.5).

Weissel's hint at the utopians may be extended further. Neurath's distribution socialism is reminiscent of Gracchus Babeuf. Babeuf outlined an economic system that rested on small communities with a centralised economy in kind: the revolution had to bring about collective happiness by abolishing inequalities. There are also parallels with his father's acquaintance, Josef Popper-Lynkeus, to whom Neurath dedicated a *laudatio* on the occasion of his 80th birthday.[158] In his main work Popper-Lynkeus argued for a social system that aimed to secure the satisfaction of the basic needs of every member.[159] He proposed a 'nutrition army' that would consist of everyone of a certain age – except sick and disabled people. The nutrition army would secure the minimum for everybody. Serving the prescribed number of years secured the right to a certain minimum of housing, nutrition and clothing until death. Popper-Lynkeus advanced the slogan 'Nutrition service instead of army service' and he envisaged the capitalist economy remaining side by side with this 'duty and obligation economy'. The capitalist economy would produce luxury articles; and the economy in kind would not extend beyond supplying the basics. Popper-Lynkeus did not envisage a 'fully socialised' economy as Neurath did, yet the regimentation of social labour required by the general draft for the nutrition service would appear to be nearly as drastic.

Babeuf argued for the rebellion of the masses to seize their rights and Popper-Lynkeus was moved by injustice and poverty to seek refuge in utopianism. Neurath appears to have been motivated by a general insight from his economic training: the free market system does not work. Neurath's socialisation model far exceeded rivals in internal

---

[157] Weissel 1976, p. 154.   [158] Neurath 1918a.   [159] Popper-Lynkeus 1912.

## In the Bavarian revolution

consistency and unity. Nevertheless it fell into oblivion with the fall of the Munich Soviet Republic.

### 1.3 IN THE BAVARIAN REVOLUTION

As we saw in section 1.1.2, already in his studies on war economy Neurath had presented approaches towards a planned economy. His socialisation model developed them in a systematic form. In the Bavarian revolution Neurath saw a chance to implement his ideas.

#### *1.3.1. The appointment*

On 8 November 1918 Kurt Eisner proclaimed the Bavarian republic in his role as president of the Workers' and Soldiers' Council in Munich.[160] The Bavarian king Ludwig III had fled the day before, and the Council appointed a government that consisted of members of the USPD, SPD and one independent, with Kurt Eisner as Prime Minister. The following day the German emperor Wilhem II resigned. In Berlin Friedrich Scheideman proclaimed the German republic. Friedrich Ebert became Reichskanzler of the government of 'people's deputies'.

On 23 January 1919 Neurath travelled to Munich for the first time. There he discussed his economic programme and problems with its political realisation with Eisner and his minister of finance, Edgar Jaffé. Jaffé proposed that Neurath present his programme to the Munich workers. Two days later Neurath gave a talk to the Workers' and Soldiers' Council on 'The Character and Course of Socialisation'.[161] He presented his socialisation model convincingly. Although the talk was not greeted with undivided approval, it played an important role in Neurath's later appointment to the presidency of the newly founded Central Economic Administration. An extensive quotation follows, from a retrospective summary of this talk which Neurath wrote during his imprisonment after the suppression of the Munich experiment:

At that time I roughly presented the following: For the near future it is not crucial to expropriate as many businesses as possible nor to nationalise as much as possible, but to organise production, import and export in the interest of the

---

[160] On the Bavarian history of this time see, e.g., the partly contrasting accounts of Mitchell 1965, Grunberger 1973 and H. Beyer 1957 and 1982. For contemporary reports by participants see, e.g., the collections Dorst 1969, Morentz and Münz 1968, Schmolze 1969, and the memoirs Neurath 1920b, Mühsam 1929, Toller 1934 and Niekisch 1958 [1974], pp. 38–101.

[161] Neurath 1919d [1973], pp. 135–49.

whole and according to a plan. The Marxist tradition talks not only about socialisation and nationalisation, but also about the arrangement of production according to a plan. Whereas the seizure of the means of production by the state figured prominently in the Social Democratic propaganda, the propaganda in favor of a planned economy occupied only relatively little space. Profiting from my long term familiarity with the concept of associations and cartels through the works of my father, building on the rather unspecific forecasts of Marx and Engels, and influenced by an appreciation of the economic plans of Ballod-Atlanticus and Popper-Lynkeus, I showed how to unite production in groups, and how to distribute housing, nutrition, clothing, education and entertainment according to plans and in accordance with basic socialist principles. Just as the economy had been made to serve the needs of war with the Hindenburg programme, one should be able to make it serve peoples' happiness.

At that time I already pointed out that the political council system did not appear appropriate to me from a social-technological point of view, that is, it did not seem to promote the functioning of the economic system. But the workers and other adherents to socialism might of course strive for a specific type of state for reasons of power, even though this type embodies social-technological disadvantages. I also pointed out that it was important to exploit the elements of organisation available already and to subject them to the central administration of the community as a whole. That means that the factories and mines in large associations, the craftsmen, and small and medium farms are to be organised into co-operatives and assigned publicly determined duties, as for instance the delivery of goods. At the same time the central administration would supply them with raw materials. In addition to industrial associations, we would need a central bank to organise credit under the control of the community as a whole. In order to be able to control money and credit transactions, it would be mandatory to introduce a moneyless payment system. This would also prevent the hoarding of money and tax evasion. The reform of the credit system would have to proceed together with socialisation.

An economic plan would have to be the basis for all measures taken by the large organisations which are to be created. It would be mandatory to trace the movements of raw materials, energy, people and machines on their way through the economy. Therefore one needs a universal statistic that provides comprehensive overviews for entire countries and even the whole world. All specific statistics have to be incorporated into it. The Centre for Calculation in Kind would have to devise economic plans which perhaps would be comparable to the financial plans already in use.

These plans would be presented to the assembly of people's representatives. The significance and realisability of each measure would be evident from the overall plan. The provision of housing, nutrition and clothing to all, i.e., the satisfaction of basic needs, would become decisive – not the profit motive. Wages in kind and barter would again become important tools on this higher level of socio-economic organisation. Controlled by the community as a whole and working together with the production associations the central bank would have to organise agriculture, mining and industry simultaneously, supply farms with industrial products and administer agricultural production. Industrial products would be awarded for agricultural productivity.

In principle socialisation must be enacted across the board. The most central associations and institutions must be created right away. Socialisation should not be simply from the bottom up; rather, one has to form the organisations from the top down since this would be the only way to secure that everything receives its appropriate position.[162]

For the time being, however, Neurath returned to Saxony, for which, together with the economist Hermann Kranold (an SPD member of the executive committee of the Saxony Farm-Workers' Council[163]) and his friend Wolfgang Schumann (also SPD), he designed another socialisation programme, later known as the 'Kranold-Neurath-Schumann programme'. Even though the Saxony Central Council took note of the programme and even kept it 'under consideration' it was not implemented because of fierce infighting in the SPD, which formed the majority in the Saxony parliament at the time together with the USPD.[164] 'When we faced opposition against our programme even in Saxony and partly within our own party, it seemed best to incorporate Bavaria into the plan because it would be easier for a planned economy to succeed if the farmlands of Bavaria were to be connected with industrial areas of Saxony.'[165]

Due in large part to a venomous press campaign against Eisner, prompted by his 'non-patriotic' stand on the question of German war guilt, the USPD lost heavily to the SPD and conservative parties in the Bavarian elections on 12 January 1919. These results were not yet official, however, since one province had delayed its vote until 4 February.[166] A period of open tension between the Workers' and Soldiers' Council and the parliamentary forces began. On 21 February Eisner was assassinated by Graf Arco-Valley on his way to the first session of the elected state parliament, at which he intended to resign.[167] Since the parliament dispersed in confusion, an action committee called the 'Central Council of the Bavarian Republic' (Zentralrat), headed by Ernst Niekisch

---

[162] Neurath 1920b, pp. 3 ff. Excerpts translated in Neurath 1973, pp. 18–28. The present quotation fits precisely between the last two sentences of the second paragraph, p. 19.
[163] See the protocol of the questioning of Hermann Kranold of 19 June 1919 at the Standgericht. State Archives Munich.   [164] Neurath 1920b, p. 6.   [165] Ibid., p. 9.
[166] Schmolze 1969, p. 199.
[167] Niekisch 1958 [1974], pp. 50, 52, suggests that Arco-Valley was at least a sympathiser of the fascist Thule Society: on the activities of this proto-Nazi organisation during the Bavarian revolution see Hillmayr 1974, pp. 32–4, 87–9.

(SPD), claimed authority in the state and appointed a government which, however, refused to serve. In early March the Central Council ceded authority to the parliament again after a compromise government had been agreed upon by the SPD and USPD. On 17 March the state parliament reconvened and elected Johannes Hoffmann (SPD) prime minister, who formed a less radical cabinet with members of the USPD, SPD, one representative of the Farmers' Alliance (Bauernbund) and two independents.[168]

Announcing the legislative programme of his government to the Central Council on 15 November 1918, Eisner said: 'We do not wish to leave anybody in doubt about our unchanged socialist goals. But we must say with all frankness that it seems to us impossible to transfer the ownership of industry to society at a time when the productive forces of our country are nearly exhausted. One cannot engage in socialisation when there is hardly anything to socialise.'[169] We saw above that Neurath's first visit in January 1919 had no immediate result. In the time between the death of Eisner and the election of Hoffmann Neurath and Schumann visited Munich again to propose joint action with Saxony to the Central Council; the latter agreed, but Saxony's government hesitated.[170] A few days after Hoffmann's government assumed office, on 21 March, the Hungarian Soviet Republic was declared under Bela Kun. To respond to the renewed radicalisation of the workers and council movement – which only on 14 March had declared itself in favour of 'full socialisation'[171] – Edgar Jaffé, authorised by trade and industry secretary Josef Simon, sent a telegram to Neurath in Leipzig inviting him back to Munich for further talks with the Socialisation Commission and the Council of Ministers. Ernst Niekisch recalled Neurath's appearance before the latter (which Niekisch attended ex officio as Head of the Central Council) as follows:

> Neurath's vitality was almost irresistible. He visited everybody whom he wanted to support his goals. He was confident that he would be able to convince those who resisted. When he sensed my scepticism he showed up at my place and attacked me with arguments for hours. He actually managed to get his proposal [for socialisation] considered by the Council of Ministers. He was invited to join the meeting to present his case. Prime Minister Hoffmann was a reserved and thoughtful man who did not want to get involved in adventures. Neurath's forceful personality made him uneasy. It was amazing how Neurath tyrannized the whole cabinet. He fought for every part of the [socialisation] bill, resisted

---

[168] Beyer 1982, p. 174.   [169] Eisner 1919, pp. 24–5.   [170] Neurath 1920b [1973], p. 20.
[171] Mitchell 1965, pp. 292–3.

every change, issued several ultimata, threatened to leave abruptly and so intimidated the ministers one by one.

He was just as tough when it came to the salaries. He made rather high demands; if I remember correctly he demanded an annual salary of roughly 24,000 marks for himself, 22,000 for Kranold and 18,000 for Schumann, sums exceeding the ministers' salaries. It was understandable that the cabinet hesitated to accept these proposals, but they forgot to reckon with Neurath's quick wit, argumentative skills and his fanaticism. He begged and pleaded, argued his points and carried on so incessantly and for so long that he finally got his way in every respect. The cabinet approved his proposal and Neurath could start organizing his office of economic planning. He did so with the same drive that he had exhibited during these preparatory stages.[172]

Josef Simon, the Minister of Trade, was clearly on Neurath's side. Together they immediately organised a wide leafleting campaign to inform the populace about the programme of total socialisation.[173] Neurath then tried to convince Sebastian Schlittenbauer, leader of the conservative Bavarian People's Party (Volkspartei). Neurath needed his support in the agricultural sector, since he felt it would be impossible to socialise Bavaria without the support of the leader of the farmers. Neurath planned to organise all farmers into co-operatives. These in turn were to be united in a central co-operative, which together with the Central Economic Administration and the people's deputies would decide what was to be cultivated, how the produce was to be distributed, and so forth. For the time being no expropriations were planned. 'The programme aimed to decouple, as much as possible, the socialisation of agriculture from that of industry to avoid unpleasant confrontations between the industrial and the agricultural population.'[174]

After another meeting with the Socialisation Commission on 25 March Neurath was appointed president of the Central Economic Administration (sometimes also called the 'Central Planning Office'[175]) and offered a six-year contract (Fig. 1.3).[176] (He reckoned at the time that it would take 'five to ten years' to implement total socialisation.[177])

Neurath envisaged the following personnel for the Central Economic Administration: head of the Centre for Organisation: Heinz Umrath (SPD, commercial director of the AEG in Chemnitz); head of the Centre for Calculation in Kind: Professor Karl Ballod (USPD, author of a utopian scheme for a 'social state' published under the pseudonym

---

[172] Niekisch 1958 [1974], p. 54.   [173] Neurath 1920b, p. 11.   [174] Ibid., p. 15.
[175] As in Neurath 1919d.
[176] Record of the questioning of Dr Georg Schmidt on 4 June 1919 at the Standgericht. State Archives Munich.   [177] Neurath 1920b [1973], p. 23.

# Das Zentralwirtschaftsamt.

Vor zwei Wochen lag der Plan, ein Zentralwirtschaftsamt für die Sozialisierung Bayerns zu gründen, noch in der Ferne. Heute ist es gegründet. Was hat es zu tun?

Das Zentralwirtschaftsamt, welches dem Minister für Handel, Gewerbe und Industrie untersteht, hat die Durchführung der **Sozialisierung auf allen Gebieten** vorzubereiten.

Das Zentralwirtschaftsamt hat sofort für die planmässige Erzeugung von Wohnung, Nahrung und Kleidung des gesamten Volkes mit allem Nachdruck Sorge zu tragen.

Dem Zentralwirtschaftsamt steht zur Seite der **Arbeiter- u. Bauern-Kontrollrat**, um gemeinsam mit den Arbeiter- und Bauernräten, Gewerkschaften und Genossenschaften die Durchführung der Sozialisierung in vollem Umfang zu **kontrollieren**.

Das Zentralwirtschaftsamt wird im Einvernehmen mit dem Ministerium über die gesamte Wirtschaft, über die Rohstoff- und Geldbewegung sowie über alle Betriebe volle Klarheit schaffen und verbreiten.

Das ganze Volk, **körperliche u. geistige Arbeiter, Handwerker, freie Klein- und Mittelbauern**, werden mitwirken und das grosse Werk kontrollieren.

**Simon**
Minister für Handel, Gewerbe und Industrie.

Zentrale für Aufklärung und Volksbildung in Bayern.

Druck von A. Waldbaur (vorm Schmidtmann), München.

Fig. 1.3 Handbill announcing the establishment of the Central Economic Administration, Munich 1919 (Monacensia-Bibliothek, Munich)

'The Central Economic Administration.

Two weeks ago, the plan to establish a Central Economic Administration for the socialisation of Bavaria pointed far into the future. Today this Administration has been founded. What are its tasks?

The Central Economic Administration reports to the Minister of Commerce, Trade and Industry and is to prepare the socialisation in all fields.

The Central Economic Administration is charged to press with all due means for the planned provision of housing, food and clothing for the entire population.

The Central Economic Administration is assisted by the Workers' and Farmers' Control Council, which, together with the workers' and farmers' councils, the trade unions and the production associations, will exercise the full control of the implementation of socialisation.

The Central Economic Administration will, in cooperation with the ministry, make fully transparent and publically known all economic facts, e.g., the movements of raw materials and money and the position of all firms.

The entire people blue-collar as well as intellectual workers, tradesmen, free small and middle-size farmers – will participate in and control the great project.

Simon
Minister of Commerce, Trade and Industry

[Published by the] Centre for Information and People's Education in Bavaria

'Atlanticus').[178] The Control Centre was meant to be split, with Kranold as head of Production Control and Rudolf Hilferding, who declined, as head of Monetary Control. Schumann was to head a Centre for Information (Aufklärungszentrale).

On 5 April a Munich newspaper published this official government announcement concerning the Central Economic Administration:

> The Central Economic Administration that is subordinated to the Ministry of Trade, Business and Industry has the task of preparing the implementation of socialisation in all fields. As of now the Central Economic Administration is entrusted with planning and securing the supply of housing, food and clothing for all. The Central Economic Administration will be supported by the Workers' and Farmers' Control Council and representatives of Workers' and Farmers' Councils, unions and co-operatives so as to ensure and control the implementation of total socialisation. In co-operation with the Ministry of Trade, Business and Industry, the Central Economic Administration will gather information about the entire economy, i.e. the movements of raw materials and money as well as all business transactions, and disseminate it.[179]

### 1.3.2 In office

The first major action taken by the Central Economic Administration was the preparation of a bill for the socialisation of the daily newspapers. Red Marut, who later acquired literary fame under the pseudonym B. Traven, and Gustav Landauer, the prolific writer, critic and anarchist visionary, were appointed members of a committee that also consisted of representatives of the press (journalists, editors, publishers, typesetters, printers and designers). This committee was to 'produce a draft bill for rendering the contents of newspapers as independent from their sources of revenue as possible, without thereby restricting free speech'.[180]

The committee's proposal justified the socialisation of the press as follows: 'It is not just the capitalist and exploitative conception of some [press] enterprises that is offensive, but rather primarily the fact that many members of the public are *forced* to read a newspaper that is politically, spiritually and intellectually alien and disgusting to them.'[181] This attempt to compensate for the advantage of well-funded newspapers

---

[178] Atlanticus 1898; cf. Rühle 1971, p. 73. In this book Ballod-Atlanticus presented an overview of the things that he believed could in 1897 be achieved in a socialist state with available methods of science and technology.    [179] *Münchner Neueste Nachrichten*, Saturday/Sunday, 5/6 April 1919.
[180] Neurath 1920b, [1973] p. 19.
[181] *Gesetz über die Sozialisierung der Tagespresse* (law on the socialisation of the daily press) p. 1. In State Archives Munich.

was not meant to limit free speech (Neurath was opposed to censorship).[182] Of course, the creation of area associations for daily newspapers was a first measure towards socialisation. These associations were to take over the administration and control of the 'economic, technical and managerial side' of the newspapers and a newspaper council, subject to the Central Economic Administration, was to control the socialisation of the press until an autonomous organisation assumed office. With its demand for editorial staff councils and other measures intended to strengthen democracy within news organsiations, this bill deserves to be called progressive even today. (Neurath had not anticipated the socialisation of the press in his theoretical works, but it was integrated into his conception without problems.)

The tense and unsettled political situation during the Hoffman government resulted in runs on banks and capital flight to foreign countries, events which threatened to destroy the Bavarian economy. As head of the Central Economic Administration, Neurath felt forced to counteract this process with the socialisation of the monetary system. The political changes seemed congenial to his plan. Members of both the SPD and the USPD, discouraged by the slow rate of progress, demanded the declaration of the Bavarian Soviet Republic (Räterepublik) in early April. On the evening of 4 April Ernst Schneppenhorst (SPD), minister for military affairs, called representatives of the SPD, the USPD, the Bauernbund, the anarchists and members of the Central Council for a meeting at which he strongly urged the declaration of a Soviet Republic as a 'means to unify the proletariat'.[183] All parties present agreed – including the pacifist poet and playwright Ernst Toller (USPD), the anarchist writers Gustav Landauer and Erich Mühsam, Joseph Simon (USPD) – except the representatives of the Central Council, notably Niekisch, who demanded that the KPD be consulted as well. In a second session in the early hours of 5 April, now also attended by Neurath, Leviné (KPD) strongly opposed the idea. Nevertheless, the parties who had previously agreed to the proposal insisted on declaring the Soviet Republic on Monday 7 April. (Schneppenhorst had argued for the delay in order to secure the agreement of the other Bavarian cities first; he left the morning after the meeting only to switch sides.) When the Soviet Republic was declared that Monday the Hoffman government withdrew to regroup their forces in Bamberg (some 150 miles to the north).

These political changes intensified the problems in the banking sector.

---

[182] Neurath 1920b [1973], p. 25.   [183] Quoted in H. Beyer 1982, p. 70.

Neurath reported: 'It was obvious that there would not have been any deposits left in Munich after a few days. So the central council closed all the banks.'[184] The decree that Neurath had 'prepared with some bank directors was now issued by the central committee without hesitation'.[185] Cash withdrawals were limited, only giro checks honoured without restrictions and payments to foreign territories were not allowed to exceed 5,000 marks.

In the meantime a Bank Council that had been formed by younger bank employees declared that it would take control. They found mass withdrawals still continuing and sensed a danger to industry payrolls. The Bank Council proposed limiting cash withdrawals to 700 Marks a week and handing out money needed for salaries, interest rates, equipment, etc., only on the condition that the respective factory councils agreed to countersign. Within the Soviet Republic payments with giro checks were to be allowed without restriction.[186]

Neurath supported this proposal because it was in the spirit of the socialisation measures. His banking decree caused deep indignation in financial circles. The Munich Chamber of Commerce even published a brochure castigating the 'destruction of Munich's economic life by the communist economy'. Economic life was certainly disturbed. Rationing of raw materials, coal and food was considered, and Neurath even thought about pawning the Bavarian state forests in order to counteract the food shortage.[187]

Even though this first Bavarian Soviet Republic lasted no longer than six days, Neurath managed to set in motion many parts of his socialisation programme. That the 'Revolutionary Central Council' supported him is shown by two credits for the Central Economic Administration (20,000 marks on 8 April and 500,000 on 11 April 1919) and a general authorisation to prepare total socialisation.[188] In addition to the measures already discussed, during the time of the first Soviet Republic the Central Economic Administration also announced the formation of Expert Councils and a Workers' Control Council and began the socialisation of the mines.

On 13 April forces of the Hoffmann government tried to regain authority. Members of the soviet government were arrested and transported to Bamberg. Neurath was arrested as well. He gave the following

---

[184] Neurath 1920b, pp. 22 f.   [185] Ibid., p. 23.   [186] Ibid.
[187] See the report of Neurath's questioning on 20 June 1919 at the Standgericht. State Archives Munich.
[188] See the indictment of the public prosecutor against Neurath of 25 June 1919. State Archives Munich.

report: 'During the struggles to restore the Hoffmann government, I was arrested in the Augustinerbräu by officials of the police headquarters and taken to the train station. Chief Constable Staime and others were present there. But they released me again due to the nature of my position, which they obviously considered unproblematic.'[189]

The first Soviet Republic had only a short life. Niekisch had resigned on 8 April and Ernst Toller succeeded him as the chairman of the Revolutionary Central Council. In the following six days there was chaos and confusion. The KPD felt compelled to save the situation by replacing the 'pretend' Soviet Republic with a 'real' one under the leadership of Leviné and Max Levien. On 13 April, with Hoffmann seeking to retake power from Bamberg, Leviné was elected as the chief executive of the second Soviet Republic.[190]

In Neurath's eyes the leaders of this second Soviet Republic 'applied Russian experiences to the German situation'.[191] He reported their first actions in these words:

The forcible dismissal of former officials and organisations was believed to be necessary. The leaders were of the opinion that the existing apparatus had to be destroyed as far as possible. Accordingly they immediately dismissed all policemen and officials and replaced them with Communists. They hired not only non-expert yet honest people, but also criminals who took advantage of the opportunity. In addition the second republic reanimated the tradition of 1789. The church bells rang alarm during the nights, often for several hours, and shots were fired without purpose – this was all intended for mass suggestion. The factory councils, which consisted of several thousand people, were assembled and their determination was appealed to! Under the influence of inciting speeches they were asked to make big decisions quickly, to elect leaders ... The result was a maximum of power for the leaders and a minimum of power for those who were being led, under the illusion of direct popular rule.[192]

Neurath remained head of the Central Economic Administration during the second Soviet Republic due to support from the workers – against the wishes of the Communist leadership.[193]

The Soviet Republic faced insurmountable difficulties. The economy, which had already suffered, was further damaged by a blockade imposed by Hoffmann's government and lack of co-operation from the farmers. In view of his own limited resources Hoffmann enlisted the help of the

---

[189] Report of Neurath's questioning at the Standgericht on 16 May 1919. State Archives Munich.
[190] Mitchell 1965.   [191] Neurath 1920b [1973], p. 27.   [192] Ibid., p. 26, translation altered.
[193] Neurath 1920b [1973], p. 27.

## In the Bavarian revolution

Free Corps – various well-equipped right-wing militias that had previously been employed by the Berlin government to suppress the Spartacist uprising. A Red Army was quickly deployed by the second Soviet Republic for the defence of Munich. Clashes with the advancing government troops began on 16 April and the fighting continued until after 2 May. That day Munich was conquered by the government troops. The ruthless suppression of the Soviet Republic resulted in more than 500 dead, 300 wounded and numerous summary executions.

### 1.3.3 On trial

Neurath was re-arrested on 16 May and accused of high treason. He was tried under the Bavarian martial law at the Standgericht Munich, a mixed court of civilians and military personnel. The indictment reads as follows: 'Dr. Otto Neurath is under sufficient suspicion of having consciously supported in word and deed the attempt to change by violent means the constitution of the federal state of Bavaria.'[194] Neurath defended himself first by pointing out that he had, in his own words, 'only decided to participate in active life after lengthy arguments with my friends and their hints at the cowardice that would reside in not doing the right thing, in not doing what one was able to do in times as serious as these'.[195] Second, Neurath maintained that his only goal throughout had been to 'implement socialism as an economic system' and that he never intended to have any active role in politics.[196] 'I had planned to avoid any involvement in politics and only desired to continue the implementation of my socialisation plan under the soviet government as it had been accepted in principle by the [previous] Ministry.'[197]

Numerous witnesses confirmed Neurath's lack of interest in politics. Niekisch testified: 'Dr. Neurath did not take any political action during the whole process.'[198] Niekisch also noted, rightly, that the Central Economic Administration was a lawful institution pre-dating the Soviet Republic, and that it was unjust to accuse Neurath of participation in treason for having continued his work as a civil servant.[199] Neurath's

---

[194] Indictment of the public prosecutor against Otto Neurath of 25 June 1919. State Archives Munich.
[195] Handwritten remarks of Neurath on the report of 30 May 1919. State Archives Munich.
[196] Ibid.
[197] The questioning of Neurath on 16 May 1919 at the Standgericht. State Archives Munich.
[198] Report of the questioning of Ernst Niekisch on Neurath's case, Eichstätt, 21 May 1919. State Archives Munich.  [199] Niekisch 1958 [1974], p. 55.

deputy in the Central Economic Administration, Hermann Kranold, further emphasised Neurath's statement about his attitude towards the different administrations. He testified that Neurath ordered him to 'remove the word "Soviet Republic" from all official stamps'.[200] Supposedly, Neurath did not appreciate the need 'to advertise the respective form of government on every letter'.[201] Rather he used different letterheads depending on the situation, one of them reading 'Central Economic Administration of the Federal State Bavaria', the other 'Central Economic Administration of the Bavarian Soviet Republic'.[202]

Neurath's lawyers secured his release on bail on 27 June 1919. The list of witnesses for the defence includes Professor Karl Ballod (Berlin), Dr Harnisch (Minister of Justice in Saxony), Dr Olga Neurath, Professor Max Weber, Dr Otto Bauer and others. Max Weber's statement to the court has not been preserved.[203] According to a newspaper report, Weber described Neurath as an extraordinarily capable scholar whose more recent work, however, lacked a sure grasp of reality: he was 'too easily carried away by a utopian scheme'.[204] Later that year he wrote to Neurath that he considered his schemes for planned economies an 'amateurish, objectively absolutely irresponsible foolishness that could discredit "socialism" for a hundred years and will tear everything that could be created now into the abyss of a stupid reaction . . . I fear you are contributing to this danger, which you greatly underestimate.'[205]

Otto Bauer sent a written report about Neurath. Bauer had known Neurath's work on war economies since they had worked together in the Austrian war ministry. He said about this work: 'At that time I already thought that Neurath's purely technical and therefore intrinsically unpolitical way of thinking constituted the most curious aspect of his personality.'[206] According to Bauer Neurath was not a politician but a social technician or engineer. 'He does not concern himself with the constellations of political power that are preconditions for the realisation of socialism, but only with the technical and organisational means for its

---

[200] Questioning of Kranold at the Standgericht, Munich, on 16 June 1919. State Archives Munich.
[201] Ibid.
[202] See the questioning of Neurath at the Standgericht on 20 June 1919. State Archives Munich.
[203] Max Weber's statement is not attached to record No. I.2139 of the State Archives Munich. He only appears on the list of witnesses. The report on Weber's statement from *Münchner Neueste Nachrichten* 24 July 1919 (evening edition) is reprinted in Weber 1988, p. 495.   [204] Ibid.
[205] Weber 1921, p. 488. Weber later complained in similar terms about Lukács and his involvement in the Hungarian Soviet Republic (see Baumgarten 1964, p. 533).
[206] Expert report attached to the letter of Otto Bauer to Neurath's defence lawyer of 30 June 1919, p. 3. State Archives Munich.

implementation.'²⁰⁷ Bauer did not see any difference between Neurath's teachings on war economics and his model for a fully planned economy and came to the following conclusion: 'At bottom [Neurath's] socialism is authoritarian. He recommends enforcing from the top down a planned order and a transformation of economic life by a government over and above society. He does not care whether this is the Austro-Hungarian army command, a democratic parliamentary government or a council dictatorship.'²⁰⁸ Neurath would accordingly be prepared 'to serve under any government if only he gets the chance to organize his "demand-satisfaction-economy".'²⁰⁹ In his report Bauer pointed to a weak spot of Neurath's conception, writing: 'A Marxist can accuse him of not understanding that all social reform is determined by the constellation of political powers under which it takes place.'²¹⁰

As we saw above (section 1.2.5), however, Neurath intentionally evaded questions concerning political practice and equally intentionally planned his socialisation model in a purely socio-technological manner. Moreover, Neurath's lack of interest in possibly correct but also very abstract considerations that try to anticipate within theory itself all conditions for future action seems to have been more fruitful than Bauer's approach, which embodied a far more detailed political analysis. In seven days under Neurath's guidance Bavaria implemented more socialisation measures than Austria managed under the leadership of Otto Bauer, despite the fact that Bauer had more time.²¹¹

Neurath was convicted on 25 July of assisting high treason and sentenced to one year and six months incarceration in a fortress (*Festungshaft*).²¹² But he did not serve his full sentence, for he was deported to Austria on an intervention by Otto Bauer, then Foreign Secretary of Austria.²¹³ Neurath's sentence was a relatively light one compared to his fellow revolutionaries.²¹⁴ Leviné received the death penalty; Mühsam received fifteen years, Toller five years, and even Niekisch two years' imprisonment in a fortress; Landauer was murdered in prison. By contrast the death-sentence for Graf Arco-Valley, the murderer of Eisner,

---

²⁰⁷ Ibid., p. 4.   ²⁰⁸ Ibid., p. 7.   ²⁰⁹ Ibid.   ²¹⁰ Ibid.   ²¹¹ See note 88 above.
²¹² This form of incarceration is distinguished from *Gefängnis* [prison] and the still more severe *Zuchthaus* by certain privileges granted to the prisoners. For some remarks on the differential granting of these privileges to revolutionary and counter-revolutionary prisoners after May 1919 see Niekisch 1958 [1974].
²¹³ Conversation with Grete Schütte-Lihotzy. See also Schütte-Lihotzky 1982, P. Neurath 1994, Dahms and Neumann 1994.
²¹⁴ See Neurath 1920b [1973], p. 27 and the list from E. Gumbel, *Zwei Jahre Mord*, Berlin, 1921, reprinted in Stadler 1982a, p. 233.

passed in January 1920, was immediately commuted to life imprisonment; he was released in 1924, still before the general amnesty intended to free Hitler in 1925, when Mühsam was also set free. The surviving KPD leaders escaped or were deported to Russia where they disappeared in the purges of the 1930s; Mühsam was rearrested by the Gestapo in 1933 and murdered in Oranienburg concentration camp; Toller committed suicide in New York after the fall of Madrid to Franco in 1939. Graf Arco-Valley meanwhile became one of the first directors of the pre-war Lufthansa.[215]

## 1.4 IN RED VIENNA

Otto Neurath returned to a changed Vienna. The First World War not only resulted in the dissolution of the monarchy but in the breakup of the multinational state into numerous individual nation-states. The remains, often referred to as *Deutsch-Österreich* (German-Austria), reconstituted itself as a republic. During this 'Austrian revolution' the labour movement obtained powerful positions. At first the demoralised bourgeoisie did not manage to oppose the Social Democratic offensive. Even though their opponents soon regained their strength, the Social Democratic party was able to maintain some power to shape society during the entire period of the first republic. Vienna gives the most striking example. In what has come to be called 'Red Vienna,' far reaching reforms in the social, health care, education and teaching sectors were achieved.

This section describes some of the ideological foundations of Austrian socialism from 1919 to 1934. During this period the Austrian Social Democrats embarked on a socio-cultural programme to turn Vienna into a model of the future 'socialised humanity'. At the same time Neurath expanded and began to publish his views on unity of science; these efforts were to culminate in his founding of the Unity of Science movement. We want to show how Neurath's ideals of social change fitted into the dominant political milieu of Red Vienna and how his philosophical ideas continued to develop in tandem with his political activities.

### 1.4.1 People's education

Austria, a republic since November 1918, became a federation with Vienna as one of its independent states. In the municipal elections of

---

[215] Herz and Halfbrodt 1988, p. 322.

## In Red Vienna

May 1919 the Social Democrats gained sole control of the Viennese government. These circumstances made Red Vienna possible. The Austrian Social Democratic Party was unique among European socialist parties, not only in the amount of power it held but also in its policies, which were based on Max Adler's dictum that the future of democracy lay not in politics but in pedagogy. This emphasis on education (*Bildung*) came to define the widely admired programme of *Bildungspolitik*, designed to reform the whole of working-class life through policies affecting everything from housing to higher education.

An overview of the unique achievement of *Bildungspolitik* is given by A. Rabinbach in his study of Austrian socialism:

> The designers of the new Vienna understood the concept of *Bildung* in the broadest sense, from the sweeping educational reforms of Otto Glöckel to the creation of a 'new socialist individual' . . . by the end of 1933 the city had built more than 61,175 new apartments, mostly in the form of large housing blocks, the *Wiener Höfe*, with parks, swimming areas, schools, kindergartens, gymnasia, health facilities and community centers. . . . Moreover, the municipality had introduced an extensive program of adult education that supplemented the party libraries, bookstores, and *Bildungs*-commissions that were set up in all Vienna districts. But most important, the construction of a socialist Vienna was oriented towards the realization of the ideal of a 'socialised humanity', not at the level of economic expropriation but at the level of cultural appropriation.[216]

The construction of a new order of life was undertaken with an optimism that is reflected in a remark by Neurath, recalled by one of his students at that time: 'What will we do when all the world's energy will come from such a small centre?'[217]

The emphasis on education in general and science in particular had long been common in socialism. Since Ferdinand Lassalle had proclaimed the 'alliance between science and the worker,' the task of the 'enlightenment of the worker' had occupied a prominent place among the activities of the socialist parties. But only the Austrian labour movement adopted the slogan 'education is power'. Four different factors can be identified that led to the adoption of *Bildungspolitik* as the central political strategy in the construction of Red Vienna.

(1) The failure of Bauer's socialisation plan. In January 1919 Bauer presented his plan for the socialisation of the economy of the new republic. In March the central government created the National Socialisation Commission, with Bauer presiding. The plan called for gradual socialisa-

---

[216] Rabinbach 1983, p. 28.   [217] Zeisel 1985, p. 123.

tion without nationalisation, starting with key industries such as forestry, coal mining, iron and steel production, and armaments. The work of the commission was prompted by the developments in Hungary and Bavaria, since it was feared that the Communists might attempt a Bolshevik revolution with the support of the increasingly radical workers.[218] It was especially important to socialise the factories manufacturing army supplies, which were critical for maintaining control in case of an uprising. But the pace of socialisation had to be slowed down because of the desperate condition of the Austrian economy. By the summer the collapse of both the Bavarian and Hungarian revolutions put a virtual end to the attempts at socialisation in Austria.[219] Deprived of the possibility of implementing radical economic reforms, the Social Democrats focused on social and cultural policies.

(2) The constitution of the Party. The Austrian Social Democratic Party, founded in 1889, had grown out of a number of cultural societies (*Bildungsvereine*) that were formed in Austria during the period of political liberalisation in the 1860s. These societies stressed the virtue of self-improvement and workers' education. Education was promoted as a mechanism for enacting social equality and as a tool in the struggle against the obscurantist absolutism of the church and the aristocracy. This cultural function of the party was preserved and explicitly acknowledged in its constitutional platform at its founding congress in Hainfeld:

The Austrian Social Democratic Party, working for the whole people without distinction of nation, race, or sex, strives to free people from intellectual atrophy ... to organise the proletariat politically, and to fill them with a consciousness of their position and their task, to make them ready physically and mentally for battle and to maintain this readiness.[220]

(3) Education for unification. An underlying motive of Austro-Marxism, particularly the doctrines of Max Adler, was to bring together the Marxist tradition with German culture, especially classical German philosophy.[221] For Bauer the promotion of traditional German culture served a clear political function: it provided the only authentic bond among the diverse nationalities that composed Austria-Hungary and Germany. In his major work of 1907 on the question of nationalities Bauer argued that if the empire could not become a nation-state it could nevertheless become a community of education: 'People tried to

---

[218] O. Bauer, 1923, chapter 4.  [219] See Gulick 1948, vol. I, part 2.
[220] *Verhandlungen des Parteitages der Österreichischen Sozialdemokratie in Hainfeld* (1889), quoted in Blum 1985, p. 108.  [221] See Bottomore and Goode 1978.

discover the nation in our present class society . . . while the growth of the new community of education has not yet been able to unite these small groups into a national whole.'²²² As Rabinbach points out in a discussion of Bauer's political programme: 'Bauer's attitude was completely consistent with [Victor] Adler's view that the party's role was to "keep the proletariat alive, to enlighten it, and to bring it forward, to educate it" – a civilising mission as much indebted to the traditions of classical German philosophy as to Marxism.'²²³ In the same year Bauer began to publish in the newly founded journal *Der Kampf* a number of articles touching on different aspects of education of the proletariat.²²⁴

(4) Max Adler's revisionism. Austro-Marxist revisionism opposed orthodox and 'vulgar' Marxism, which taught that economic factors were the sole determinants of psychological and intellectual developments, thus asserting an absolute determinism unalterable by personal efforts. One of the intellectual leaders of the Austrian Social Democrats was Max Adler. With his neo-Kantian idealism he advanced an anti-reductionist interpretation of Marx and Engels that assigned an active role for the mind and for ideological and cultural factors: 'The superstructure is just as real as the base because both are parts of a real building' and '[e]conomy and ideology are certainly different, but at the same time they are parts of a single social-cultural system of human life.'²²⁵ Adler stressed the political importance of improving the cultural life of the proletariat: the labour movement 'can achieve its own class interests after the abolition of enslavement of labour only if general cultural interests triumph'.²²⁶ Notably Adler, like Neurath, regarded science as a crucial component of the new culture, as the means for understanding situations and taking action to change them. He wrote:

The proletariat, in contradistinction to [other political groups], acts through clear, scientific knowledge of the existing situation . . . . The procedure of the socialist party must be determined by scientific knowledge and from such a basis Social Democracy – for it is a scientific socialism – stands over all parties, as is always the case with a science.²²⁷

Neurath's work during this period clearly reflects the social and cultural atmosphere of Red Vienna. As he put it in its final years: 'How to organise human life socially – that is the great question which people are asking today with ever greater insistence.'²²⁸ Neurath with his long-

---

²²² O. Bauer 1907a.   ²²³ Rabinbach 1983, p. 16.   ²²⁴ E.g. O. Bauer, 1907b, 1907c, 1910.
²²⁵ M. Adler 1930–2, vol. 1, p. 73.   ²²⁶ M. Adler 1910a, p. 52.
²²⁷ Quoted in Blum 1985, p. 101.   ²²⁸ Neurath 1933d [1973], p. 220.

standing interest in practical reforms must have considered Vienna as a challenge, especially given his defeat in Munich (which he did not consider to be a defeat of his ideas). Before he began work in Vienna Neurath accepted an offer from the Central Committee of the German Trade Union Association in Czechoslovakia to head its first training institute for teachers of factory councillors.[229] This task may be considered a continuation of his Munich plans for the organisation of factory councils.

### 1.4.2 The Housing Movement

In Vienna Neurath established contact with the leaders of the Austrian socialisation movement. He soon became General Secretary of the Research Institute for Social Economy (Forschungsinstitut für Gemeinwirtschaft), founded in 1920. This institute was financed by a society of the same name, whose aim was to lend theoretical and practical support to the social economy. 'Among others the society has the following tasks: to create and support research positions; build up a source collection, archives, libraries and study halls; to establish an information office and centre where reports of experiments in social economy can be exchanged and international contacts be fostered; and to organise a news service.'[230] The attempts of the Research Institute to support the Austrian co-operative housing movement (Siedlerbewegung) 'in the spirit of social economy'[231] had a special practical relevance.

Soon Neurath was dedicating his work entirely to the co-operative housing movement. He played a decisive role in organising the Austrian Co-operative Housing and Allotment Association in 1921, of which he became general secretary.[232] The Association was to serve as an umbrella organisation for all other housing co-operatives and societies. It provided Neurath with ample opportunity to formulate and propagate his general critique of the capitalist way of life and his ideas for improving the standard of living in the concrete case of housing. He objected that houses were not built for people's desires and needs, but solely for profit; houses were usually built in competition between several people instead of proceeding co-operatively. Neurath favoured simple terraced houses with

---

[229] Neurath 1920d, p. 15.   [230] Neurath 1921d, p. 280.   [231] Ibid., p. 281.
[232] See Feldbauer and Hösl 1978, for information on the housing shortage and the beginnings of the housing co-operative movement.

connecting gardens and a communal house in the middle to serve as centre and meeting point. He envisaged the end of private kitchens and planned 'mini-kitchens' or kitchens-cum-living-rooms as a transitional stage. He also proposed communal raising of children, youth care and communal cultural and educational facilities. He urged solidarity in the fight against the poverty and bad housing that encouraged diseases and proposed specific measures against tuberculosis and alcoholism. These ideas were an essential part of the Social Democratic reform movement of the time; as such they resulted in numerous sub-organisations of the party, including the Workers' Samaritans (Arbeitersamariterbund), the Workers' Temperance Association (Arbeiterabstinenzlerbund).

The co-operative housing movement was also a self-help organisation. Many of its members came from the allotment movement. During the First World War this movement made an important contribution to the alleviation of food shortages by turning fallow land around Vienna into temporary vegetable gardens. After the war they proposed the building of a belt of small detached inexpensive family houses with kitchen gardens or allotments around Vienna. In response to a demonstration in front of the town hall, the mayor promised financial support for the movement. The housing bill of 15 April 1921 decreed that co-operatives had to pay only 10 per cent of the construction expenses – in labour or money – the rest was to be paid by the city of Vienna.[233] In contrast apartments in multi-storey communal buildings were completely financed by the city of Vienna. The houses of the co-operative did not become private property but the 'settlers' received the right of hereditary tenancy.[234]

The Co-operative Housing and Allotment Association was responsible for the planning and construction of houses. It provided building materials, supplied information on interior furnishing and helped with the furniture as well, if required. The Association also organised lectures on, for instance, gardening, small-animal husbandry, beekeeping, house construction and hygiene. The Association was administered under the aegis of the municipal housing and allotment department. Adolf Loos functioned both as architect responsible for the housing movement and as advising architect for the city's housing administration.

Undoubtedly one of Neurath's major achievements was to get distinguished modern architects like Loos, Josef Frank (strongly committed to the system of terraced housing and equally strongly opposed to building

[233] See Hoffmann 1978.  [234] Neurath 1922b, p. 87.

large blocks of flats), Josef Hoffmann, Oskar Strand, Grete Schütte-Lihotzky and others to work with the housing movement. Neurath tried to obtain commissions for them from the city of Vienna. The houses were planned as extendable units. Because of financial limitations, they were very small (40–60 square metres), and it was thus proposed that the tiny apartments should open on to the garden. The interior design was also planned to be easily upgradeable. Neurath insisted on reducing construction costs by standardising building materials (a measure earlier proposed in his studies on war economies). The first model village was built at the Heuberg in Vienna.[235] One could also count the Werkbund village as a continuation of the efforts of the housing movement even though it was built in 1930–3 after the movement had already been dissolved. In addition to the architects named above, Oskar Wlach, Heinrich Kulka and Karl A. Bieber, to name but a few, participated in building this village consisting of small houses that were easy to reproduce elsewhere as terraced buildings.

'The happiness of the inhabitants has to be the measure for housing politics,' wrote Neurath, indicating his approval of the city's housing policies.[236] The projects mentioned so far accounted for only a small part of the famous housing policy of the city of Vienna in this period, most of which consisted of building blocks of flats or small apartments. These estates included communal facilities, such as advisory services for mothers, playgrounds, sports facilities, laundries, swimming pools, study and lecture halls and commercial facilities like theatres that could also be used by the community at large. In 1923 the city of Vienna began to support the construction of such multi-storey apartment blocks.

In 1923 the Housing and Allotment Association organised an open-air exhibition in the centre of Vienna meant to inform the population about its activities. Neurath described the following scene: 'Real, completely furnished houses were built on the square in front of the town hall. A selection of vegetables was exhibited and large charts displayed the achievements of the Association.'[237] Later, with the support of the Vienna city administration, this exhibition was enlarged and became permanent, evolving into the Museum for Housing and Urban Development in Vienna, located at Parkring 12.[238] The incumbent mayor of Vienna became president of the museum. In this period

---

[235] Conversation with Grete Schütte-Lihotzky.   [236] Neurath 1931e, p. 110.
[237] Neurath 'Visual Autobiography', quoted in Kinross 1979, p. 17.
[238] See the *Arbeiterzeitung* of 11 March 1925.

*In Red Vienna* 63

Neurath also took up lecturing again, dealing not only with housing and economic problems, notably socialisation, but also with women's rights.[239]

### 1.4.3 The Museum of Economy and Society

In 1924, when financial problems began to signal the end of the housing co-operative movement,[240] Neurath concentrated his activities on the museum. He proposed to Hugo Breitner, the city councillor for finances, that the housing museum be transformed into a social and economic museum. In October 1924 Neurath went all over Vienna by taxi collecting signatures for a society that was to run the Museum of Economy and Society in Vienna, first from mayor Karl Seitz, then from Dr Julius Deutsch and others.[241] Before the end of the year the city council granted a subsidy of 20,000 schillings to this society with Neurath as its head.[242] The museum consisted of three departments, called 'Labour and Organisation', 'Culture and Life', and 'Housing and Urban Development'. It began operations on 1 January 1925. In the beginning the museum rented an office from the housing association where Neurath was still employed. Josef Jodlbauer and Marie Reidemeister were the first permanent staff. (Marie Reidemeister had first heard of Otto Neurath and his plans for the museum in 1924 from her brother, the mathematician Kurt Reidemeister. She had then already decided to join Neurath's project, but she first finished her studies of mathematics and physics in Göttingen.) In the beginning other employees were hired on an hourly basis only, for example, the architect Rosl Weiser. The museum at the Parkring was remodelled and reopened as the Museum of Economy and Society.

Neurath conceived of the museum as an institution for public education and social information. Its primary task was the development of methods to inform the public about the results of sociological and economical research. Since 'highly developed industry and modern administration require a certain minimum of education of all citizens',[243] a method was needed to put basic social knowledge across to the people. Such a method would also help to enforce hygienic and social measures

---
[239] See the announcements for his talks in *Arbeiterwillen*, e.g., on 16 January 1923, 14 August 1923, 7 November 1923, 3 April 1924, 5 September 1924, 10 October 1924 and 14 December 1924.
[240] Conversation with Grete Schütte-Lihotzky.
[241] Ibid. Grete Schütte-Lihotzky accompanied Neurath in the taxi.
[242] See *Arbeiterzeitung* of 11 March 1924.   [243] Neurath 1933b, p. 25.

## 64     *A life between science and politics*

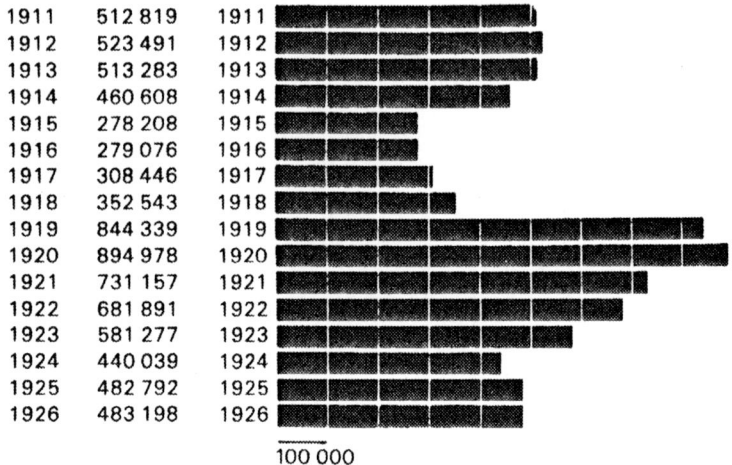

Fig. 1.4 Marriages in Germany per year (Neurath 1936b [1991], p. 382)

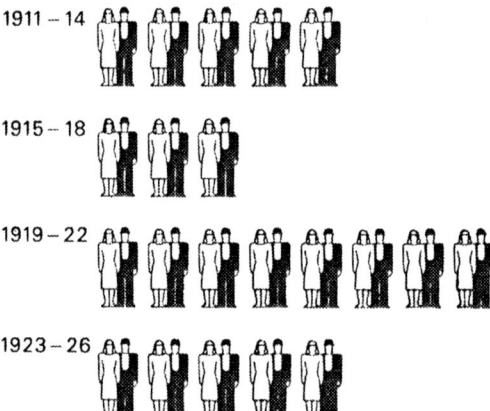

Fig. 1.5 Marriages in Germany per year. Each symbol represents 100,000 marriages (Neurath 1936b [1991], p. 383)

and spread other information. Neurath developed his method of visual education – picture statistics – for this reason, hoping to ensure that even 'passers-by . . . can acquaint themselves with the latest sociological and economical facts at a glance'.[244] Picture statistics were meant to transmit their contents to children as well as to adults, irrespective of language barriers or educational limitations. Neurath's motto was: 'Words divide, Pictures unite.'[245] Over the years Neurath and his employees continued to develop the idea of visual education. The result came to be known internationally as 'the Vienna method of picture statistics'.

The Museum of Economy and Society prepared its first graphical displays for a Viennese hygiene exhibition in May and June 1925. Soon display boards were also ordered by the city of Vienna, unions and the Viennese Council for Workmen and White Collar Workers which, together with the city and the social security insurance funds, paid for most of the running costs of the museum. The permanent showcases and models built for the hygiene exhibition later became permanent exhibits in the museum where they were used in the new department of 'Living Standards and Culture, Public Health'. In 1926 the illustrator Erwin Bernath, the book-binder Josef Scheerer, the graphic artist Bruno Zuckermann and several others joined the staff. That year the museum also designed displays for an exhibition for health, social care and sport (GESOLEI) in Düsseldorf, which had been commissioned by the city of Vienna and the social security insurance companies. The displays were exhibited in the Austria House in Düsseldorf.

On this his first visit to Germany since 1919, Neurath met the graphic artist Gerd Arntz, who was also exhibiting in Düsseldorf at the time, and hired him for the Vienna museum. When Arntz arrived in Vienna in 1928 he exerted an important influence on the design of the symbols used in picture statistics. The exhibition in Düsseldorf was the first foreign show of the museum and the Viennese method of picture statistics created something of a sensation.[246] Neurath was invited to Dessau for the reopening of the Bauhaus and subsequently received many return invitations for lectures and discussions.[247] In 1926 the museum designed display boards for the International Urban Development Exhibition in Vienna and in 1927 for an exhibition on Vienna and the Viennese at the Burgenland Fair in Eisenstadt. There were also commissions from

---

[244] Ibid., p. 33.   [245] For example in Neurath 1933c, p. 211.   [246] Neurath 1925c, pp. 3 f.
[247] Conversations with Marie Neurath.

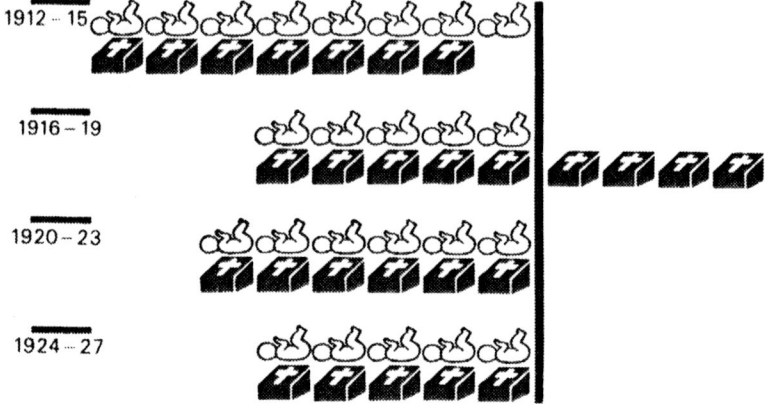

Fig. 1.6 Births and deaths in Vienna. Each child represents 20,000 live births; each coffin represents 20,000 deaths (Neurath 1933e [1991] p. 320)

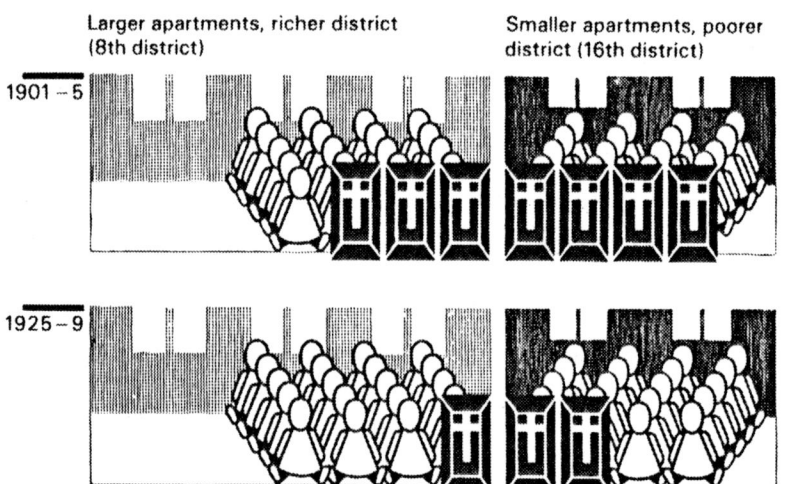

Fig. 1.7 Infant mortality and social conditions in Vienna. Each coffin represents 1 death in the first year of life per 20 live births (Neurath 1933e [1991], p. 311)

foreign countries, including displays for youth exhibitions in Amsterdam and Berlin and a hygiene exhibition in Calau.[248] The Vienna method of picture statistics had begun to be an international success.

The basic units of picture statistics are pictorial symbols, signs with forms as simple and abstract as possible and with heavily emphasised margins that contrast sharply with the white background. The same type of object is always represented by the same type of symbol, so that the symbols engrave themselves in the viewer's memory. The basic principle of picture statistics is that 'a larger number of objects . . . is to be represented by a larger number of symbols'.[249] Rows or linear arrangements of objects were held to be easier to compare with one another than volumes.[250] For Neurath 'it is preferable to memorise simplified quantitative pictures than to forget exact numbers'.[251]

Every sign was a member of a series and it had to be possible to connect it with all the others. No perspectival representation was used. Colours were also given a special meaning, and specific colours were correlated with specific objects. 'For instance, following the example of Egyptian wall paintings, the Museum of Economy and Society always pictured men darker than women.'[252] Besides their form and their colour, the spatial arrangement of symbols (for example along the vertical axis) played an important role. From the start the museum began to compile a dictionary of symbols but it was never completed. The pictures were designed to 'facilitate quick recognition and easy recall'.[253] According to Neurath the simplicity of the picture language promoted clear thinking. This language helped to visualise connections and enabled comparisons of economic, geographic, historical and social interrelations. Neurath urged: 'What you can show with a picture you ought not to say with words.'[254] (Fig. 1.4 through 1.13 give examples of Neurath's Vienna method of picture statistics.)

In addition to picture statistics the museum team worked to develop new forms of representation such as 'new types of maps with unusual foci (projections) that were advantageous from an educational point of view; technical pictures, especially to represent problems of rational organisation; large wooden models . . .; many-levelled models of ground plans of buildings made of transparent materials; and magnetised display boards for the representation of changing quantities. Photography was also used quite often for various purposes.'[255]

[248] Kinross 1979, p. 202. [249] Neurath 1931f, p. 153. [250] Neurath 1927, p. 130.
[251] Neurath 1931f, p. 153. [252] Neurath 1927, p. 132. [253] Neurath 1931g, p. 125.
[254] Neurath 1933c, p. 212. [255] Neurath 1931g, p. 125.

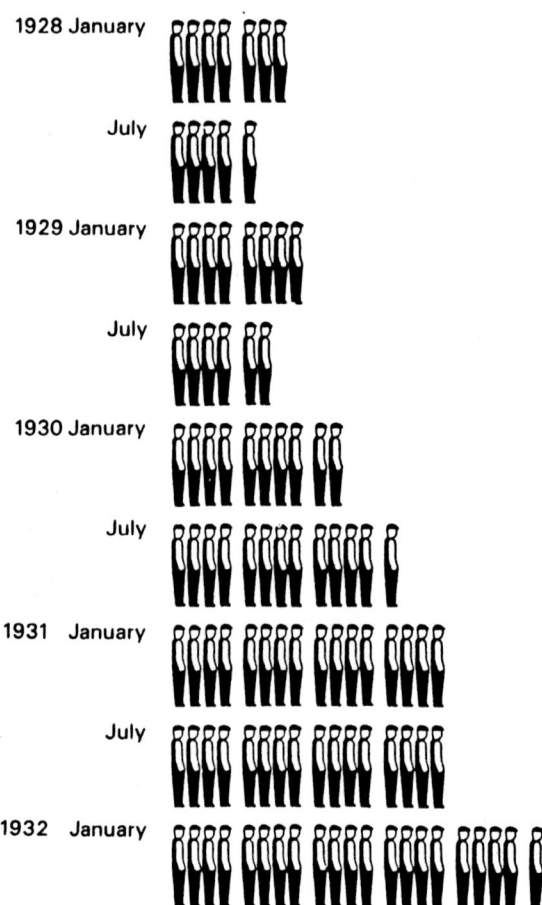

Fig. 1.8 Unemployment in Berlin. Each symbol represents 25,000 unemployed (Neurath 1933b [1991], p. 233)

On 7 December 1927 the permanent main exhibition of the Museum of Economy and Society opened in the new town hall in rooms designed by the architect Josef Frank. This permanent exhibition dealt with the world economy, Germany and Austria, the labour movement, population topics, and issues concerning the city of Vienna.[256] The old exhibition hall of the housing movement at the Parkring became a branch of the museum dedicated to matters of health and social security. Another

---

[256] Ibid., p. 127.

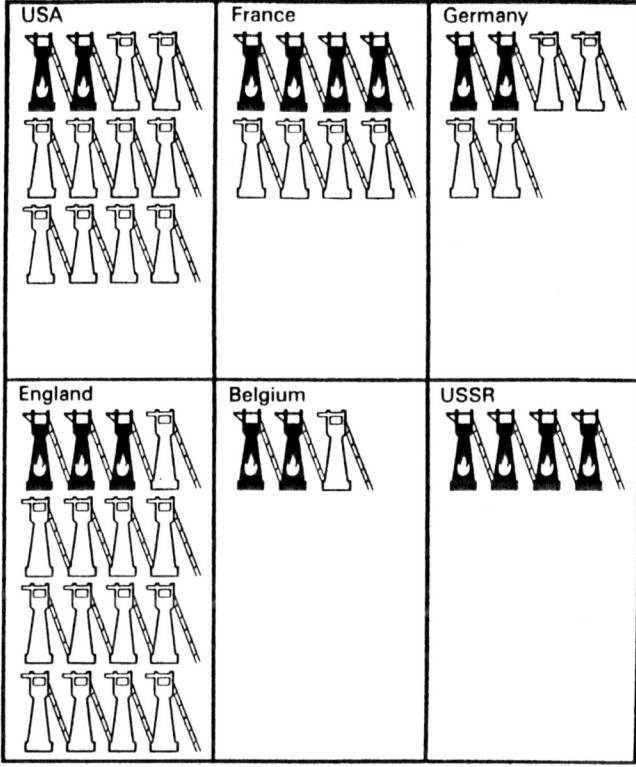

Fig. 1.9 Economic crises and industrial furnaces in March 1932. Each symbol represents 25 industrial furnaces: black = operational, white = inoperative (Neurath 1933b [1991], p. 238)

branch that dealt with the topic of the world economy was opened a couple of years later on a council estate at Am Fuchsfelde.[257]

In 1928 the museum designed the Viennese exhibition 'Woman and Child'. Other displays were produced for a congress on housing and urban development in Paris and work began on the museum's own publications. In 1929 the picture book *Die bunte Welt* (*The colourful world*) was published with illustrations by Gerd Arntz. Under his influence the symbols used became even clearer and still more simple and abstract.[258] The museum designed touring exhibitions as well, sending material to the social museums in Berlin, Zagreb and Klagenfurt, to a new school of further education in Mannheim, to the 'peace exhibition' in The Hague

[257] Ibid.  [258] See Kinross 1979, p. 29.

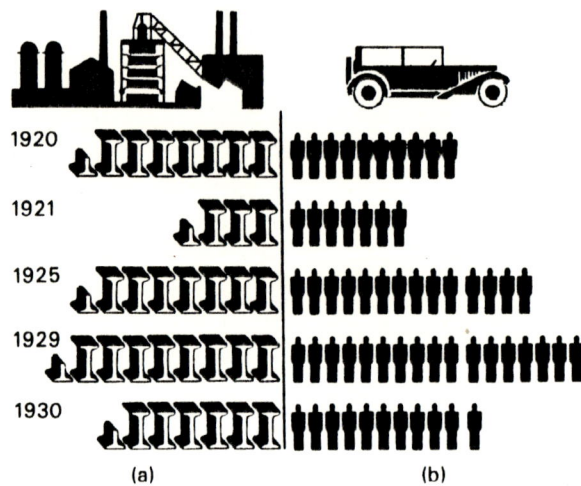

Fig. 1.10 Pigiron production and employment in the US automobile industry. (a) Each symbol represents 5,000,000 tons; (b) each symbol represents 25,000 workers (Neurath 1933b [1991], p. 234)

and to the Museum of Science and Industry in Chicago.[259] The displays of the touring exhibitions were always returned, so the stock of the Museum of Economy and Society grew rapidly. A special archive for visual education was added to the museum. The method of picture statistics was also used in schools. 'The Viennese city school council decided to undertake large-scale testing of picture statistics in a pilot school.'[260]

Neurath did not consider the museum in Vienna to be unique but rather a model that could be copied any time. In fact he expected a steep rise in the number of social museums in the near future. To promote this development, he founded the Mundaneum Institute in Vienna, which was to disseminate the method of picture statistics, found new museums, organise exhibitions, advise schools and produce teaching materials.[261] Branch offices were opened in Berlin, The Hague and London. Neurath gave talks on visual education in Sweden, Norway and Greece and strengthened his contacts in the USA on a journey there in 1933.[262]

In 1931 the Soviet embassy in Vienna invited Neurath to found a similar museum in Moscow, and Neurath accepted. According to the contract five members of the Museum of Economy and Society were to

---

[259] See Neurath 1931g.  [260] Neurath 1933c, p. 210.  [261] See Neurath 1933c, p. 210.
[262] Conversations with Marie Neurath.

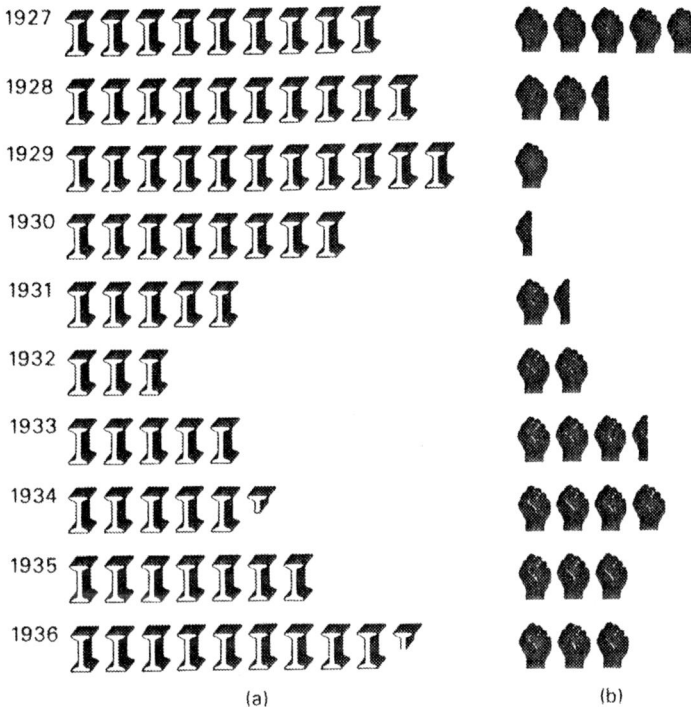

Fig. 1.11 (a) Steel production in the USA. Each symbol represents 5,000,000 tons. (b) Strikes in the USA. Each symbol represents 5 lost days (Neurath 1939 [1991], p. 531)

live in Moscow permanently. Neurath was required to stay there for sixty days a year and seems to have done so until 1934, after which he did not return.[263] Neurath said about the office in Moscow: 'Our method met with exceptional success in the Soviet Union. In 1931 the Council of People's Commissars decreed that "all public and co-operative organisations, unions and schools are directed to use picture statistics according to the method of Dr. Neurath".'[264]

Little remains by way of records to determine Neurath's attitude towards the Soviet system, in the service of which he was now training local staff in the Vienna method and creating charts that informed people about the second five-year plan and related enterprises. Certain things are clear, however. The first is that given Neurath's long-established stance as a social engineer (see section 1.3), his position in

[263] Conversation with Grete Schütte-Lihotzky.  [264] Ibid.

Moscow does not indicate full agreement with the political system he worked for. Still, why should an Austrian democratic socialist support the dictatorship of the proletariat in the USSR? At first it may seem that Neurath simply suspended judgement on the internal politics of the Soviet Union; perhaps he did not wish to impose his Western experiences on the East, as the KPD had done in reverse in Munich (see section 1.3). But early on Neurath also evidenced enthusiasm for the Moscow job. Thus he wrote to Carnap during his first longer stay: 'It is a relief to be able to be active, to create forceful pictures, and not to be part of the decay that is so depressing even if one has confidence in the new that is to come.'[265] Initially Neurath was deeply impressed by the fact that economic development in the USSR far outstripped that of the West, which was in the grip of the Great Depression. (Of the atmosphere in Vienna Neurath had remarked in the previous year 'it smells of putrefaction.')[266]

Yet this enthusiam did not last. After Hitler's ascension to power and the forced closure of his Berlin branch, Neurath typically looked westwards for a new base for his museum – not towards Moscow.[267] The changing political climate there may very well have been responsible. Russian staff simply disappeared between one day and the next.[268] In addition the official, increasingly forceful demand for 'realistic' representation in art was hardly likely to make Neurath's team of graphic designers feel welcome as time went on.[269] Finally we may note that a Viennese friend, who had settled in Moscow and visited Neurath in The Hague, suggested that by 1937 he had become very critical of Soviet policy,[270] while another old friend reported that after the Hitler–Stalin pact of 1938 Neurath 'had totally given up on the idea of collaborating with Communists'.[271] It seems fair to conclude that once the nature of Stalin's reign became clear Neurath suspended his previous suspension of judgement about Soviet Communism.

### 1.4.4 The Vienna Circle

When Neurath returned to Vienna after completing his studies in 1906 he resumed contacts with his former fellow-students Hans Hahn, Olga Hahn and Philipp Frank. Rudolf Haller has named this group the 'First

---

[265] Neurath to Carnap, 25 January 1932 (RC 029-12-69 ASP).
[266] Neurath to Carnap, 22 July 1931 (RC 029-13-08 ASP).
[267] Neurath to Carnap 13 March 1933 (RC 029-11-20 ASP) and 18 June 1933 (RC 029-11-14 ASP).
[268] Broos 1982, p. 217, Bool 1982, p. 223.   [269] Stadler 1984.   [270] Schütte-Lihotzky 1982, p. 42.
[271] Neider 1977, p. 41.

# In Red Vienna

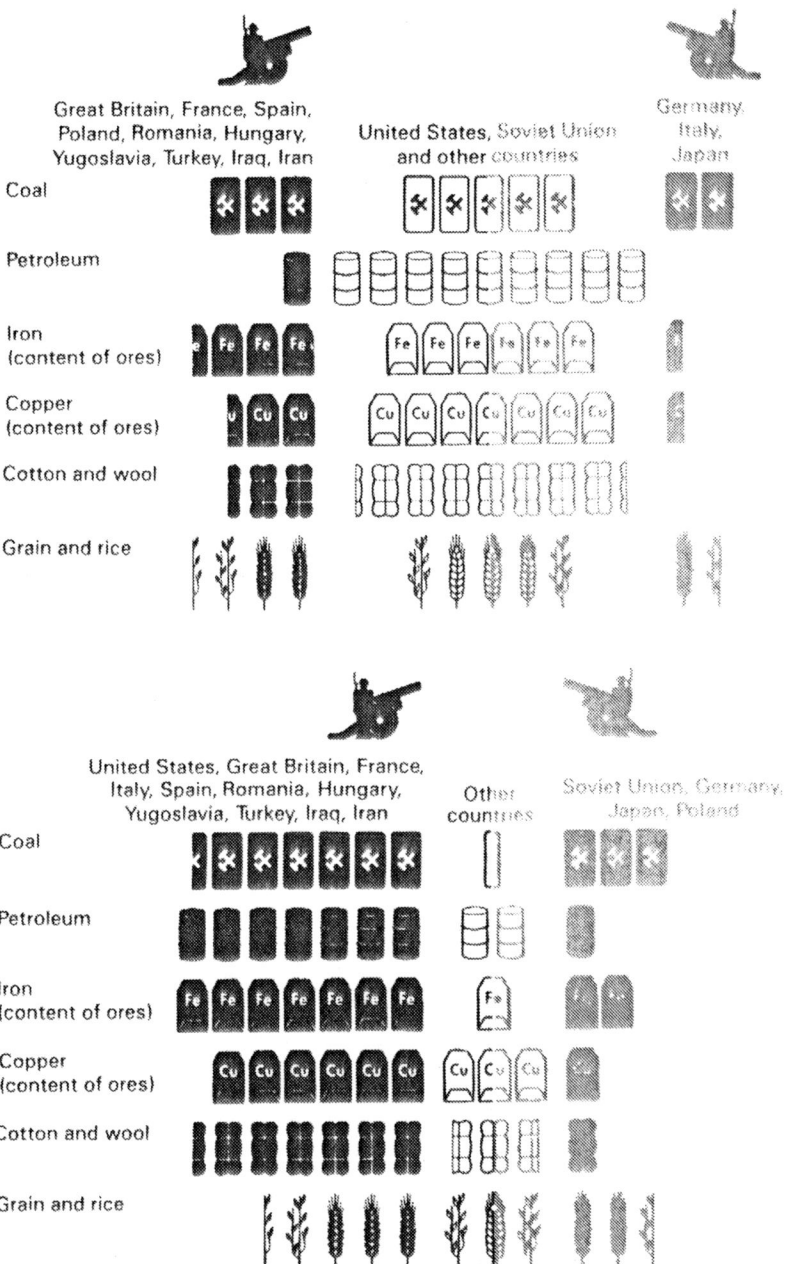

Fig. 1.12 Silhouettes of war economy. Each symbol represents 10% of world production (Neurath 1939 [1991], p. 519)

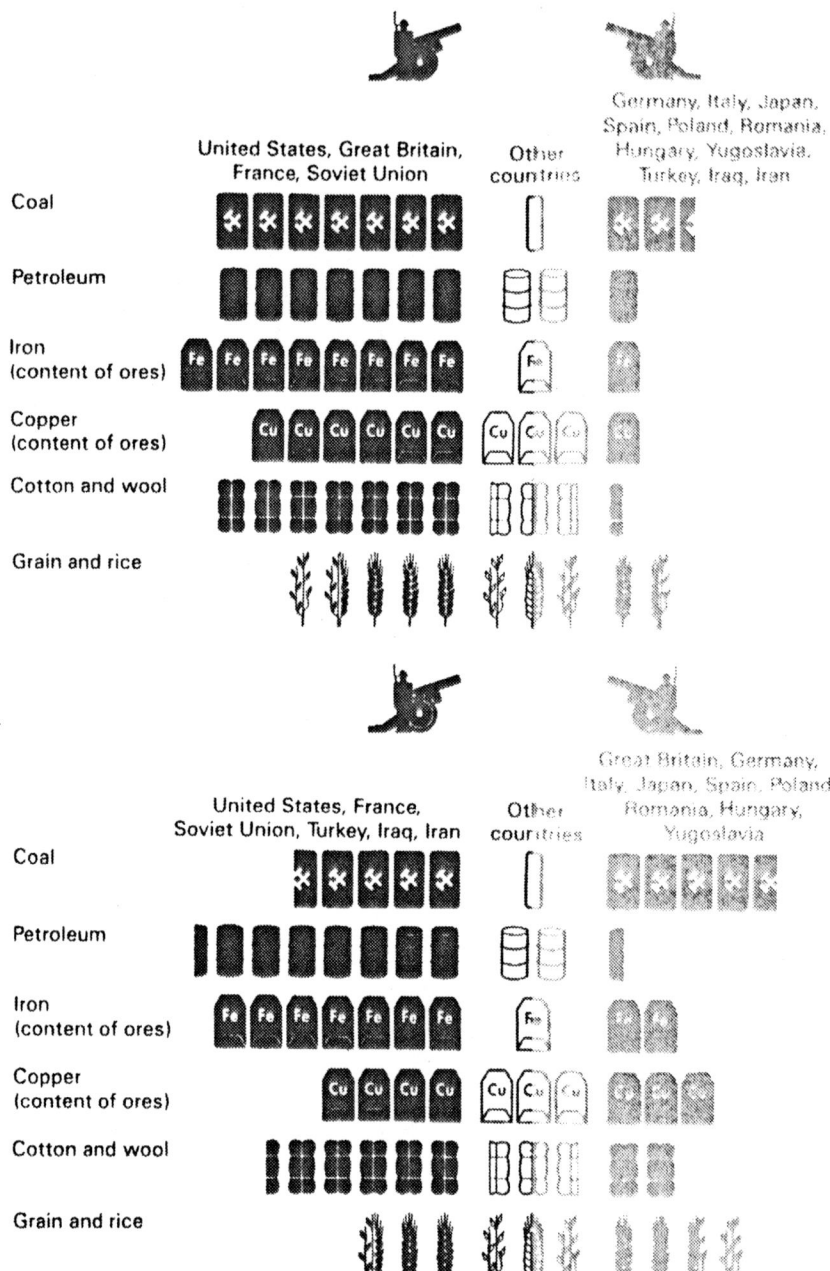

Fig. 1.13 Silhouettes of war economy. Each symbol represents 10% of world production (Neurath 1939 [1991], p. 520)

## In Red Vienna

Vienna Circle'.[272] They discussed the new trends in theoretical physics, mathematics and logic. Neurath wrote that their shared 'empiricist and anti-metaphysical attitude' was influenced by 'the French Conventionalists, by Pragmatism, by Mach and Einstein, by the modern logicians, by empirical sociology'.[273] Further topics are indicated in the programmatic 1929 pamphlet of the Vienna Circle: 'Above all these were epistemological and methodological problems of physics, for instance Poincaré's conventionalism and Duhem's conception of the aim and structure of physical theories . . .; also questions about the foundations of mathematics, problems of axiomatics, and the like.'[274]

While still in Berlin Neurath had learned about the new logic from Gregorius Itelson, an 'empirical rationalist'[275] and anti-metaphysician, and some of his early philosophical – as opposed to economic – essays deal with logical problems. But it is his review essay on Wilhelm Wundt's multi-volume *Logik* that gives the best idea of the philosophical position he held when he was twenty-eight years old. Neurath was interested in foundational research. He noted that Comte had argued that the interrelations between scientific disciplines should be displayed in 'a fundamental systematisation of the sciences' which lays out the principles common to all sciences.[276] The 'unity of human knowledge' could only be achieved by such a 'universal science'. Even though he was less apriorist in his views on unity than Comte, Neurath was interested in this project because if such a system could be achieved, there would no longer be just a loose conglomerate of highly specific disciplines, but instead the partial investigations of many would unite to a 'common work'. A degree of success would guarantee that:

> the division of science into individual disciplines will no longer result in the isolation of the researcher. Instead research of a more general summarising type will determine the common principles and thus elucidate the idea that one can present an overview of the system of the sciences, whereas nowadays it faces chaos. What is common to all sciences will be formulated pointedly and so it will also be possible to organise scientific work.[277]

According to Neurath, Wilhelm Wundt was a pioneer of this approach, despite Wundt's division of the sciences into the humanities and the natural sciences, which Neurath rejects as 'in fact not useful'.[278] Twenty years later Neurath would return to and elaborate the idea of a unified

---

[272] Haller 1985. [273] Neurath 1936a [1981], p. 695.
[274] Carnap, Hahn and Neurath 1929 [1973], p. 303. [275] Neurath 1936a [1981], p. 694.
[276] Neurath 1910a [1981], p. 39. [277] Ibid., p. 46. [278] Ibid., p. 26.

science. His work in this area is one of the major themes of part 3, where we will trace how Neurath's views on unity are first expressed in the essay on Wundt shift and become more radical after Neurath's work in the Bavarian revolution. Before that, in part 2, we will return to the First Vienna Circle and Neurath's Wundt essay for a detailed analysis of his philosophical development up to his participation in the Vienna Circle proper.

With Frank's and Hahn's appointments in 1912 far from Vienna, in Prague and Cernowitz, the contacts between the members of the first Vienna Circle became less frequent. Neurath developed his ideas for a planned economy, worked for the war economy department of the Austrian ministry of war and headed the war economy museum in Leipzig. He also underwent an enormous politicisation. In 1919 while awaiting his trial in Bavaria Neurath wrote *Anti-Spengler*. He did so in order to protect the young from metaphysics. 'We want to liberate from Spengler, from the kind of mind which, tempting and violating, destroys clarity of judgement and precision of reasoning and distorts feeling and perception.'[279] For Neurath the question of how one might use scientific findings became increasingly important. Ten years after his *Anti-Spengler* he was to write in an essay on Bertrand Russell: 'Anti-metaphysicians strengthen the force of the proletariat.'[280] Social changes are not achieved by appeals to pure reason, but by pursuing a scientific approach; by an unmetaphysical, nonreligious comprehension of the world. Political struggle is reflected in science. 'Two fronts face each other: on the one hand the bourgeois front, which for sociological reasons is necessarily half scientific and half unscientific in its attitude; on the other hand the proletarian front, which is completely scientific.'[281] The bourgeois front is 'shot through' with metaphysics and theology,[282] the church is prepared 'to support the fight against the working class!'[283] 'Bourgeoisie against the proletariat' also means 'metaphysics against science'.[284] Neurath was sure that the fight for proletarian interests is both a fight against metaphysics and a fight for a scientific approach. And vice versa.

In 1921 Hans Hahn became professor of mathematics at the University of Vienna. In him a representative of the scientific philosophy came to occupy an important position. Rudolf Haller has noted that he 'was not only the first to introduce the new mathematical logic to

---

[279] Neurath 1921a [1973], p. 163, translation amended.  [280] Neurath 1929b [1981], p. 338.
[281] Neurath 1930b [1981], p. 353.  [282] Ibid.  [283] Neurath 1924, p. 181.
[284] See Neurath 1930b [1981], p. 356.

imperial Vienna, he was also the first in Vienna to consider questions concerning the philosophy of the new logic. He came to know Wittgenstein's logical-philosophical treatise through the works of Russell, which he discussed in a seminar in the winter-semester of 1922/23.'[285] On Hahn's initiative Moritz Schlick was appointed to the former chair of Ernst Mach. Schlick had received his Ph.D. under Max Planck and in 1917 had published a philosophical defence of the theory of relativity.[286] Soon Schlick had gathered around him a group of scientists and students. He organised a Thursday-evening discussion group which met from 1924 onwards.[287] In addition to Hans Hahn, Rudolf Carnap became an important member of the Thursday group. Carnap had received his Ph.D. for a treatise comparing different conceptions of space and had written to Schlick on Hans Reichenbach's recommendation in the summer of 1922; in 1926 he habilitated in Vienna.[288] Other members of the group included at various times Gustav Bergmann, Herbert Feigl, Kurt Gödel, Felix Kaufmann, Viktor Kraft, Marcel Natkin, Heinrich Neider, Rose Rand, Friedrich Waismann and Edgar Zilsel. Otto Neurath and Olga Hahn were introduced to the Thursday group by Hans Hahn. It is well known that early on Wittgenstein had a strong influence on the group even though he did not participate in it. Neurath by this time already believed that an anti-metaphysical attitude was progressive; perhaps he saw in this group a potential synthesis of progressive science and progressive politics.

In November 1928 the group went public by participating in the founding of the Verein Ernst Mach ('Ernst Mach Society') which was dedicated to popularising what Neurath called the 'scientific world conception'. The following year Carnap, Hahn and Neurath co-authored the programmatic paper 'The Scientific World Conception: The Vienna Circle.' This paper outlined the anti-metaphysical and empiricist convictions of the Vienna Circle and formulated its goals: 'The Vienna Circle believes that in collaborating with the *Verein Ernst Mach*, it fulfills a demand of the day: we have to fashion intellectual tools for everyday life, for the daily life of the scholar but also for the daily life of all who in some way join in working at the conscious re-shaping of life.'[289] By associating 'the point from which the Circle speaks to a wider public'[290] with the name Ernst Mach the Circle consciously tried to fit itself into the positivist and scientific tradition in Vienna.[291]

[285] Haller 1977 [1979b], p. 19.   [286] Schlick 1917.   [287] See Stadler 1982b, part 2.
[288] See Carnap, 1963.   [289] Carnap, Hahn and Neurath 1929 [1973], p. 305.
[290] Ibid.   [291] On this aspect see Haller 1977, Rutte 1977 and Stadler 1982b.

According to the scientific world conception there are no enigmas for philosophy to solve. Rather, all statements can be divided in two groups: one consists of statements like those 'made by empirical sciences; their meaning can be determined by logical analysis, more precisely, through reduction to the simplest statements about the empirically given. The other statements . . . reveal themselves as empty of meaning, if one takes them in the way that metaphysicians intend.'[292] Unified science is a long-term goal of the scientific world conception. 'The endeavour is to link and harmonise the achievements of individual investigators in their various fields of science.'[293] The hope is 'to reach the goal, a unified science, by applying logical analysis to the empirical material'.[294]

Neurath worked hard to turn the Circle into a movement. He was responsible for contacting the publishing house Arthur Wolf which eventually published the pamphlet with a print run of 5,000 copies. (His own Museum of Economy and Society also published there.) The pamphlet was distributed with the help of Philipp Frank and the Society for Empirical Philosophy in Berlin. In September 1929 the Ernst Mach Society and the Society for Empirical Philosophy organised the First Conference on the Epistemology of the Exact Sciences in Prague. It was planned to coincide with the Fifth German Physics and Mathematics Congress in order to gain a broader public for their views. A second such conference was held a year later in Königsberg. Neurath suggested that the lectures and discussions of the conferences be recorded in shorthand. After the *Annalen der Philosophy* had became *Erkenntnis* in 1930, sponsored by both the Ernst Mach Society and the Society for Empirical Philosophy, the conference proceedings were published there. *Erkenntnis* became the most important publishing organ of the Vienna Circle. In addition Frank and Schlick edited the book series '*Schriften zur Wissenschaftlichen Weltauffassung*' (Writings on the Scientific World Conception) from 1929, and Neurath began the series '*Einheitswissenschaft*' (Unified Science) in 1933.

Neurath's political engagement was not shared by all members of the Vienna Circle. He called himself a Marxist and often gave talks on Marxism, for example on Marxism and Unified Science. Schlick did not attend these talks. As Heiner Rutte notes in his history of positivism, Schlick was a 'rather unpolitical person and ideologically he was best considered as a "liberal-bourgeois".'[295] Neurath believed that it was not

---

[292] Carnap, Hahn and Neurath 1929 [1973], pp. 306–7.  [293] Ibid., p. 306.  [294] Ibid., p. 309.
[295] Rutte 1977, p. 51.

just his socialist conviction that brought him into disfavour with some members of the Vienna Circle. He was the only leading member of the group who did not hold a university position; still worse, he had been dismissed from his university lectureship. Neurath always aimed to create unity within the movement by citing the works of the others and stressing the goals of the group and their common task. But he considered himself diminished in the estimation of others, especially by Schlick. At most he felt he was part of a smaller group consisting of Carnap, Frank and Hahn.[296] (Neurath's support for the Vienna Circle can be seen for example from the fact that he added a department for unified science to the Mundaneum. This department granted research stipends and employed Marie Jahoda-Lazarsfeld, Walter Hollitscher and Rose Rand in 1932.)

The first work that Neurath published in *Erkenntnis* was the opening lecture of the Prague conference. This paper traces the historical roots of the scientific world conception. For Neurath 'changes in the way of thinking . . . are closely connected with the concrete technical and social changes of mankind'.[297] He noted: 'What characterises the modern scientific conception of the world . . . is the interconnection of individual empirical facts with systematic testing by experiment, the integration of the particular in the texture of all sequences of events and the uniform logical treatment of all trains of thought, in order to create a unified science that can successfully serve all transforming activity.'[298] Science is a system of sentences that enables us to make predictions. To arrive at predictions, science tries to lay out correlations that hold within the spatial and temporal order, an order not constructed by us. 'That knowledge of the world is possible rests not on human reason impressing its form on the material, but on the material being ordered in a certain way. The kind and degree of this order cannot be known beforehand.'[299]

All statements within the scientific system are either tautologies, like mathematical statements, or factual statements, that is, statements that can in principle be controlled by observation statements. Other statements are fake sentences that can be rejected as metaphysical. In these respects Neurath's conception does not differ from that of the Vienna Circle, except perhaps that he did not so naturally express this view as one about meaning. The first difference emerges in connection with his conception of truth. Neurath is usually said to have adhered to a

---

[296] Neurath to Carnap, 18 June 1933, RC 029-11-14 ASP.   [297] Neurath 1930a [1983], p. 33.
[298] Ibid., p. 42, translation amended.   [299] Carnap, Hahn and Neurath 1929 [1973], pp. 312–13.

coherence theory of truth. But it would be more accurate to say that his theory was an account not of truth, but of acceptability, since it was a theory explicitly about the conditions of admissibility of sentences.[300] He expressed no view about what made the sentences true or false, and indeed would probably have found that a senseless metaphysical undertaking. According to Neurath it makes no sense to check the truth of statements against reality. 'Thus statements are always compared to statements, certainly not to some "reality", nor to "things", as the Vienna Circle also thought up till now.'[301] Neurath's views stood in marked contrast to Schlick, who seemed to be grasping for 'The Foundation of Knowledge'.[302]

If a statement is made, it is to be confronted with the totality of existing statements. If it agrees with them, it is joined to them; if it does not agree it is called untrue and rejected; or the existing complex of statements of science is modified so that the new statement can be incorporated; the latter decision is mostly taken with hesitation. There can not be another concept of 'truth' for science.[303]

There is a special group among the scientific statements: the protocol statements, which record our observation claims. 'With the help of hypothesis one can proceed from protocol statements to predictions that are in turn ultimately checked against protocol sentences.'[304] We will later discuss in detail Neurath's views on the empirical basis of science (section 2.5). But as a start towards understanding Neurath's positive doctrine it is important to realise that protocol sentences do not serve as a firm basis of knowledge. The acceptability of a protocol sentence is always decided relative to the totality of sentences otherwise accepted. Neurath presents a novel picture by now well known to philosophers: 'We are like sailors who have to rebuild their ship on the open sea, without ever being able to dismantle it in dry dock and reconstruct it from the best components.'[305] Nor are protocol sentences simply a record of the facts. 'There is no way to establish fully secured, neat protocol statements as starting points of the sciences.'[306] Science is ambiguous and often there are several totalities of sentences – containing differing protocol sentences – that contradict each other even though each is internally coherent. We must choose among these competing totalities. Choice and decision play a crucial role in Neurath's conception, as we shall see in parts 2 and 3.

---

[300] See Uebel 1991b and 1992a, pp. 134–6.  [301] Neurath 1931b [1983], p. 53.
[302] This is the title of Schlick 1934.  [303] Neurath 1931b [1983], p. 53.
[304] Neurath 1934 [1983], p. 102, translation altered.  [305] Ibid.
[306] Neurath 1932d [1983], p. 92.

Neurath's conception of physicalism has often been misunderstood. Once he said that physicalism was a more neutral label for materialism.[307] His basic idea was that all meaningful statements of the empirical sciences 'contain references to the spatio-temporal order, the order that we know from physics'.[308] But this did not mean that the physicalist language was the language of physics: the unified physicalist language is used alike in physics, psychology and sociology. Even though the label 'physicalist language' suggests precision, the physicalist language will also contain imprecise terms. 'The unified language of unified science is not a sum of precise formulations, but a kind of universal slang, which contains also all those imprecise terms, "agglomerations" (*Ballungen*), that cannot be replaced by more precise ones.'[309] Neurath says: 'One can speak in strict physicalist terms and yet with perfect naturalness speaking in the spirit of unified science, use traditional words. It is a mistake to think that we can use only very precise and complicated turns of phrase. One only has to be able to relate everything back to protocol sentences.'[310]

With the help of the method of logical analysis and a common physicalist unified language Neurath hoped to achieve the goal of the scientific world conception: unified science. Unified science was supposed to contain all legitimate sentences of the different scientific disciplines. The idea was not to provide a single picture of the world, but rather to collect the sciences together to enable them to be unified at the point of action. Neurath was sure that we will never arrive at *one* system of science that is then confronted with 'the real world'. The diversity, indetermination and multiplicity of predictions that build upon the multiplicity of possible protocol sentences remain. 'We know that "everything flows" and that multiplicity and indetermination exist in all sciences and that there is no *tabula rasa* for us that we could use as a safe foundation upon which to heap layers upon layers.'[311] We will explain in more detail in part 3 what unity means for Neurath.

In contrast with most other members of the Vienna Circle, Neurath also considered questions of theory formation within the social sciences. In his 1931 monograph on empirical sociology he tried to show how one could pursue economics, history and sociology in the same way as any other empirical science. Sociological problems can be formulated in a physicalist language since everything that one cannot formulate in, or

---

[307] Neurath 1931d [1981], p. 411.  [308] Neurath 1931b [1983], p. 54.
[309] Neurath 1933a [1987], p. 3.  [310] Ibid., p. 8, translation altered.
[311] Neurath 1935a [1983], p. 118, translation altered.

relate back to, statements about spatio-temporal matters is empty and metaphysical. Neurath says: 'Empathy, understanding and the like may help the research worker, but they enter the totality of scientific statements as little as does a good coffee which also furthers a scholar in his work.'[312] Yet Neurath also acknowledged distinctions between the methods and the conceptual structures of physics and sociology. Predictions advanced in empirical sociology differ in character from those that are advanced in physics since there are, for example, no experiments in sociology. Sociology does not start with laws but with generalisations. 'One extends experiences that were collected in connection with a relatively restricted set of examples to a set of further cases that is also relatively restricted.'[313] According to Neurath sciences like sociology may not allow precise predictions of the kind provided by physics. It is in this sense that his physicalism should not be understood as reduction to physics.

### 1.5 EXILE IN THE HAGUE AND OXFORD

Despite the fact that both the Vienna Circle and Neurath's own picture statistics were gaining recognition around the world it became increasingly difficult for Neurath to maintain his position and his museum in Vienna. Thoughout the first Austrian Republic the Social Democrats continued to lose power to the bourgeoisie. By June 1920 they were excluded from the federal government and after the events following the fire in the Palace of Justice in Vienna in 1927 they could no longer hope to exert a decisive influence on a national scale. Despite these developments, however, the Social Democratic administration of Vienna enjoyed sufficient support among the voters to maintain its politics of reform there.

In 1933 a session of the national assembly was interrupted for formal reasons and, in the confusion that followed, Chancellor Engelbert Dollfuss managed to suspend the powers of parliament. From then on he governed by emergency decrees. Over the following months Dollfuss gradually dissolved all the organisations and institutions of the labour movement and its press. On 12 February 1934, when the situation was already hopeless, the Republikanischer Schutzbund (the paramilitary organisation of the Austrian Social Democrats) engaged in armed resis-

---

[312] Neurath 1931c [1973], p. 357.
[313] Neurath 1937e [1981], p. 788. Compare also his 1932a [1983], p. 75.

tance to further unconstitutional encroachments on their rights. The civil-war-like clashes ended after two days with victory for the Dollfuss regime. The defeat of the labour movement was sealed with a ban on all Social Democratic and Communist organisations, the imprisonment of their functionaries and a purge of the civil service and all public institutions including educational ones.

Following these February events the Ernst Mach Society was dissolved by the authorities, who accused it of having disseminated Social Democratic propaganda. Neurath's Museum of Economy and Society was also affected by these events. The permanent exhibition in the town hall was closed; the museum's offices were searched by the police and Neurath's papers confiscated. Neurath himself was in Moscow at the time and a telegram from Marie Reidemeister warned him not to return to Vienna. Instead he went to Philipp Frank in Prague.[314] The museum was closed on 5 April 1934 and replaced by a new institute for Austrian picture statistics headed by the Heimwehr (a right-wing paramilitary organisation).[315] Reidemeister and Mary L. Fledderus, the president of the International Foundation for Visual Education (the Dutch branch of the museum), managed to rescue a few exhibition pieces. They were able to move them to a hastily rented room at the door of which they installed the sign of the Dutch museum. Thus they succeeded in saving this material from the police.[316] Meanwhile Neurath travelled from Prague to The Hague in Holland via Poland and Denmark. There he was joined by his wife Olga, Marie Reidemeister and several collaborators. Their new beginning in the Netherlands was arduous, since they received very few commissions.

Despite these difficulties Neurath still found time to organise philosophical congresses. In 1934 from exile in the Netherlands he organised the Preparatory Conference for the First International Congress for the Unity of Science. The Congress itself was scheduled for September 1935 in Paris. The Preparatory Conference took place on 31 August and 1 September 1934 in Prague. Despite stringent financial conditions Neurath used the facilities of his Dutch museum to organise the Paris Congress as well. These conferences launched the international Unity of Science movement. Surely part of the motive for organising them was the desire to keep contacts among the members of the Vienna Circle intact. This had become increasingly difficult and would not get easier.

---

[314] Conversation with Marie Neurath.   [315] Conversation with Franz Rauscher.
[316] Conversation with Marie Neurath. Today this material is at the University of Reading, Great Britain. It is the only material left from this time.

In 1931 Feigl had gone to the USA and Carnap had accepted an offer to work in Prague. In 1934 shortly after Neurath's emigration Hahn died unexpectedly. In June 1936 Schlick was murdered in the University of Vienna. Neurath records: 'Even though the murder resulted from the personal motives of an unsuccessful scholar who suffered from paranoid jealousy, part of the public of the Catholic corporate state considered his death to be the result of the negative effects of the teaching of Schlick and his Circle.'[317] Later that year Carnap went to the USA, partly for political reasons, and Frank followed him there in 1938.

The Paris Congress was followed by the Second International Congress for the Unity of Science in Copenhagen in 1936. Further conferences took place in Cambridge, England, in 1937, Paris in 1938 and in Cambridge, Massachusetts, at Harvard, in 1939. During these years Neurath concentrated on his longstanding plan to edit a large encyclopedia of science on the model of the French encyclopedia compiled by D'Alembert and Diderot. He said about this project: 'The fundamental idea of the new Encyclopedia is to display the logical framework of our modern science, and to do this in such a way that attention is drawn to gaps, difficulties and points of discussion, thereby avoiding the false impression one wants to replace the speculative systems [of the past] by "the system of the sciences".'[318] The general project of the Encyclopedia was discussed at the First International Congress for the Unity of Science in Paris and its approval was voted. In 1936 Neurath created 'The Unity of Science Institute' as a department of the Mundaneum Institute in The Hague. Within this department, renamed the 'International Institute for the Unity of Science', Neurath set up in 1937 the 'Organisational Committee of the International Encyclopedia of Unified Science'. This committee was composed of Neurath, Carnap, Frank, Charles Morris, Jørgen Jørgenson and Louis Rougier and its task was the production of the Encyclopedia. The first monograph of the 'International Encyclopedia of Unified Science' series was published in 1938 by the University of Chicago Press. Rudolf Carnap and Charles Morris acted as co-editors of the Encyclopedia. Neurath's own monograph in that series, 'Foundations of the Social Sciences', was published in 1944, having been re-written from scratch after he had to leave behind the first nearly complete version when he fled the Nazis in Holland.

Neurath's work in visual education continued. In 1936 he published a book titled *International Picture Language*.[319] It contains a comprehensive

---

[317] Neurath 1937c [1983], pp. 193-4.   [318] Rutte 1977, p. 51.   [319] Neurath 1936b.

description of the Vienna method of picture statistics – now rechristened ISOTYPE – as it was developed by Neurath and his staff during the 1930s. The term 'ISOTYPE' (International System of Typograhical Picture Education) had been suggested by Marie Reidemeister in 1937 to replace the too 'local' 'Vienna method' of picture statistics. Later the same year Neurath also travelled to the USA. His journey received financial support from the National Tuberculosis Association of America and the Oberlaender Trust.[320] In the USA he was commissioned to organise an exhibition on the prevention of tuberculosis. The exhibition was duplicated 5,000 times – made possible by the development of new printing techniques – and presented all over the USA. Neurath also visited Carnap in Chicago, both to join him in negotiations with the prospective publishers of the Encyclopedia and for long discussions about the development of their philosophical positions. Carnap reported to Olga:

[Neurath] celebrated big triumphs here and in America generally. He finds much recognition and interest, already has innumerable connections everywhere, of late even with China, and is considered a great international man. Because of this his mood is much better than last year in Paris and even more than in Prague [before that]. Once again he sees before him lots of opportunities for activity, in fact more than even he could handle.[321]

In 1937 Olga Neurath died from the effects of an operation.[322] In the same year, during a lecture tour in Mexico, Neurath and Marie Reidemeister participated in founding the Museum for Science and Industry in Mexico City. [323] A large exhibition called *Rund um Rembrandt* (Around Rembrandt) was the last work by Neurath and his staff in the Netherlands. The Fourth and Fifth International Congresses for the Unity of Science took place in Cambridge, England (in July 1938), and in Cambridge, Massachussetts (in September 1939), respectively. Neurath was in Cambridge, Massachussetts, when the war broke out and despite the British blockade, he managed to return to The Hague.

In 1940 Neurath had to flee at the first signs of the German invasion. Joseph Scheerer was drafted and Gerd Arntz decided to stay in the Netherlands. On 13 May Rotterdam was burning. On 14 May Neurath and Maria Reidemeister headed off to the Scheveringen harbour. They were the last of the fifty-one passengers on the *Seaman's Hope*, the official lifeboat of the province of South Holland that had been seized by four

---

[320] See Kinross 1979, p. 41.   [321] Carnap to Olga Neurath, 31 October 1936 (RC 102-52-16 ASP).
[322] Marie Neurath 1973b, p. 63.   [323] Conversation with Marie Neurath.

students. The boat departed when night fell, with full moon, low fog and calm waters. On the afternoon of the 15th the overloaded and by then troubled *Seaman's Hope* was intercepted by a destroyer of the British Navy and its passengers rescued and carried to Dover. There Neurath was taken away by the police first to Pentonville Prison, then an to internment camp on the Isle of Man. Marie Reidemeister was taken first to Fulham Institution and finally to the Holloway Prison in London. Neurath was not released until 9 February 1941. Marie Reidemeister had been released just the day before. They were immediately offered hospitality in Oxford by the Marxist historiographer G.D.H. Cole. Neurath and Marie Reidemeister were married in Oxford on 26 February.

In Oxford Otto and Marie Neurath promptly resumed their work, at the usual feverish pace. Neurath lectured (on logical empiricism and the social sciences) at All Souls College for two terms; and the ISOTYPE Institute was founded in Oxford for the continuing international promotion of visual education, with Otto and Marie Neurath as directors.

In a time of worldwide crises Neurath's work responded to his lifelong social concerns, well expressed in 'Personal Life and Class Struggle' (1928): 'Men here on earth who flee sorrow and pain and wish to be kind to each other, happiness, friendship, life as it is really lived on earth, these are our concerns; speculation concerns us only so far as it helps to shape life and to make it happy.' The promotion of visual education was perceived by Neurath as a neutral common ground for the international achievement of human happiness through the establishment of participatory democracy, peace, freedom and higher living standards.

The ISOTYPE Institute organised exhibitions and produced film documentaries (in collaboration with the filmmaker Paul Rotha) and popular books on subjects such as food production, health, women at work and democracy. In 1948 the institute moved to London. Its records are now held in the archives of the University of Reading. In a continuing involvement with housing projects, in 1945 Neurath's office was requested by the Borough Council of Bilston to collaborate on the planning of new community houses for slum dwellers. A.V. Williams, town clerk at that time, mentions in a memorandum of Neurath's visit to the town in July the proposal of an exhibition with the emblematic subject of 'housing and happiness'.[324]

Neurath loved England. He liked his small house on Headington hill, and he was delighted when he could find another in the same neighbour-

---

[324] Cf. Neurath 1973, p. 78.

## Exile in The Hague and Oxford

hood; he liked his garden and his landlady; he was grateful to those who were hospitable to him; he was amused by the Oxford College System.[325] Most of all he liked the freedom and diversity that English life allowed, which he did not see as in any way opposed to the goals of planning he had always striven for. In a letter to Carnap in 1942 he argued:

I disagree totally with all people who think of sacrifice of personal freedom because they want planning. Planning avoids destruction of raw materials, avoids unemployment, provides a society with raw materials, horse power, etc., and now it depends on how we go on with the distribution. The most terrible thing would be if people got power to bully other people.

Even Carnap, he urged, was too Prussian, too formal, too concerned with 'testing and cutter efficiency'. 'As for myself', says Neurath, 'I PREFER OUR BRITISH MUDDLE.'[326]

Otto Neurath died in Oxford in his house in Divinity Road on 22 December 1945. Marie Neurath's recollections from that day are worth quoting in full:

The day before had been a very busy day for Otto in London. He had a good long sleep and in the morning, still in bed, he read Moebius' book on Goethe's medical biography. He told me about the many illnesses Goethe had and compared our lives with his and stressed how lucky we were never to need a doctor. After breakfast we decided to do our Christmas shopping at Blackwell's, the big bookshop. In the afternoon Mr Hung, the Chinese philosopher, came to work with Otto on a publication about Schlick. In the evening we were invited to visit friends who lived within walking distance. We had a nice supper and vivid talks. The conversation was much more personal and relaxed than it had ever been with these friends. It started with some remarks about 'happiness' – an article on Otto had just appeared in the *News Chronicle* entitled 'Man with a Load of Happiness': 'What does happiness mean? I can tell you what I mean by happiness. If Godfather would come to me and say, "Dear Otto, I make you an offer. As you are living now on your hill, with your books and your work and your wife, you can live forever and ever, but you will never be more prosperous than you are now – will you accept?" I would answer, "Yes, dear Godfather, I gladly accept."'

We walked quietly home, nothing hurrying us. The hill is rather steep, and we took our time. We were looking forward to two days entirely for ourselves. When I went into the study to get the newspaper, Otto came in too, and sat down at his desk. Looking over the letters he had written in the afternoon, he gave me one of these to read to him – it would make me laugh. It was in verses based on Goethe's *Iphigenie*. When I hesitated at a certain point which was not quite clear

---

[325] See Neurath's letters to Carnap throughout 1941–2.
[326] Neurath to Carnap, 29 July 1942, Vienna Circle Archives (capitals in original).

88 *A life between science and politics*

Fig. 1.14 Otto Neurath, 21 December 1945

to me, he said, '*so heißt es in der Iphigenie*'. When I read on I heard him laugh – but it sounded strange and I looked up. His head lay on the desk before him. His hand had not yet touched the volume of Goethe that lay to his left. There was no more answer, no more pressure from his fingers. The doctor who came had never seen him alive.[327]

[327] Marie Neurath 1973c, pp. 79–80.

PART 2

# On Neurath's Boat

## 2.1 THE BOAT: NEURATH'S IMAGE OF KNOWLEDGE

Otto Neurath's philosophy – or better: his 'anti-philosophy'[1] – found its characteristic expression in the simile named after him, 'Neurath's Boat':

> We are like sailors who have to rebuild their ship on the open sea, without ever being able to dismantle it in dry dock and reconstruct it from the best components.[2]

In this part we will trace the history of Neurath's Boat. The formulation just quoted is the one most commonly referred to and dates from the heights of the Vienna Circle's protocol sentence debate. With it Neurath cast into dramatic form his insight that knowledge has no secure foundations, that it all depends on how we 'make it up'. This was not the first time that Neurath employed the simile to this end; it was, in fact, the third of five. The Boat represents a recurring motif in Neurath's work. The 'first Boat' dates to 1913.

Neurath's Boat is a simile, a literary figure. Indications are that it is original with Neurath, even though he fashioned it from rhetorical driftwood of his time. The image of the gradual replacement of a ship's components figures in Plutarch's Ship of Theseus, long employed as an example of the problems of identity. It is clear, however, that Theseus' Ship is not at sea, but that its repair took place in the harbour of Athens. Equally only partially to the point is a passage from Plato's *Phaedo*:

> if a man can find neither the truth by exercise of his own faculties, nor through the help of another, then having chosen that which is at all events the best and most irrefragable of Human Doctrines, he ought to embark thereon, like a mariner going to sea on a raft (in default of a better conveyance), and sail through life's voyage ... unless it were possible to proceed on one's way more

---

[1] Neurath 1931a [1983], p. 48.   [2] Neurath 1932d [1983], p. 92, our italics.

securely and with less danger on some firmer vessel, or on some Divine Doctrine.'³

Neurath's Boat combines the image of permanent reconstruction from Plutarch with that of being at sea from Plato, suggesting a qualitatively new situation: there is no 'better conveyance', 'firmer vessel' or 'Divine Doctrine' on which to fall back. Nor can Neurath's Boat simply be read back into Peirce's adoption of Plato's metaphor in 'The Fixation of Belief':

A book might be written to signalise all the most important of these guiding principles of reasoning... Let a man venture into an unfamiliar field, or where his results are not continually checked by experience, and... he is like a ship on the open sea, with no one on board who understands the rules of navigation. And in such a case some general study of the guiding principles of reasoning would be sure to be found useful.⁴

Neurath's Boat does not carry an eternal manual of reason and even where there are experimental checks we are still 'at sea'.⁵

The constraints on the interpretation of Neurath's Boat as a literary figure are looser than those on the interpretation of a logical argument, but they are there nonetheless. The interpretation of the Boat must take into account its central position in the corpus of the texts produced by Neurath. Following its adoption by Quine, Neurath's Boat has been widely understood as an emblem of *naturalism*, the metaphilosophical view that denies the autonomy of philosophy. It needs argument to show that this was the meaning of the simile for Neurath too.⁶ This is especially true given Quine's opposition to the views commonly associated with the Vienna Circle, to which Neurath belonged. To accomplish this task and yet also to distinguish Neurath's naturalism sufficiently from Quine's, a careful explication of the Vienna Circle's idiosyncratic

---

³ Plato, *Phaedo* 85D. W. D. Geddes, whose translation from the notes to his edition of the Greek text (Macmillan, London, 1885) is here used, remarks that 'there are few passages of ancient literature which in modern times have been the subject of so frequent reference as this' (ibid., p. 281)!

⁴ Peirce 1877 [1970], p. 65. This source is suggested by Mohn 1978, p. 57, who first noted Neurath's first Boat, and seconded by Hofmann-Grüneberg 1988.

⁵ For a history of seafaring metaphors in philosophy and literature see Blumenberg 1979; for remarks on Neurath's third Boat see ibid., pp. 73–4; on Boats 2 to 5 (he misses the first) his 1987, pp. 125–9. Blumenberg's reading stays within a falsely presumed positivistic horizon. For closer attention to the potential of the Boat metaphor (with little concern for Neurath's own use) from the perspective of contemporary analytical philosophy see comments in Cherniak 1986, Hookway 1988, 1990.

⁶ For the gradual unravelling of Neurath's naturalism see Haller 1979a, 1979c, 1982a; Rutte 1982a, 1982b; Hempel 1982; Heidelberger 1985; Koppelberg 1987, 1990; Uebel 1991b, 1992a, 1993a.

nomenclatures is essential. It is even more important to understand the historical development of Neurath's views, for this will allow an appreciation of the practical nature of his thought that cannot be provided by considering the Vienna Circle ambience alone.[7]

Neurath's 'anti-philosophy' was a 'philosophy of practice', the Boat his guiding *image of knowledge*.[8] Neurath's first Boat is found buried halfway through a long article on 'Problems of War Economics' in a German political science journal. There Neurath discussed general problems of scientific concept formation, albeit from a special point of view. The specific topic was the founding of the new sub-discipline of economics that was to provide pilot studies for experiments in economic planning. Neurath's second Boat appears in his *Anti-Spengler* of 1921, where with acerbic wit he devastated the pretensions of a German doomsayer who found a wide echo in Central Europe after the First World War (and even impressed Wittgenstein). Neurath objected that Spengler's work was pseudo-scientific. The 1932 deployment occurs during his intricate argument with Carnap about the social base of scientific knowledge, which was published in *Erkenntnis*. The Boat is again invoked in a 1937 contribution to the American journal *Philosophy of Science*. In this article Neurath advertised his very own conception of 'Unified Science and its Encyclopedia'. There he aligned himself with 'the Commonsensism of Peirce, the "natural world-view" of Avenarius and the theory of experience of Dewey' against Cartesianisms and Kantianisms old and new. Neurath's fifth and last Boat concludes his 1944 contribution to the *International Encyclopedia of Unified Science*. Here, under the title 'Foundations of the Social Sciences', he argued against oversystematisation and for reflexivity in unified science without philosophy.[9]

The figure of the Boat constitutes a thread that runs through Neurath's work. From 1913 to 1944 his views developed, of course, yet the tenor of the Boat remained constant throughout. It was used to argue against all forms of 'pseudo-rationalism' – a term of Neurath's for doctrines that suggest determined reason where there is but contingency

---

[7] On various aspects of the pre-Vienna Circle Neurath, besides part 1 above, compare the documents and memoirs in Neurath 1973, ch. 1. See also Cohen 1967; Ay 1969; Glaser 1976; Mohn 1978, 1985; Hegselmann 1979; Haller 1982a, 1985, 1993; Nemeth 1981, 1982a, 1982b, 1991, 1992; Paul Neurath 1982, 1994; Stadler 1982b, 1982c, 1989; Köhler 1982; K.H. Müller 1982; Bergström 1982; W.R. Beyer 1983; D'Acconti 1986; Zolo 1986; Koppelberg 1987; Martinez-Alier 1987; Soulez 1988; Hofmann-Grüneberg 1988; Freudenthal 1989; Nyiri 1989; Chaloupek 1990; Uebel 1993b, 1995b, Reisch 1994, Dahms and Neumann 1994.
[8] On the concept of 'image of knowledge' see Elkana 1981.
[9] Neurath 1913a, p. 457; 1921a [1973], p. 199; 1932d [1983], p. 92; 1937a [1983], p. 181; 1944, p. 47.

of volition – and for the exercise of the constitutive power of convention. 'Pseudo-rationalism', as we shall see, was a special enemy for Neurath. The audience of the Boat changed as well. At first Neurath wrote for the liberal-progressive academics of pre-First World War Austria-Hungary and Germany; next for the post-war youth in the first-ever republics of both countries; then for his colleagues in the Vienna Circle. In the end his appeal was directed to the international Unity of Science movement, first at its moment of expansion into Anglo-American academia, then again in the middle of the Second World War, as a loyally critical voice in anticipation of the role this movement might play in the post-war reconstruction of life not only in Europe but also elsewhere.

What was it that Neurath sought to impart to these different audiences? Consider in addition Neurath's engagements in the Verein für Sozialpolitik (Social Policy Association), at the First World War Austro-Hungarian Ministry of War and in the short-lived Bavarian revolution, then in the housing movement and the adult-education programmes of Red Vienna, finally his practical work in museology and the development of an international picture language. What gave unity to these disparate involvements? In all of them Neurath acted as a reformer, an organiser and an educator. He was a man of the Enlightenment, fired by an idea simple in conception, yet difficult to realise: to develop and employ a conception of knowledge as an instrument of emancipation. This *project* unites his disparate involvements. It is essential therefore to recognise that it was not the dry and technical conception of anti-foundationalism as a discovery of the nature of knowledge that demanded dissemination. What propelled Neurath was an *idea*: the idea not simply that our stock of knowledge claims keeps on changing forever, but that a decisive revision of our concept of knowledge is required if reason is to fulfil its Enlightenment promise. This idea and its attendant project lay behind Neurath's mature *theory* of science. This theory was the result of working out the implications of his guiding image of knowledge – in the context of his practical concerns. It would be wrong to think of the relationship between his idea (and later theory) and his practice as a one-way relation. Practice enters theory itself in such a way that the theorising can only be understood if it is seen as embedded in a context of practices.

Neurath's practice was unified by his theory of practice, but his theory must itself be understood pragmatically. We will argue that the view that theory and practice stand in a reciprocal relationship constitutes the third and most complex of three apparent 'holisms' portrayed by

Neurath's Boat. Ultimately it is this view of the relation of theory and practice that accounts for Neurath's distinctive stance. The first holism, the holism of theory in the face of data and the consequent thesis of underdetermination, Neurath already shared with the Conventionalists, Poincaré and Duhem, and with his colleagues in the Vienna Circle. The second, the dependence of thought on antecedent concept formation, Quine would come to share with him. Together these three 'holisms' (or, better, 'entanglements of thought') backed up the attack on foundations that Neurath mounted on all fronts, an attack that spelt, as we shall see, descriptive and normative, first- and second-order anti-foundationalism. The three 'holisms' also map out the different dimensions of scientific thought that Neurath's anti-philosophy tried to comprehend – and in which it itself must be understood. Neurath's theory of scientific discourse, which the debates in the Vienna Circle prompted him to articulate, was a consistent, though not inevitable, development of the ideas he adopted as early as 1913.

Neurath's Boat addressed a pressing problem for theory and practice: how could we make our conventional determinations of order responsive to experience? Neurath's answer: negotiation. All knowledge claims must be conceived in such a way that people can discuss them by reference to intersubjective evidence. If they cannot reach reasoned agreement, they should at least be able to determine the intelligible, if not mutually endorsed, grounds for their disagreement. The transparency of argument, insofar as it is achievable, could not derive from a base in personal experience but from its embeddedness in a multi-dimensional context of discursive practices. Philosophy – epistemology as a foundational enterprise – was to be replaced by the explication of the constraints that operate in these contexts, by the functional assessment of these constraints and by attempts at a partial reordering of the contexts. Neurath's Boats teach that in place of ultimate justification there can only be spot checks, and that these legitimating procedures themselves needed legitimation which in turn cannot be foundational either. Scientific knowledge is a communal project that has to hold itself in place.

The Boat pictures the basic insights that led Neurath to his theory of epistemic practice. Rather than chase the pipe-dream of apodictic grounds for human reason, Neurath urged us to explore the embedding of reason in what is not reason, to render intelligible the actual workings of reason and, where possible, expose them to conscious intervention. In the Circle's manifesto Neurath displayed a remarkable optimism: 'The scientific world-conception serves life, and life receives it.' If one is tempted to quip that life has not exactly been grateful so far, one must also add that it has not proved entirely malicious either. What still sounds refreshing in our self-consciously 'post-modern' times is Neurath's view that we are not only condemned to conceptual contingency, but should be encouraged by the sober recognition of this fact consciously to refashion our concepts and our self-image as epistemic agents.

The whens, whys and hows of Neurath's voluntarism make up the story of Neurath's Boats. Told in full, this story would take us from turn-of-the-century debates about the foundations of the sciences to a surprisingly 'contemporary' metaepistemological proposal – in little more than three decades. Clearly the full story of Neurath's Boat is too long to be told here. In this part we will not deal extensively with the later Neurath nor intervene in his voice in current debates. Instead we will deal mostly with the early Neurath and his path to his controversial Vienna Circle position.

Even so delimited, however, we must treat our topic selectively. There are intriguing stories to be told about Neurath's personal relation with and intellectual background in the Viennese Enlightenment of Ernst Mach and Josef Popper-Lynkeus, Friedrich Jodl, Wilhelm Jerusalem and his own father Wilhelm Neurath, and about his early academic background in and personal relations with the protagonists of the numerous debates of (mostly German) social science before the First World War: his revered mentor Ferdinand Tönnies; his dissertation advisers Gustav Schmoller and Eduard Meyer; somewhat more distantly, Georg Simmel and Max Weber as well as the Austrian value theorists von Philippovich and Böhm-Bawerk.[10] But we must quickly turn to the systematic question of what considerations prompted Neurath's first Boat of 1913. We will explore the 'deep' background of Neurath's thinking insofar as it pertains to his central argumentative moves, but we will not deal with it in its own right. Nevertheless this background figures importantly in the story of Neurath's Boat. We will begin the investigation of the development of Neurath's theory of science by stating three hypotheses which delineate the specific direction his unusual background gave to his

---

[10] Needless to say, the sketch of Neurath's development given here is not complete. In addition to matters pertaining to Neurath's 'deep' background, there is also the question how far the early Neurath may have been influenced by the widespread interest in pragmatism of American and home-spun varieties. To be considered are Vaihinger (see Fine 1993) and especially James (whose *Pragmatism* was translated into German by Wilhelm Jerusalem who debated its virtues with Gregorius Itelson at the 3rd International Congress for Philosophy in Heidelberg in 1908 (Elsenhans 1909, p. 91)). Compare Neurath 1936a [1981], p. 695 (quoted in section 1.4.4).

*In the first Vienna Circle* 95

thinking and to the dialectic of its development. It provided him with a unique set of problems for which his developed theory of science was to furnish the solution.

## 2.2 IN THE FIRST VIENNA CIRCLE

Before considering the development from 1910 to 1913 of Neurath's theory of science (or scientific metatheory), it is important to set the scene for the early Neurath. We will begin by pointing out two crucial continuities between Neurath's later and early views. They show that Neurath's rather unorthodox version of 'logical empiricism' has very early roots. These roots provide the distinctive perspective with which Neurath approached the discussions in the first Vienna Circle. Of Neurath's 1910 programme for 'universal science' we will then ask: how far does it fall short of the radical anti-foundationalism of his Boat? What ideas still had to change before 1913? (How they changed is the topic of section 2.3.)

### 2.2.1 *Three hypotheses*

Characteristic for the Neurath of the second Vienna Circle is a conception of science as a distinctive form of historically located discourse. 'Philosophy' was to be replaced by an interdisciplinary inquiry into the conditions of this form of discourse: physical, biological, sociological, psychological, linguistic and logical. 'The possibility of science becomes apparent in science itself.'[11] Epistemology as a discipline that provides legitimation for knowledge claims from beyond science was discarded by Neurath alongside metaphysics. 'Metaphysics' was a category Neurath used idiosyncratically: 'I like to use the word', he once explained to an exasperated Carnap, 'whenever I am confronted with a view which is supported by the tendency to formulate uncontrollable [assertions].'[12] (We will retain Neurath's term *kontrollierbar*, for it usefully stands equally distant from verification and falsification, confirmation and disconfirmation, merely asserting the controllability of assertions.)

Neurath's concern with controllability spelt the renunciation of the autonomy of philosophy. We saw in section 1.4.4 that Neurath's 'physicalism' did not require the reduction of different forms of scientific discourse to that of physics, but only that they be so formulated as to allow

---

[11] Neurath 1932a [1983], p. 61.   [12] Letter to Carnap, 20 February 1935, RC 029-09-80 ASP.

an actual or possible 'disconfirmation' by events in space and time. Hence his replacement of philosophy did not have to be a reductive scientism. Neurath's theory of science allowed different disciplines to engage in autonomous concept formation without reducing them to physics. Since, further, the 'linguistic turn' of the Vienna Circle dictated that all philosophy should concern itself with ways of representing rather than with the nature of what was represented, what could be more 'natural' for Neurath than to develop a theory of science that was also non-reductive with regard to the idea of representation and to develop a theory of science as a many-dimensional discourse?

With this position Neurath stood virtually alone in the Circle, in contrast to his concern for the controllability of assertions. Yet even the latter reached back to his student days in Vienna and Berlin when, he once recalled, 'this whole business of criticising language was in the air'.[13] Gregorius Itelson taught him 'to be cautious in the use of expressions' and his father's acquaintance Popper-Lynkeus impressed him by being 'open in his criticisms of the tendency to conceal the unpleasantness of historical events by using certain fashionable vague terms and phrases that sound good'.[14] Of course Neurath's replacement of traditional epistemology emerged fully only in the Circle's protocol sentence debate as a critical alternative to what appeared as Carnap's reworking of Machian ideas in his *Aufbau*. Yet already with his Boat of 1913 Neurath had embraced an anti-foundationalist position that 'cautioned' against holding any knowledge claim to be certain.

Another distinctive feature of Neurath's Vienna Circle views also goes back to the early days. In the Circle Neurath affirmed the political partisanship of scientific metatheory. Was his emphasis on the political dimension merely a reflex of the Marxist convictions he had acquired in the final days of the Austrian and German empires? Having participated in the discussions of the Verein für Sozialpolitik on value judgements in science, Neurath always took care to distinguish the elaboration of models for practical reform from prescriptions for their implementation. It was in a capacity different from that of the scientist that he endorsed sweeping socio-economic changes. Did Neurath finally conflate these roles when he affirmed the political partisanship of the 'scientific world-conception'?

Neurath's view may be taken to represent a careful qualification of

---

[13] Neurath 1941a [1983], p. 217. The implicit reference is to Mauthner 1901 (2nd edn 1906).
[14] Ibid. About Itelson see Buek 1926 and Bergmann 1971; about Popper-Lynkeus see Wachtel 1955 and Belke 1978.

the ethical non-cognitivism he shared with his Vienna Circle colleagues. Neurath regarded ultimate value judgements as expressions of 'will', as Carnap did of 'character'.[15] Yet, sensitised by his long interest in the sociology of knowledge, he also denied that any bodies of doctrine and socially shared beliefs enjoyed 'extraterritoriality'[16] – even if they were as seemingly esoteric in nature as the scientific world-conception itself. These doctrines always stand in more or less well-defined relation to other elements of social life that are always open to investigation. Thus when Neurath emphasised the need for a sociology of sociology in the 1930s, he was reacting not only to the discussions of Mannheim's *Ideology and Utopia* (in which he participated as one of the sharpest critics on the left), but was also reflecting on old lessons.[17] That, for instance, 'those theories of money always and everywhere appear correct which correspond to the needs of those countries which possess preponderance in the world market' was something he had already learnt from his father.[18]

Clearly some of Neurath's mature Vienna Circle views have deep roots. Yet the image of knowledge represented by Neurath's Boat differs significantly from the views common in his early intellectual environment. The Boat marks a break with the tradition. To understand it we must understand what prompted this break.

Neurath was the only trained social scientist at the core of the Circle. Not surprisingly he wrote papers and monographs on the metatheory of social science during his association with the group. But he did not merely apply the Circle's – nor his own – logical empiricism to his original field of scientific interest. Much has rightly been made of the importance of the 'new physics' of relativity theory for the philosophy of the Vienna Circle and the associated Berlin group. The *first hypothesis* defended here is that this was not the only source of the logical empiricists' new theories of scientific knowledge. The case of Neurath shows that developments in social science before the First World War complemented the challenge of the new physics and provided additional independent impetus. Unlike the new physics, of course, the pre-war social sciences did not provide exemplary knowledge requiring merely a new metatheory; rather, their inconclusive state demanded a metatheory that

---

[15] Neurath 1913b [1983] p. 10, 1920a, pp. 44–5, 1920b, p. 1; Carnap 1934b, 1963, pp. 999–1017.
[16] Neurath 1931c [1973], p. 406; cf. Paul Neurath 1982.
[17] E.g., Neurath 1931c [1981], p. 426; 1944, pp. 42–3. For Neurath's critique of Mannheim see Neurath 1930b. The debate over Mannheim is documented in Meja and Stehr 1982 (see also editors' discussion there and Ringer 1968, §7.3). [18] Wilhelm Neurath 1880a, p. 503.

could bring their promise to fruition. Neurath wanted to provide such a metatheory.

Even a cursory glance at Neurath's biography supports this hypothesis. As a student Neurath had become acquainted with the protagonists of the most important scholarly disputes in social science at the time: Tönnies in the battles against Social Darwinists and Kantian value apriorists; Meyer in the methodological turn-of-the-century dispute among historians (*Historikerstreit*); Schmoller in the earlier economists' dispute about method with Carl Menger (*Methodenstreit*) and in the pre-war dispute in the Verein für Sozialpolitik about the propriety of issuing scientific policy recommendations (*Werturteilsstreit*); and Simmel (whose sociology seminar he attended), together with Tönnies and Weber, on the other side of the latter dispute.[19] We will consider various aspects of these disputes in greater detail below; for now it suffices to note that the first two concerned the role of universal laws in history and in economics, respectively, and the third the role of value judgements in social science.[20]

Generally Neurath's sympathies lay with the classic theorists of German social science, Tönnies, Simmel and Weber. Much of their work concerned what we would now call the problem of 'modernity'.[21] In particular Neurath applauded their conception of the value-neutrality of science. But he could not content himself with Tönnies' and Weber's implict cultural pessimism. Pursuing leads ultimately stemming from his father's economic theories about 'the institution of credit',[22] Neurath studied antiquity with comparative intentions. While he struggled with the problems of categorisation and periodisation in historiography, he learnt from Tönnies, Simmel and Weber that social science too requires an abstract conceptual apparatus with which to express its theories and principles of explanation or description. What remains problematic is the 'ground' of this abstract apparatus and thereby the

---

[19] Attendance of Simmel's seminar is noted in Neurath's letter to Tönnies, 31 October 1903; Neurath met Weber at the 1909 Social Policy Association meeting in Vienna at the latest.

[20] For a discussion of the *Streite* in early twentieth-century German social science from an idiosyncratic, phenomenologically influenced Viennese position, see Kaufmann 1936, Part II. For more recent discussions of the *Methodenstreit* see Hansen 1968; Ringer 1968, § 3.2; Hennis 1987; Schön 1987. On the *Werturteilsstreit* see Lindenlaub 1967, ch. 6; Ringer 1968, §§ 3.3 and 6.2; Gorges 1980, ch. 6; Turner and Factor 1984; Krüger 1987. On the *Historikerstreit* see Ringer 1968, § 5.4; Iggers 1969; Lepenies 1985 [1988], pp. 252–3; Whimster 1987; Schleier 1988. On German sociology from the 1910s to 1933 see Käsler 1983, 1984; Stölting 1986. On the history of Austrian sociology see Christian Fleck 1990.

[21] See, e.g., Ringer 1968; Dahme and Rammstedt 1984; Frisby 1986; Liebersohn 1988; Rammstedt 1988; Bickel 1991.  [22] See Paul Neurath 1982 and Uebel 1995b.

'ground' on which overarching conceptions of history (like stage theories of law-like development) are erected.[23] Abstract concept formation in the sciences was thus a problem that Neurath had encountered even before reading the French Conventionalists in the first Circle.

The *second hypothesis* about Neurath's development is that Neurath was prompted to adopt his radical position in part by his drive to legitimate the pursuit of a social science oriented towards emancipation. The partisanship of the scientific world-conception was not a consequence of his later Marxism, but was consistent with his efforts all along.

Again Neurath's biography furnishes support. Even before his days at university he was strongly influenced by the ideas of his father. Wilhelm Neurath's thought, like that of Popper-Lynkeus, was characterised by a strong social commitment and a decidedly utopian orientation. Both presented far-reaching plans for the reorganisation of the economy of Austria-Hungary. Yet unlike Popper-Lynkeus, Wilhelm Neurath backed his reformist endeavours with a philosophical idealism that was in opposition to Mach's positivism, which abjured all teleological reasoning in science. Since the son tended towards Machianism, he could not but find the father's reliance on ideal values problematical.[24] Otto Neurath's own work as an economist was dedicated to the construction of new conceptual frameworks as tools for the formulation of alternative economic schemes, and so was also marked by the practical concerns and theoretical needs of a reformer. Were his reform economics any more scientific than his father's spiritualist pan-cartelism? Indeed, what was the status of engaged social science? Clearly these questions – closely related to the issues of the dispute over value judgements – cannot have failed to provide Neurath with a pressing impetus to develop his own general conception of science.

These two hypotheses characterise distinctive interests that Neurath pursued in the first Circle around Hahn. The third hypothesis spans Neurath's development as a whole. The questions of the 'ground' of abstract concept formation in social science and the legitimation of a research orientation gained unsuspected topicality, albeit in a new guise, in the first Circle's discussions of the challenges to Mach's positivism, especially those from the French Conventionalists. Strict empiricists like Mach were unable to account for the abstract principles employed by science. Yet the Conventionalists also failed to provide a satisfactory

---

[23] For Neurath on stage theories of history (such as Breisig's) see Neurath 1903b, 1904b and passing comments in 1906a, 1906/7, 1909, 1910a, and the discussion in 1921a [1973], pp. 166–73 [1981], pp. 148–61. [24] Cf. Uebel 1993b and 1995b.

answer. Short of returning to the Kantian *synthetic a priori* there seemed to be only one option: embrace a biological theory of knowledge and declare innate proclivities to be constitutive principles of theory building. This was Simmel's interpretation of the *synthetic a priori*, apparently followed independently by the Conventionalists and Boltzmann.[25]

Here Neurath's social scientific training interacted with his metascientific interests. Familiar with the early opposition of his father and Popper-Lynkeus to Social Darwinism and Tönnies' critiques of Ammon and Schallmeyer,[26] Neurath could not in the end accept the biologising of social science. Given that he also accepted the unity-of-science thesis and rejected the categorical separation of science and its philosophy, this means that he also could not accept the biologising of the metatheory of natural science. This left the question: if high-level scientific principles could not be legitimated by a positivistic reduction to experience, how were they to be justified? Indeed, how could 'convention' justify? Social and natural science seemed to share the same metatheoretical problems.

The *third hypothesis* is that the conception of science Neurath developed in the Vienna Circle proper represents his solution to the problem of the status of conventions in science. Neurath's answer was to dispose of traditional foundational justification altogether, to turn pragmatic and preserve Mach's naturalistic impulse, albeit in a non-reductionist setting. Neurath 'naturalised' conventions by transposing Mach's pragmatism from the evolutionary dimension to the domain of social practice. To avoid the trappings of biologistic shortcuts and instead represent conventions as decisions taken by theorists, naturalism had to reject the reduction of cognitive to biological norms. The conventions of science were to be 'naturalised' in a discourse theory of science.

The Boat accompanied the development of this non-reductive naturalism from its inception. As early as 1913 Neurath held knowledge claims to be legitimated only to the extent that they satisfy epistemic norms which are in turn legitimated by their instrumental utility in science. Even though there is a radical difference between cognition and conation, there is no radical distinction between theoretical and practical science. All theory is located within a practical context and it is only with reference to this context of interests, needs and projects that the regulation of theory can be understood. Neurath demanded that the conventions of science be subjected to scrutiny in just these respects.

---

[25] Simmel 1895; cf. Capek 1968.
[26] Wilhelm Neurath 1878; Popper-Lynkeus 1878, 1912; Tönnies 1904, 1905–9; cf. Uebel 1993b.

Answers to questions about concept formation must similarly be understood as instrumental decisions taken by scientists about how to structure their professional discourse. The alternative to both irrationalism and 'pseudo-rationalism' is a rationality controlled by constantly scrutinising one's conventions.

It is a consequence of this general conception of science that the pursuit of an emancipatory social science requires no more theoretical justification than any other science. (What does such an emancipatory orientation imply but a certain choice of subjects for investigation and, ideally, of the organisation of science itself? All other choices of subject or organisation are no less pragmatic.) Basic standards must be met by any pretender to the title 'science'. A possible event somewhere, sometime, should make a difference to the acceptability of the statements of these discourses. The emancipatory orientation does not unlock a world of value hitherto unknown, but simply restructures the way we deal with the world we have been confronting all along.

Needless to say, the conventionalism Neurath arrived at with his first Boat was not unproblematic: the constraints on concept formation remained seriously underspecified. Neurath's development toward his own version of logical empiricism – his non-reductive naturalism – is marked by his attempts to answer outstanding questions and shore up his early anti-foundationalist and pragmatist insights. The third Boat marks the first articulation of his final answer: a theory of science as discourse. Only by then had the Marxist lessons of the Munich episodes been digested. Social and economic theory, even scientific metatheory, are not formations of free-floating intelligence, but are part of the socio-economic, indeed ideological, struggles being fought. In a word Mach had to meet Marx and take the linguistic turn before Neurath could view science fully as a socially and historically conditioned, even partly politically and economically determined, social practice – a discursive formation with emancipatory potential.

This is the story we will now begin to tell by considering Mach and his importance to the first Vienna Circle in general and to Neurath in particular. Before we can join the actual plot we must come to appreciate what it meant for Neurath to say that 'Many of us . . . have been brought up in a Machian tradition.'[27] The arc of the third hypothesis extends from Mach via the Conventionalists to the Neurath of the Vienna

---

[27] Neurath 1946a [1983], p. 230; cf. his letters to Mach in Thiele 1979, pp. 99–101. On Mach see, e.g., Cohen and Seeger 1979; Stadler 1982b; Haller and Stadler 1988; Blackmore 1992; Holton 1992.

Circle, with the Boats marking different stages of the development of his theory of science. The first part of this arc (up to the first Boat) would appear wholly traditional for a member of the Vienna Circle, yet Mach played a special role for the young Neurath.

### 2.2.2 Mach's legacy

Philipp Frank's remarks about the legacy of Mach, their not uncritically viewed hero, characterise the intellectual situation of the first Vienna Circle and illuminate how the group understood itself. Frank located Mach, and with him the first Circle, in the tradition of the Enlightenment. An 'essential characteristic' of this Enlightenment was its 'protest against the misuse of merely auxiliary concepts in general philosophical proofs'.[28] Frank even discerned what he called a 'tragic feature' in this philosophy:

> It destroys the old system of concepts, but while it is constructing a new system, it is already laying the foundation for new misuse. For there is no theory without auxiliary concepts, and every such concept is necessarily misused in the course of time.[29]

'Hence', Frank concluded, 'in every period a new enlightenment is required in order to abolish this misuse.' Frank's characterisation suggests a dialectical conception even though he himself spoke of 'eternal circles'. Thus Mach was reported as proposing that 'the task of our age is not to fight against the Enlightenment of the 18th century but rather to continue its work'. In this sense it was the task of the first Vienna Circle to study and criticise Mach and what his theories had become.[30]

Before turning to Mach's specific importance for Neurath, let us briefly consider his theory of science and its reception. Mach held that the principle that regulates all thought and thus all scientific activity is the 'economy of thought'. The pursuit and point of science is continuous with common everyday cognition as an aid to survival. Thought is governed by two processes, the 'adaptation of ideas to facts and to each other'.[31] In Mach's theory of science the former is reflected in the demand for the ultimate reducibility of theory to experience and the second in the postulation of the unity of science. As his own 'historical-critical' studies had shown, this unity is not only a practical requirement

---

[28] P. Frank 1917 [1949b], p. 73.   [29] Ibid., p. 78.
[30] Ibid., pp. 73, 78, 75, respectively; for the characterisation of the then current task, see ibid., pp. 70, 76.   [31] Mach 1905 [1976], p. 120.

of scientific progress but represents the 'ideal of a monistic view of the world which alone is compatible with the economy of a sound mind'.[32]

Both adaptations to facts and to other ideas should observe considerations of maximum economy. Mach demanded that all genuine, i.e., non-auxiliary, scientific concepts be reducible to expressions denoting elements of experience alone. Theories where this does not hold are incomplete. (Mach realised, of course, that the universal sentences of science could not be justified by experience and, long before Popper, he counted on their refutability to provide the methodological delimitation of science: 'Where neither confirmation nor refutation is possible, science is not concerned.'[33]) The unity of the object domain of science was similarly prescribed by Mach's theory of neutral elements of nature. The sense data to which singular empirical propositions are to be reduced constitute such elements – albeit under one particular description (besides the psychological, there were the physical descriptions).[34] It was Mach's idea that the systematisations of science be represented in a 'universal character' or 'international universal script'. Mach gestured towards a language by which to capture the elemental composition of phenomena: 'To read it would be to understand it.'[35]

Scientific concepts thus express only a certain kind of ordering of the elements of experience.[36] What affords thought its economy is the fact that its representational tools are universals. Language 'is itself an economical contrivance. Experiences are analysed, or broken up, into simpler and more familiar experiences, and then symbolised at some sacrifice of precision.'[37] Concepts shape what they collect. First, the symbols for recurrent elements denote what are in truth strict particulars.[38] Second, as economical expedients, scientific concepts reflect the cognitive interest of their disciplines.[39] It follows from these 'sacrifices in precision' that it is not the 'pictorial content' of scientific concepts that renders them serviceable.[40] There remains a level of language, however,

---

[32] Mach 1883 [1902], p. 465.  [33] Ibid., p. 490; cf. ibid., p. 237.
[34] Mach's was a neutral monism: 'it is only in the connexion and relation in question, only in their functional dependence, that the elements are sensations. In other functional relations they are at the same time physical objects.' (Mach 1886 [1959], §8.)
[35] Mach 1883 [1902], pp. 481–2, transl. altered.
[36] E.g., 'Space and time are well-ordered systems of series of sensations'. (Ibid., p. 506.)
[37] Ibid., p. 481.
[38] E.g., 'nature simply is' (ibid., p. 483), a thing is 'an abstraction . . . for a compound of elements whose changes we disregard' (ibid., p. 482).  [39] Ibid., p. 507.
[40] Thus his warning that the mechanists 'should beware lest the intellectual machinery employed in the representation of the world on the stage of thought be regarded as the basis of the real world.' (Ibid., p. 505.) This is also one of the themes of Neurath 1916.

that stands in a more direct representational relation to the elements: the familiar sensation terms for colours, pressures, etc. Mach's positive science was built on measurable experience alone. Experience provides both the object and the method of investigation. The point of his theory was to understand science on this austere basis. To recognise that 'science itself . . . may be regarded as a minimum problem consisting of the greatest possible representation of facts with the least possible expenditure of thought'[41] meant simultaneously to focus on the composition of scientific concepts from the elements of experience and to discount their pictorial content which led to unwarranted hypostatisations. Thus metaphysical temptation was to be avoided and the self-sufficiency of positive science established.

For Mach the unity of the object domain of science was discernible ultimately by an analysis of the concepts of the scientific language. It was found in the homogeneity of the bare elements of nature, complexes of which are picked out by concepts. The 'international' intelligibility of Mach's 'universal script' derives from, and depends on, its relation to what Mach considered the objective foundation of human knowledge. Its non-primitive terms exhibit their reductive relation to the experiential given by some (not further specified) formal property. Mach's foundations for science rest in the co-ordination of primitive terms with measurable elements of experience. The unity of science was an ideal of theory building that mirrored an objective fact of nature.

Poincaré taught the first Circle how Mach's reductionist project failed. Mach held that 'with respect to the space in which we live, only experience can decide whether it is finite, whether parallel lines intersect it, or the like'.[42] Poincaré showed that the concept of space was not 'brought to us' by sensations but rather by the study of 'the laws by which . . . sensations succeed one another'.[43] The properties of actual space cannot be determined by experience alone. Hypotheses are required which themselves cannot be tested by experience. This leaves the notion of space irreducible to what is given in experience. The first Vienna Circle sought 'to retain the most essential points of Mach's positivism' (i.e., its antimetaphysical stance), but where it opposed the latest developments in science, 'they planned a reconstruction of his doctrines.'[44]

Mach was criticised for failing to see that the mathematical expression of symbolic systems (e.g., the axiomatic representation of geometry as a

---

[11] Mach 1883 [1902], p. 490.  [12] Ibid., p. 493n.  [13] Poincaré 1902 [1952], p. 59.
[44] P. Frank 1941a, p. 7.

theory of space) allows for the formulation of high-level hypotheses that cannot be experimentally tested. In addition he was criticised both for supposing that the primitive terms of scientific language correspond neatly with the elements of nature as well as for his opposition to atomic theory:

> We tried to supplement Mach's ideas by those of the French philosophy of science of Henri Poincaré and Pierre Duhem, and also to connect them with the investigations in logic of authors [such] as Couturat, Schröder, Hilbert, etc. The attitude towards the atomic theory was indicated to us by the ideas first of L. Boltzmann and then of Einstein.[45]

If Mach's entire 'historical-critical' approach to science was not to fossilise into dogma, his theory of elements had to be given up. The reductionism of his theory of elements could not be upheld. But that opened an old problem:

> We admitted that the gap between the descriptions of facts and the general principles of science was not fully bridged by Mach, but we could not agree with Kant, who built this bridge by forms or patterns of experience that could not change with the advance of science.[46]

The first Circle was united in this conclusion. The problem that both it and the second Vienna Circle faced was to account for the creativity of scientific work without jeopardising its empirical character.

Roughly speaking, there were two answers to this problem in the second Circle. The first is the well-known and well-developed bipartite conception of the language of scientific theories and the hierarchical conception of unified science (later enshrined in the 'received view'). The second is Neurath's less well-known and less developed conception. The Boat may serve as its guiding image.

Had this alternative already been considered in the first Vienna Circle? This need not be decided here. It suffices to note that it was clearly recognised by Neurath in his first Boat. It marks the beginning of his distinctive answer to the general philosophical problem that the Circles confronted. To show this, we return to Frank. How did the development proceed once Mach's mistakes were identified?

> According to Mach the general principles of science are abbreviated descriptions of observed facts; according to Poincaré they are free creations of the human mind which do not tell anything about observed facts. The attempt to integrate the two concepts into one coherent system was the origin of what was later called logical empiricism.[47]

---

[45] Ibid.   [46] P. Frank 1949a, p. 8.   [47] Ibid., pp. 11–12.

In Frank's narrative the confrontation between Mach and the French Conventionalists led to the (second) Circle's orthodox conception of scientific theories. According to this conception the formulation of scientific theories requires two languages, an 'observation language' in which observable spatio-temporal processes are recorded, and a 'theoretical language', which was held to be not fully interpreted because only select terms are tied to the observational language by partial operational definitions (correspondence rules). This orthodox conception could, if one were so inclined, support a moderate anti-realism, an instrumentalism towards the theory proper which could be seen as a merely convenient systematisation of experiential regularities. Observation statements remain unaffected. By contrast Neurath's position seems to entail a much more radical anti-realism. Just how radical he became we will see fully only in part 3, but as in his Boat of 1932, so also already in 1913 Neurath denied that there was available a *tabula rasa* upon which a 'clean' structure of scientific concepts and hypotheses could be erected. There are no secure foundations, neither descriptive nor normative, neither first- nor second-order. No observation statements are beyond doubt. The challenge was global, the response holistic.

It is clear then that Neurath dissented from what was to become the orthodox solution in the Vienna Circle to the problem of the creativity of scientific theorising at a very early date, at the latest by 1913. What prompted this dissent? The three hypotheses announced in the previous section indicate our general answer to these questions. Yet why did Neurath seek to preserve Mach's naturalism, his rejection of philosophy over and above science? Neurath discerned in it a distinctive feature of Mach's own renewal of Enlightenment thought. This aspect of Mach's thinking was well described by Robert Musil (a neutral observer for our purposes) in his dissertation of 1908:

[Mach was] the first person to take seriously the assertion that his (positivist) convictions were solely obtained by applying views which had proven themselves in the natural sciences and that they are no more than a result of the development of exact research. Mach, therefore, makes good in his own person what had hitherto been only a more or less empty claim and so makes it possible to find out whether positivism lives up to one of its most dazzling and appealing promises, the claim that it is merely the backwardness of philosophers which explains their failure to recognise the extent to which exact and fruitful science is already following the tracks of positivist philosophy.[48]

---

[48] Musil 1908 [1980], p. 18.

*In the first Vienna Circle* 107

Neurath sought to preserve Mach's naturalistic impulse – establishing the self-sufficiency of positive science – in a non-reductive setting (the plurality of the sciences did not reduce to physics and cognitive norms did not reduce to evolutionary givens). This required that a more circumspect form of reasoning replace reduction as the ideal of metascience.[49] Neurath's Boat gives a clear expression to this new problem situation. In his mature theory, Neurath's concern was reflected in his demand that metatheoretical concepts be themselves explicable scientifically. Metatheory was not to be exempted from the control to which it subjected first-order theories.[50] This alone promised to render scientific reason scrutinisable and intersubjectively intelligible in the absence of foundations. But already with his first Boat, Neurath forswore reliance on extra-scientific insight, for how was it to be checked in turn? Neurath's unorthodox solution to the problem of scientific creativity sought to renew Enlightenment thinking by insisting on the controllability in principle of every step of scientific and metatheoretical reasoning.

The story of Neurath's Boats traces the steps of this development. We will now join the plot with Neurath's 1910 programme for 'universal science', midway through this period of the first Circle. Once an 'overprecocious Viennese youth',[51] then three years past his dissertation without a university position in sight, Neurath used the occasion of his review of Wundt's *Logik* to develop his own conception of scientific metatheory and of economics. This review marks certain ambiguities in Neurath's early programme for a general theory of science. Well on his way to the Boat, crucial steps remained to be taken.

### 2.2.3 The 1910 programme

As we have seen in section 1.4.4, one Machian theme stands out in this review: the unity of science. Neurath hedged with qualifiers like 'seems to', but nevertheless dismisses the distinction between the natural sciences and the human or cultural sciences (*Geisteswissenschaften*) as 'in fact not useful' and thinks it 'very doubtful' whether a better categorisation of the sciences than by single disciplines would be possible. Neurath also

---

[49] In a broadly related vein (but unconnected to the first Boat) Zolo speaks of Neurath's 'reflective recognition of the situation of circularity' (1986 [1989], p. xvii). That Zolo deems 'anti-naturalistic' the need for reflexivity – which he rightly stresses as of great importance to Neurath – turns on a restrictive reading of the possibilities for naturalistic theories (Uebel 1992a, ch. 12).
[50] Cf. Uebel 1991b, 1992a, ch. 10.
[51] Tönnies to Paulsen, 6 March 1904, repr. in Klose, Jacoby and Fischer 1961, p. 375.

dismisses attempts to develop a 'fundamental systematisation' (*grundsätzliche Systematik*) and rejects as unsupported the Comtean idea of a hierarchical ordering of the sciences with those at the top investigating the least determined objects. For him the exact form of the unity of the science was to emerge from a close investigation of all of the individual sciences.[52]

Even though the demarcation of the individual sciences is not a settled matter (to say nothing of the demarcation of what he called the 'section social science'), Neurath stated some definite theses concerning first-order theories.[53] What is most immediately striking is that any scientific hypothesis can be judged for its correctness only holistically: theories are judged in their entirety.[54] Neurath also followed the Conventionalists' stress on the formal and abstract nature of scientific theories. Scientific theories search out their objects in their historical reality, yet consider them not only in their actual configuration but also in their possible ones: modal reasoning was an integral part of science.[55] It follows that the distinction of practical and theoretical science is a false one. All science is theoretical.[56] That theory construction and choice must be understood pragmatically was yet another matter. (The availability of purely 'theoretical' conceptual constructions could be of paramount importance for scientific and social practice.)

The 'understanding that science is a unity'[57] is to be promoted by a 'universal science' which would investigate 'the foundations of the sciences in general'.[58] What is the nature of this 'foundational' inquiry?

> When the roughly conceived unity of human knowledge dissolved into individual sciences, one tried in a conscious fashion to unify these anew in an orderly system. Such a universal science would not be the sum of individual sciences, because it would contain assertions concerning connections which could not be dealt with in any single one of them and because it would unify the principles which are common to all of these individual sciences.[59]

Apparently Neurath's version of universal science was intended to have a unique domain. What exactly was it? Neurath distinguished object and metalevels of investigation but his description of universal science is ambiguous in this respect. And still another distinction must be noted. Neurath advised universal scientists to proceed in their descriptive work with the historical perspective pioneered by Mach and Duhem: abstract

---

[52] Neurath 1910a [1981], pp. 25–6.
[53] For Neurath's 1910 views on value statements and formal calculi, see section 2.3.1 below.
[54] Neurath 1910a [1981], pp. 44–5.   [55] Ibid., p. 29.   [56] Ibid.   [57] Ibid., p. 45.
[58] Ibid., p. 24.   [59] Ibid., pp. 24–25.

## In the first Vienna Circle

concepts and principles must be investigated in their development from the problems in response to which they were originally devised.[60] Yet these studies were expected to yield normative consequences. Universal science would be aided by this descriptive work to decide whether, say, the continued use of given concepts was still appropriate: 'A more general comprehensive research will set down the common principles.'[61]

To determine the domain of universal science then, distinctions must be drawn between first- and second-order inquiries and between descriptive and normative inquiries. Descriptive first-order inquiries concern the relations between the objects of the individual sciences; descriptive second-order inquiries concern the relations between the individual sciences and the scientific enterprise in its historically conditioned states of development. Normative first-order inquiries concern questions of the good life; normative second-order inquiries, questions of scientific methodology. Neurath's ambiguity can be resolved as follows. 'Universal science' is not a traditional philosophy of nature. Unlike his later 'unified science', however, Neurath's 'universal science' takes only scientific theories as its proper domain (in 1915 Neurath was to speak of a 'theory to classify theories').[62] Neurath's later form of the unity of science as unity at the point of action or the even later form of unity as orchestration were still to emerge. In the meantime the general idea of the unity of science functioned as a regulative ideal. Universal science was both a descriptive and a normative inquiry of second order. It could not of course be pursued in isolation from the first-order sciences. In other words, universal science corresponds to the metatheoretical part of his later 'unified science' which includes both first- and second-order theories.

Already in his 1910 programme, scientific metatheory bore a distinctly Neurathian cast: universal science was a historically located, collective enterprise with a practical intent. The practical point of the unity of science thesis had two aspects. Intra-scientifically it would lead to overcoming the exaggerated separation of the sciences and to a 'true and fruitful division of labour'. Extra-scientifically it would greatly facilitate the popularisation of scientific knowledge. Then universal science would 'allow everybody to have a part in the greatest achievements of the human spirit, the world to stand before us as a whole again.'[63]

We must leave open the question whether Neurath's concern to

---

[60] Ibid., p. 27.  [61] Ibid., p. 45.  [62] Neurath 1916 [1983], p. 31.
[63] Neurath 1910a [1981], p. 46.

restore wholeness was an echo of complaints voiced by German academics of his time when faced with the rationalisation of modern life, or whether it represents an early move towards the 'democratisation of science' which he promoted in his Vienna Circle days.[64] Indications are that it was not yet the latter. Neurath did not start out in the trenches of class warfare.[65] In part 3 we shall describe how in later years the necessities for political change set a still more practical programme for unification at the heart of his concerns. We may also note that in 1910 Neurath was willing to speak of '*the* comprehensive system of the sciences'.[66] Apparently, Neurath had some distance to cover before reaching his pluralist 'encyclopedism' of the 1930s.

Already, however, Neurath's endorsement of Duhem's holism betokened his rejection of Mach's reductionism. But Neurath also returned to the broad scope of Mach's thesis of the unity of science (affirmed by the Conventionalists only for the physical sciences) and extended Duhem's holism from theoretical physics to all sciences.

> It is a great task to comprehend as far as possible the complete order of life and to reduce as many relations as possible to simpler principles. Without doubt one cannot always proceed step by step, often one will have to try to apply a whole system of relations immediately, just as in the case of physics; there one does not hold true one theory after another and so increase the knowledge already gained; rather, the system itself is often called into question. The biggest difficulty consists in isolating individual inquiries while at the same time not losing sight of the remaining connections. One must always know how long it is useful to hold on to an entire system of theory and to explain a certain fact by auxiliary hypotheses, and when it is, on the contrary, more useful to remodel the entire system.[67]

Several important questions arise at this point, for which the text does not provide answers. What were Neurath's reasons for extending Duhem's conclusions to all sciences? How radical was Neurath's antifoundationalism at this early stage – was he prepared, as he was three years later, to reject observation reports that contradicted a theory's predictions instead of changing the theory? To answer these questions we shall need to trace the emergence of radical anti-foundationalism in Neurath's views of first- and second-order, descriptive and normative inquiries. Thus we shall ask: how did Neurath deal with the normative

---

[64] On the former see Ringer 1968; on the latter Dvorak 1982, Stadler 1989.
[65] One of his earliest publications (1903a) sounded a distinctly paternalistic tone and exaggerated the importance of the merely cultural improvement of the lower classes.
[66] Neurath 1910a [1981], p. 45, our italics.  [67] Ibid., p. 44/5.

*From the Duhem thesis to the Neurath principle*

element of universal science? What prompted him to extend the holism of theoretical physics to all sciences? What prompted his truly radical anti-foundationalism? Having surveyed in section 2.3 what the Boat denies – and how the denial is carried through – we shall turn to the positive moral in section 2.4.

## 2.3 FROM THE DUHEM THESIS TO THE NEURATH PRINCIPLE

We shall begin to explore Neurath's holism by considering his normative anti-foundationalism and then turn to his descriptive and metatheoretical anti-foundationalisms and the steps of reasoning required for them. The latter two mark his separate path of development: these anti-foundationalisms also took longer to find articulation. They will lead us to the first Boat.

### *2.3.1 Normative anti-foundationalism*

Neurath's social science background is most evident in his theory of value. It is often overlooked that questions of value held a special interest for Neurath, the historian, economist and sociologist. These disciplines had a long history of discussions of the scientific propriety of value statements. As early as 1892 Simmel had declared unconditional value statements unscientific and Neurath himself took part in the relevant debates in 1909 and 1913.[68] It is also important to see that Neurath did not disparage scientific concern with value statements altogether and that he accepted them as perfectly respectable in their proper form. It is only against this background that it is possible to appreciate both his denial of first- and second-order normative foundations and his recourse to practical interests in providing non-foundational orientations for practical and metatheoretical reasoning.

Neurath had a theory of value of his own. Not only did it encompass many features associated with the later non-cognitivisms of the Vienna Circle, but it also stood in the tradition of the views of the classical modernists Tönnies, Simmel and Weber. These writers responded in various ways, on the one hand, to neo-Kantian value philosophy and its claim to legislate the form and general content of historical and social science, and, on the other, to the legislative pretensions of the paternalistically reformist social scientists long associated with the Verein für

---

[68] Neurath 1910b and 1913c; Simmel 1892a.

Sozialpolitik. Whether there was room for a philosophy of value or not – and here already the classical theorists seem to have differed – they agreed that there was no room for unconditional value statements in social science.[69] Neurath also agreed with the thesis of the value freedom of science. Most succinctly put, its advocates hold that social science can investigate what is in fact held valuable by people, but it cannot posit values as binding.

As we learnt in section 1.2.3 Neurath's social Epicureanism centred on the concept of a standard of living. Neurath's theory of living standards followed Aristotle in holding that economics deals with relations of exchange in so far as they determine wealth and poverty. These relations are to be investigated with the help of abstract symbolic calculi. Neurath constructed 'value calculi' that did not require measuring wealth or poverty in terms of quantities of some unit, say money.[70] A mathematically exact treatment of a subject matter was possible not only if it could be conceptualised in terms of measurable quantities, but also if merely comparative magnitudes were employed. By means of ordinal rankings of outcomes of commodity transfers under different conditions Neurath's calculi sought to render perspicuous which economic exchanges of commodities and which distributions of resources across a population fulfilled given desiderata better than others. Economics itself did not prescribe these desiderata.

Neurath followed the Austrian school in holding value to arise from the relation between human want and available quantities of commodities.[71] This conception is subjectivist in that it does not derive value from an inherent quality of the commodity. In his comparative calculi Neurath equated the wealth of an individual with the (comparative) quantity of an individual's 'pleasurable . . . sensations and feelings' and poverty with the (comparative) quantity of unpleasurable sensations and feelings.[72] As was typical for his time, Neurath did not distinguish between psychological and abstract conceptions of utility, even though his rejection of unitary measures distanced him from naive psychologism. But it would be wrong to attribute to Neurath a reductionist view of value. What people do find pleasurable, i.e., valuable, is not restricted by this conception. Thus his calculi were designed to represent also people's beliefs about the desirability of certain economic exchanges,

---

[69] Simmel 1892a; Weber 1904, 1918; Tönnies 1911.
[70] This calculus was programmatically announced in his 1910a, presented in detail in his 1911 and figured centrally in his economic thinking throughout his life; see, e.g., his 1917a, 1919a, 1925a, 1935d, 1937b, 1944.    [71] Neurath 1910d, p. 1.    [72] Neurath 1910a [1981], p. 41.

even altruistic beliefs, i.e., beliefs that take the desirability of economic exchanges for other people into account.[73] For Neurath talk of value was acceptable if valuations were understood to express subjective states of happiness. When he asserts that 'Every value statement is also an empirical statement' Neurath is best read as focusing on the expressive functions of value statements, not on the truth-conditions of valuations.[74] The expressivist reading provides empirical sense for value statements. Note that this subjectivist interpretation of value did not commit Neurath to a subjectivist view of ethics, for he doubted whether systematising such value statements would ever amount to a proper discipline, issuing in 'oughts' like ethics.

Neurath distinguished between unconditional and conditional value statements and assigned to them different grades of scientific respectability. Unconditional value statements can be taken as empirical statements once they are understood as expressions of subjective opinions, which his calculus accepted as data. Conditional value statements are equally empirical, but unlike the former do not merely serve as data but also figure in scientific arguments. They represent testable assertions concerning the appropriateness of proceeding along one or another path of action, given a specific goal (and often further preferences). 'Ought' locutions in their traditional form, by contrast, were barred from the language of science for they exhibit what amounts to a misleading grammatical form: 'The imperative is not a statement at all; it is but a means of suggestion.'[75] Imperatives can only be investigated in terms of their causal role in the production of behaviour. Understood this way, economic values are empirical:

> If science must describe facts, then it must also describe those effects which we designate as pleasurable or unpleasurable sensations or feelings. If someone says: 'This is pleasurable for me', then this assertion is of the same type as 'This is red', or 'This is sweet', or 'This is painful'. Whether I feel displeasure because of the social order or because a tooth nerve is stimulated comes to the same thing.[76]

Statements about somebody's displeasure or pain are as much empirical statements as those about an organism's spatio-temporal location. Whether the displeasure is derived from bodily sensations or from the violation of ethical convictions does not amount to a categorical differ-

---

[73] Neurath 1911, pp. 70f.
[74] Neurath 1910a [1981], p. 42; compare the distinction between 'characterising' and 'appraising' value-statements in Nagel 1961, pp. 492–5.
[75] Ibid., p. 41; cf. p. 43. These views were amplified in 1913c.   [76] Neurath 1910a [1981], p. 42.

ence for scientific value talk, which can handle either. Similarly Neurath held that 'Whether we approve of wealth or not has nothing to do with the investigation of [its] causal context.'[77] Economics deals with the maximisation of wealth. The practical interest of its results does not spoil its descriptive, non-normative character. Which of two rules for economic exchanges would bring more pleasure is an 'objectively answerable question'.[78] (This does not mean that such comparisons are always decidable.) Economics is best defined as the science of how certain economic arrangements affect the total wealth of a certain group of individuals.[79]

Consider now the bearing of Neurath's comparative calculus on the doctrine of utilitarianism. Does it follow from the fact that outcomes of commodity transfers can be ranked that it is possible to determine by these calculi 'the right' form of socio-economic organisation? As noted in section 1.2.3, Neurath provided a negative answer in 1912. He started from the observation that in the absence of a unitary measure maxima of pleasures cannot be calculated in all cases. To resolve the question of how the principle of maximum happiness is to be employed – for instance, whether the distribution should be egalitarian or not – one needs what modern decision theorists call an 'arbiter'.[80] This does not mean that Neurath's calculi are useless: they remain adequate for economic theory once a certain ranking of principles is presupposed (e.g., the evaluation of the comparative ability to produce wealth of two economic systems, or for the theoretical investigation of certain utopian projects).[81] But this is not enough to satisfy the desire to establish a definite answer for practical socio-economic questions in all cases.

Utilitarianism can be viewed as an attempt to make talk of moral values objective: one ordering of social preferences is objectively distinguishable as best by the concept of the pleasure maximum. Given his own results, however, Neurath concluded that acceptance of a utilitarian conception of the moral good does not provide the desired objectivity for unconditional value judgements: '[A] moral demand can never be "proved"'. With moral demands 'unprovable', socio-economic and political demands become 'unprovable' too.[82] The problem of coordination brings the utilitarian pretensions to a fall. It is also clear that on this point social Epicureanism does not fare better either. The possibility that the moral presuppositions of individuals engaged in finding a

---

[77] Neurath 1911, p. 92.  [78] Ibid., p. 95.  [79] Ibid.  [80] Cf. Köhler 1982.
[81] Neurath 1912 [1973], p. 118, 1913d, p. 84.  [82] Neurath 1912 [1973], p. 119.

socio-political consensus may differ and that consensus may not be established terminates any hope that the norms embodied in utilitarian or social Epicurean social theories can be given objective foundations.

In general it is not possible to create an order of life which takes equal account of different views as to the best distribution of pleasures, as would have to be the case with the pleasures of each in a purely utilitarian world. One cannot determine in general how these contradictions will solve themselves. Perhaps struggle will decide which view about the best order of life will be victorious; perhaps preference will be given to one order out of those in question, and the choice will be made with the help of an inadequate metaphysical theory or in some other way; tossing coins would be much more honest.[83]

There are no normative first-order foundations. Neurath's comparative calculi bear out his pronouncements on value talk in science. Human happiness may be comparatively assessable by means of certain indices, but that does not mean that any substantive conclusions about how happiness ought to be pursued can follow. There is no way to settle in empirically non-question-begging terms which distribution of preference satisfaction is optimal. Viewed as a normative principle, the rule of pleasure maximisation cannot be scientifically established. Utilitarian philosophy cannot provide an empirical basis for values in social science.

### 2.3.2 Radical descriptive anti-foundationalism

Neurath's Boat both extended Duhem's holism and radicalised his anti-foundationalism. Neurath's holism is often regarded as an instance of Duhem's. Yet Neurath himself claimed:

Poincaré, Duhem and others have adequately shown that even if we have agreed on the protocol statements, there is an unlimited number of equally applicable, possible systems of hypotheses. We have extended this tenet of the indetermination of hypotheses to all statements, including protocol statements that are alterable in principle.[84]

It is very important for understanding Neurath's holism to understand why he was right to make this claim. We will consider in section 3.4.1 what this claim implied when Neurath made it in 1934. Here we want to point out that Neurath's claim is also right as concerns the radical anti-foundationalism which found expression in his first Boat of 1913. His

---

[83] Ibid., p. 122; cf. his earlier discussion 1911, pp. 38ff.
[84] Neurath 1934 [1983], p. 105, transl. of *Unbestimmtheit* changed from 'uncertainty' to 'indetermination' and 'not a limited number' to 'unlimited number'.

claim marks the path from the Duhem thesis to the Neurath principle. To see how Neurath extended Duhem's conclusion to all sciences and radicalised one often neglected step of his argument, we must briefly consider Duhem's conventionalism.

Duhem and Poincaré proclaimed the irreducibility of high-level postulates and hypotheses and opposed reductionism with their holistic conception of scientific theories. Duhem insisted that high-level hypotheses have to stand the test of experience, even though there are no crucial experiments whose outcome will force the rejection of a hypothesis. Duhemian holism, as endorsed by Neurath, consists in the combination of the thesis of nonseparability with that of underdetermination. The former thesis holds that scientists can never submit an isolated theoretical hypothesis to control by experiment, but only a whole set of them which includes statements about the instruments used in the experiment; the latter thesis says that an infinite number of different such sets are equally admissible given the same experimental data.[85]

Duhem's holism depends on his view of the language of science. Duhem emphasised the mathematisation of physical science and investigated its consequences. 'A physical theory ... is a system of mathematical propositions, deduced from a small number of principles, which aim to represent as simply, as completely, and as exactly as possible a set of experimental laws.'[86] Duhem distinguished three levels of abstraction from everyday observation and two types of cases in which the determination of scientific theory by 'the evidence' fails. Physical theory correlates with everyday observation ('practical facts'), first, a level of mathematical formulations in terms of measurable quantities ('theoretical facts'), second, a level on which these 'theoretical facts' are systematised by equally mathematically formulated laws ('experimental laws', often called 'phenomenal laws'), and, third, a level on which these experimental laws themselves are systematised and unified in a theory encompassing various sub-fields of inquiry (what he and Poincaré called 'hypotheses'). The two cases of the failure of determination are 'the indetermination of theoretical facts and practical facts',[87] also called 'symbolic indetermination' by Duhem, and the underdetermination of a theory's hypotheses by experimental laws. The first concerns the logic of scientific language, the second the logic of theory testing. Somewhat crudely, both cases exemplify failures of reduction. Definitionally, claims

---

[85] On Duhem see, e.g., the papers in Harding 1976, and Schäfer 1974, Giedymin 1976, Cartwright 1983 and Ariew 1984.   [86] Duhem 1906 [1962], p. 19.   [87] Ibid., pp. 144ff.

in the language of theory do not reduce to claims in the language of observation and, evidentially, even certified contrary evidence does not single out one particular hypothesis.

The first failure of determination of theory is due to a difference in the linguistic frameworks within which practical and theoretical facts are comprehended. For Duhem the (mutual) indetermination of practical fact and theoretical fact results from the fact that intuitive and mathematical conceptualisations are each embedded in different types of representational frameworks: one defines terms individually, the second logically by axioms. Nor can the meanings of the terms of the scientific language proper be established by postulating a determinate correspondence between them and an isolatable feature of experience, because scientific terms are far more precise than the vague terms of practical observation, 'everyday testimony'. This leaves the very meanings of scientific terms empirically underspecified.

The underdetermination of theory that holds for the hypotheses that unify experimental laws is different. For Duhem testing is governed by the principle that 'the physicist who carries out an experiment, or gives a report of one, implicitly recognises the accuracy of a whole group of theories'.[88] Duhem did not deny that in cases where an experimental result contradicts the prediction, it is possible to reject a hypothesis with the help of which the prediction was derived, but he pointed out that the scientist is not compelled to do so. Nothing in the logical structure of the theory stops scientists from shifting the weight of the experimental contradiction away from the disputed hypothesis to another one also employed. Duhem assigned to 'good sense' (*bon sens*) the task of dealing with the great latitude left by the rules of logic. Logic tells us not to hold hypotheses that contradict each other and to reject 'mercilessly' any theory whose consequences are 'in plain contradiction with an observed law'.[89] Beyond that, good sense must decide when a theory is satisfactory.

Clearly physical theory proper in Duhem's sense (the hypotheses unifying the experimental laws) is *under*determined. Were it not for the phenomenon of *in*determination of practical and theoretical fact, we might even think the bottom level of the language of physics well founded. But the descriptive base of the language of theory also fails to provide for foundations due to the phenomenon of indetermination. It may appear then that Duhem too held that 'any statement can be held true come what may, if we make drastic enough adjustments elsewhere in the

---

[88] Ibid., p. 183.   [89] Duhem 1906 [1962], p. 220.

system'.[90] Yet it is not clear whether Duhem gave up on all and any foundations for knowledge. What speaks against the view that Duhem denied foundations to all scientific knowledge is first that he made his holistic claims only for theoretical physics.[91] More importantly, Duhem did not cast doubt on the well-foundedness of natural language. Indeed the certainty he ascribed to everyday testimony[92] suggests that he saw physical theory as a special case of uncertainty. We will not pursue Duhem's answer further here. It is enough to see that for close readers like those in Hahn's circle, question of the extent and depth of the foundationlessness in science did arise.

Given this sketch of Duhem's argument we can now ask what enabled Neurath both to extend Duhem's holism and to radicalise his anti-foundationalism. We begin with Neurath's horizontal extension of Duhem's holism to all sciences. Like his normative anti-foundationalism, this extension reflects his deep background in German and Austrian social science before the First World War, which led him to realise that all scientific theories make use of an abstract conceptual apparatus.[93] Tönnies' sociology provides one example. It was divided into a 'pure' and an 'applied' sociology. His *Gemeinschaft und Gesellschaft* was an example of pure sociology in which he attempted to provide an abstract reconstruction of social relations.[94] Social relations are imperceivable by the senses. The task is to analyse them into their component parts and to 'represent these elements by concepts, irrespective of whether their pure form ever attains reality'.[95] Tönnies started from a psychological standpoint and employed two pairs of theoretical constructs as his guiding notions. The pair 'natural organic will' and 'artificial reflective will' organise the motivations and legitimations entering into the two broad categories of social relations, relations of 'community' (*Gemeinschaft*) and 'society' (*Gesellschaft*), and their various mixed modes. By means of these abstract concepts Tönnies sought to describe and explain the process Weber later called the 'rationalisation' of life. This pure sociology needs to be complemented by empirical studies: applied sociology. In these studies the concepts of pure sociology are applied to order the mass of data.

Simmel's work provides the second example. His 'formal' sociology

---

[90] Quine 1951, p. 43.   [91] Harding 1976, Ariew 1984.
[92] Duhem 1906 [1962], p. 163; cf. Rey 1906 [1908], pp. v–vi, 290–1, 322–3.
[93] For general surveys of early-twentieth-century German sociology (apart from the *Streite* (see note 20 above), see, e.g., Freund 1978 and Liebersohn 1988. (On Weber's ideal type theory see Oaks 1988.)   [94] Tönnies 1887. On Tönnies see, e.g., Jacoby 1971, Cahnman 1973, Bickel 1991.
[95] Tönnies 1907, pp. 8–10

consists in the analysis of social forms, be they those of specific configurations (like institutions of church and state) or of general forms of social interaction (such as competition or imitation).[96] Like Tönnies, Simmel held that social relations are not 'intuitive'.[97] Simmel suggested that society be thought of as a structure of reciprocal relations and effects (*Wechselwirkung*) that does not simply happen in, but constitutes society and its socialised individuals through the process of 'sociation' (*Vergesellschaftung*). Simmel distinguished between the 'form' and 'content' of these sociations. 'Content' is whatever is required to realise a social relation, i.e., persons, their interests, etc. Their reciprocal relation creates and sustains the categorically distinct 'form' by a process of 'compression' or 'solidification' (*Verdichtung*).[98] The forms of sociation – 'the embodiment of social energy in structures which exist and develop beyond the individual'[99] – are defined in terms that abstract from the particular individuals who enter into the relations. These abstract terms allow for the classification of apparently disparate social phenomena.

Whatever the details and, for Neurath, objectionable features of their versions of ideal-type theories, Tönnies' and Simmel's (and Weber's) sociologies demonstrate a need for abstract concepts in the social sciences not unsimilar to that in the natural sciences. Neurath concluded that all sciences possess (or can possess) an abstract vocabulary and that therefore all sciences exhibit the holism – and the underdetermination – attributed to physical theory by Duhem. The task set for the philosophy of social science is not different in kind from that of the philosophy of physics.

That tells us about the breadth of underdetermination and holism. But how deep was Neurath's anti-foundationalism? Suppose we accept that all sciences employ abstract calculi containing hypotheses that are underdetermined by the evidence and tested only holistically. We could

---

[96] E.g., Simmel 1890, 1908. On Simmel see, e.g., Wolff 1959, Frisby 1981, Dahme and Rammstedt 1984, Rammstedt 1988. [97] Simmel 1908 [1992], p. 37.
[98] Simmel 1890 [1989], p. 134, cf. 1898 [1992], p. 315 and 1908 [1992], pp. 604, 608. For different uses of *Verdichtung* see 1892a [1989], pp. 28–9 – giving it both a biological and social meaning – and 1892b [1989], p. 318 – implicating it in the understanding of other minds. Later Jerusalem employed the term, e.g., in 1909 and 1924. Jerusalem starts with examples employed by Simmel in a passage (1908 [1992], p. 15) where he discusses *Verdichtung* without using the term; Jerusalem goes on to make *Verdichtung* the generative principle of everything intentional and social. The term *Verdichtung* would seem to go back to the *Völkerspychologie* of Lazarus and Steinthal. Translated as 'condensation', *Verdichtung* is also used by Freud, of course, as denoting one of the essential modes of unconscious processes – 'a sole idea represents several associative chains at whose point of intersection it is located' (Laplanche and Pontalis 1967 [1973], p. 82) – first (published) in Freud 1900, chs. 6 section A and 7 section E. [99] Simmel 1908 [1992], p. 15.

still treat observation statements realistically and merely consider the unifying hypotheses as instruments. But suppose we also accept Duhem's argument for the indetermination of theoretical by practical fact. The only conceivable way now to provide any kind of foundation for our means of cognition would be to hold that commonsense language is well founded in the required sense or to hold that, somehow or other, the operational definitions for theoretical terms are tied directly to experience without any symbolisation interfering.

Both strategies fail. The first does so because just as the language of science is unable to disclose bare reality, so the language of common sense fails to provide anything but makeshift conceptualisations of the world of experience. Already the very need for a special scientific language demonstrates the futility of thinking natural language well founded. Moreover since all conceptualisations are interest-relative, as Mach had pointed out, the terms of 'everyday observation' that are embedded in customary practical ways of life are surely no less interest-laden than those of scientific languages. The second strategy for avoiding all-out anti-foundationalism presupposes *per impossibile* that we can get at experience in the raw. If these two points are accepted, then indetermination and uncertainty pervade science both at its abstract hypothetical and its concrete observational end.

Was Neurath by 1910 prepared to reject observation reports that contradict a theory's predictions instead of changing them? Did he already assert then what Rudolf Haller has called 'the Neurath principle', that confronted with a recalcitrant observation sentence we must chose either 'to change the sentence to be integrated or to change the system'?[100] This is the view Neurath adopted in 1913:

Whoever wants to create a world-picture or a scientific system must operate with doubtful premisses. Each attempt to create a world-picture by starting from a *tabula rasa* and making a series of statements starting with ones recognised as definitely true is necessarily full of trickeries.[101]

Here Neurath was clearly committed to radical anti-foundationalism: if no sentences are certain, then observation sentences are not certain and unrevisable either! This means that the Neurath principle was endorsed by 1913. In 1910 Neurath conceived of the choices differently (as we saw in section 2.2.3). In the face of recalcitrant experience we can either choose to amend the system as it is or to remodel it in its entirety. Both

---

[100] Haller 1982a. [101] Neurath 1913b [1983], p. 3; translation altered.

choices, it seems, leave the observation unchallenged. Neurath's view was closer to Duhem's than his radical position of 1913. (For the role of the Neurath principle in his mature thought see section 3.3.3.)

There is a difference between the Duhem thesis and the Neurath principle: Duhem left 'practical facts' intact, whereas Neurath questioned them. What then is needed over and above Duhem's thesis to legitimate the Neurath principle? The answer is: anti-foundationalism at the level of practical fact. Duhem's rejection of crucial experiments springs, as a matter of logic, from the underdetermination thesis. Yet it would be a mistake to think that the Neurath principle instead simply springs from the indetermination thesis. Indetermination denies the unequivocal match of theory and observation but does not yet place both upon the same level of uncertainty. That observation claims can be rejected, just as theoretical claims can be, follows once it is accepted that none of our observation reports represent raw data. This presupposes that knowledge be thought of as something essentially linguistic and not irreducibly mappable onto the experiential.

Neurath denied that natural language can provide foundations for knowledge claims; he also believed that there is no other starting point possible besides natural language. Neurath was supported in these convictions by Tönnies' conception of language as presented in his Welby Prize essay, which Neurath adopted as he continued to work on foundational problems of the sciences.[102] Tönnies' semiotics incorporates the Machian economy of thought and the 'economy of language'.[103] All thought depends on signs; common and scientific concepts are symbolising instruments. 'Thought . . . is for the main part recollection of signs, and by means of signs of other things which are denoted.'[104] Given our present concern with the descriptive foundations of language, we must ask how well these symbols are grounded in reality.

Tönnies distinguished between 'natural' and 'artificial' signs. The relations that natural signs bear to their objects are either found in

---

[102] Tönnies 1899–1900. Late in his career, Neurath recollected its importance to his longstanding preoccupation with questions of the language of scientific theories (1941a [1983], p. 217). It first acquainted him with 'terminology' which he regarded as primary even to the disciplines of syntax, semantics and pragmatics, because 'it introduces the terms which are used in these disciplines' (1941b [1981], p. 919). In a letter of 29 January 1922 to Tönnies Neurath recounted rereading his writings.

[103] Tönnies 1899–1900, p. 312. Tönnies endorsed the economy of thought as a regulative principle of thought and justified by its means the contention that 'all pure science refers exclusively to [abstract] objects of thought' (1887: p. xxiii), i.e., ideal types.

[104] Tönnies 1899/1900, pp. 293. Signs were defined functionally: 'That is a sign which acts as a sign.' (Ibid., p. 295.)

nature or made by humans (involuntarily and voluntarily). An important class of natural signs is 'human expressive movements'. They are for the most part (with exceptions for voluntarily made natural signs) 'involuntary signs of the psychical states expressed in them'. Such natural human signs are understood (their 'indwelling' meaning was discerned) by 'sympathy'.[105] The relevant sympathetic dispositions, the relevant understanding, is 'conditioned by similarity of organs, and facilitated by social feelings and habitual living together'.[106] Then there are the 'artificial signs':

> [O]ut of articulate sounds arise almost exclusively the completely different genus of signs which we oppose to natural signs as being artificial signs. Here there is no longer any natural relation or bond between the sign and that which it signifies; it is the human will alone which produces the relation of ideal association through which the word becomes sign of the thing, as also the relation through which writing becomes sign of the word, and the letter-unit becomes sign of the sound-unit.[107]

To ask whether for Tönnies thought requires language is to ask whether 'ideas' – which include perceptions, recollections and feelings – can be furnished from natural signs. There is no indication that he believed that complex thought could proceed in natural signs unaided by language, that is, without artificial sign systems. 'Signs are themselves ideas, and their connexion with the ideas signified is that which must be forthcoming to make an understanding possible. When other than natural signs are to be understood this connection can only be gained by learning.' [108] Insofar as thought requires learned conceptualisations, a 'natural' relation no longer obtains between sign and signified, but only a 'willed' one.

Social life consists in the habitual shared use of artificial signs: '[T]he social will ... expresses itself in them and settles and gives to them their meaning.'[109] Social will is formed by custom or by rational deliberation.[110] Mutual understanding is 'a kind of constructive effort' on the part of interpreters: '[F]or mutual understanding a common idea-system is as necessary as a common sign-system.'[111] Artificial, that is, linguistic sign systems are either shaped and preserved by tradition or conventionally agreed upon or determined. Natural languages are forms of the former; artificial languages forms of the latter. Artificial signs must be distinguished as either private or social signs, depending on whether

---

[105] Ibid., pp. 293–6.    [106] Ibid., p. 308.    [107] Ibid., p. 297.    [108] Ibid., p. 300.
[109] Ibid., pp. 297–8.    [110] Ibid., p. 305.    [111] Ibid., pp. 299–300.

they possess their meaning 'according to the will of one or more persons'. A private sign is understood by one person alone, a social sign is also understood by others. The meaning of a private sign is willed in contrast to, or in absence of, previous customary determination. 'But all such systems of private signs, like writing itself, presuppose an existing language, and refer to it, so that they represent signs of signs.'[112] No sign could be essentially private for Tönnies. Any private or 'subjective' meaning of artificial sign systems is 'essentially conditioned . . . by the meaning which they have in regular usage'.[113] Natural language is a social sign system; its meaning is learned in a social context. The need for a shared background of understanding against which any non-customary meanings can be determined is also brought out by the case of scientific languages. Science depends on the possession of a conventionally determined system of artificial signs. It 'forms its concepts, exclusively for its own ends, as mere things of thought'.[114] The determination of the meaning of scientific concepts is not to be confused with the unearthing of customarily accepted meanings.[115] Its meanings are conventionally determined by agreed definition.

Tönnies' semiotics did not provide thought with any recourse that would enable it to deal with private signs alone. Natural language use is something that humans grow into and that does not, in that process, receive critical scrutiny. From Tönnies' own 'terminological' efforts it is plain that natural language provides no foundations for knowledge on its own – it itself stands in need of clarification. Insofar as science creates its concepts anew, it either has to find its own foundations (if there are any) or rely on the foundationless natural language. From Tönnies' semiotics Neurath learnt that all thought (of any relevance to science) is regarded as symbolic and linguistically based and that language is social: 'Words are essentially and according to the law of their development social signs'; 'private signs . . . presuppose an existing language'.[116] To hold that language is social means that thinking, insofar as it relies on language, depends on traditional concept formations. Neurath concluded that observation cannot provide raw data but only conceptualised ones. Theories are never tested against bare observations. Observation statements can be rejected in principle for they are not rock bottom either.

We already made note in the Introduction of Neurath's view that

---

[112] Ibid., p. 315.    [113] Ibid., p. 326.    [114] Ibid., p. 319.    [115] Ibid., p. 321.
[116] Ibid., pp. 297–8 and 315.

'the phenomena we encounter are so much interconnected that they cannot be described by a one-dimensional chain of statements.'[117] Neurath's talk of multi-dimensionality holds the key to the following passage:

> The correctness of each statement is related to that of all others. It is absolutely impossible to formulate a single statement about the world without at the same time making tacit use of countless others. Nor can we express any statement without applying all our preceding concept formation. On the one hand we must state the connection of each statement dealing with the world with all other statements that deal with it, and on the other hand we must state the connection of each train of thought with all our earlier trains of thought. We can vary the world of concepts within us, but we cannot discard it. Each attempt to renew it from the bottom up is by its very nature a child of the concepts at hand.[118]

Duhem's holism is clearly alluded to, but Neurath has added a simple but powerful psychological observation about thinking. Philosophers are familiar nowadays with the insight that we cannot call into question all our knowledge at once. Neurath formulated much the same thought in pointing to the need for reliance on preceding concept formation. Scientific thinking can only be understood by recognising its temporal dependence. Any thinking depends upon at least some concepts not subject to scrutiny at the time; hence knowledge is twice over unfounded. Not only can theories only be confirmed as wholes at any one time, but our thinking at any one time also depends on the thinking that came before.

Taken together, the considerations concerning the nature of language and the diachronic dependence of thought frustrate all foundationalist ambitions. The recognition of the dependency of all thought on prior thought – we shall call it 'historical conditioning' – placed the cultural determinants of thinking in the foreground. Neurath's move from the Duhem thesis to the Neurath principle trades on the recognition of historical conditioning. 'Everyday testimony' cannot provide foundations for knowledge claims. Nor can natural language be bypassed: private languages contravene the concept of language as an essentially social sign-system. Thus the absolute authority of everyday testimony is challenged and the leap from the Duhem thesis to the Neurath principle is achieved.

---

[117] Neurath 1913b [1983], p. 3.  [118] Ibid., transl. altered.

### 2.3.3 Metatheoretical anti-foundationalism

As we saw in part 1 for several years up to 1913 Neurath had been publishing works exploring the methodology and conceptualisation required for scientific studies of economic reforms. The Boat made its first appearance in one such review of the 'Problems of War Economics'. Neurath wanted to justify the conceptual apparatus underlying the new field of war economics. He faced pressure from two fronts: from various ideal-type theorists and from biologising Conventionalists. The claims of neither camp to be providing foundations were acceptable to Neurath.

Neurath was deeply impressed by Tönnies' and Simmel's sociologies, as we have seen, but one potentially problematic feature did not escape him. Like Weber's sociology, both employed so-called 'ideal types' ('normal forms' for Tönnies). Consider Tönnies' concepts of society and community as examples. They are not the product of inductive generalisation or typification from empirical data. They were thought constructs that abstracted particular features from the phenomena at issue and thereby allowed for explanation by laws or understanding of meaningfully related complexes (depending on who employs them). Some neo-Kantians (like Rickert, who influenced Weber) regarded them as constitutive forms of the understanding that render intelligible the data upon which they are imposed, as it were, as the *synthetic apriori* of the historical sciences.

Here we touch on the *Methodenstreit* of the 1880s between Schmoller and Menger.[119] Broadly put, this dispute concerned the question whether economics should search for universal laws or instead was restricted to generalisations for particular societies during particular historical periods. Menger argued for abstract deductive theory and distinguished it sharply from inductive generalisation, whereas Schmoller required theory to be directly responsive to empirical phenomena. If economics is to aim for abstract theory in Menger's sense, it must face the question of how its conceptual tools can be legitimated. Neither the answers provided by Menger's Platonism nor by neo-Kantianism were satisfactory.

Neurath was highly critical of ideal types in the 1930s, but already in the early 1910s he considered them problematical. As *synthetic a prioris* ideal types violate empiricist sensibilities, but even on a minimalist reading they present methodological problems similar to the high-level

---

[119] Menger 1883, Schmoller 1883; cf. Cartwright 1994.

hypotheses of mathematical physics: what legitimates their use? Neurath raised the conflict between proponents of inductivist and abstract-deductivist methodologies for social science in his review of 1910.

> Some claim that for the construction of a system fitting with reality it is only required that some elementary phenomena be known; a detailed knowledge of reality is not required until one wants to prove the agreement between the system and reality. Others take the view that only the detailed knowledge of reality allows for the construction of a system of sentences which can themselves be applied again to reality in virtue of the predictions they allow.[120]

At this time Neurath failed to specify a univocal position on this dispute. On the one hand, he spoke in favour of the development of abstract theory. On the other, he criticised abstract economics built around the construct of *homo economicus*, the self-interested individual endowed with economic goods and the profit motive.[121] The *homo economicus* of neo-classical economics focuses on one aspect of economic life, ignoring among other things altruistic acts. Forbidding the interdependence between the values of different individuals, this abstraction does not have an equivalent in experience. By contrast the ranking of commodity transfers in Neurath's own economic calculi (which could represent altruistic valuations, as noted) is more realistic.

Just as Neurath did not abjure abstract calculi and modal reasoning, so he also thought that any fruitful theory must build on a reasonable knowledge of the historical conditions of its objects of inquiry: 'Only historical experience made abstract theory possible.'[122]

> Scientific progress in economics obtains once empirical complexes prompt abstractions which lead to new combinations whose reality or realisability one can investigate... In our investigations only the elements and relations between the elements are as a matter of principle empirical, the resultant complicated organisations are only partially realised.[123]

Neurath's 'elements', of course, are abstract concepts 'prompted by empirical complexes'. Social science requires their use because the relations it investigates are 'not intuitive'. What then distinguishes the abstract elements of Neurath's value calculi from the ideal types he objected to? They are 'as a matter of principle empirical'. Menger's insistence that his 'abstract' types need have no independent comple-

---

[120] Neurath 1910a [1981], p. 29.   [121] E.g., ibid., p. 33.   [122] Ibid., p. 43.
[123] Neurath 1911, p. 82. Next he insisted on the propriety of the procedure: 'Mechanics tells the machine builder also about machines which have never yet been built, once its elementary parts are known.' (Ibid., p. 83.)

ments in empirical reality flaunts this requirement.[124] In this respect Austrian neo-classical economics failed to meet Neurath's standard of empiricism, for it is committed to explanations in terms of unanalysable 'elements' that are 'as a matter of principle' unempirical – like *homo economicus*. For Neurath, ideal types 'found' metatheoretical talk as little as common sense does theory.

Consider now the problem conventionalists had in justifying the use of what they had shown were 'free creations of the human mind'.[125] What constrains their conventions; what determines Duhem's *bon sens*, for instance? Mach declined to recognise the alternative of 'the renunciation of the unification of science or the introduction of metaphysical propositions into science'. He insisted on 'the unification of science by means of the elimination of metaphysics'.[126] Yet the failure of his reductionism meant that the unity of science could no longer be taken as a fact disclosed by analysis. What then justifies the unity of science and the principle of economy?

For Poincaré the guide for the adoption of constitutive postulates in the sciences was experience, though not individual but rather 'ancestral experience'. It is 'by natural selection [that] our mind has adapted itself to the conditions of the external world, that it has adopted the geometry most advantageous to the species; or in other words, the most convenient'.[127] For Duhem the economy of thought means different things for different theorists. The desire for a systematic and abstract unification of physics (to which he and Poincaré had reduced the unity thesis) 'results from an innate feeling of ours and cannot be justified by purely logical considerations'.[128] The relevant 'type of mind' is determined by hereditary qualities.[129] For Poincaré and Duhem then, 'biologically determined convention' was not a contradiction in terms, just as an 'instinct of classification' was not for Boltzmann[130] nor a 'biologically determined synthetic apriori' for Simmel. For Simmel the very dualism of noumenon and phenomenon was 'resolved in that the forms of thought that create the world as concept are determined by the practical effects and counter-effects fashioned by our mental constitution – nothing other than the body – according to its evolutionary needs.'[131]

Neurath does not seem to have rejected such evolutionist theorising

---

[124] Menger 1883 [1963], p. 62.  [125] P. Frank 1949a, p. 11.
[126] P. Frank 1938 [1949b], pp. 88–9.  [127] Poincaré 1902 [1946], p. 91.
[128] Duhem 1906 [1962], p. 102.  [129] Ibid., p. 93.  [130] Boltzmann 1905.
[131] Simmel 1895 [1982], p. 71. The process in question was earlier called *Verdichtung*; cf. 1892a [1989], p. 29.

out of hand, as his prefaces to his edition of Wolfram's *Faust* (see section 1.1.1) and his translation of (the second edition of) Galton's *Hereditary Genius* show.[132] Yet by 1911 he criticised the argument that a monetary economy was superior to the moneyless on the grounds that 'a people without money miss the standard by which its national wealth could be measured' and cannot feel the 'stimulation . . . to increase our usable property'. 'The monetary form of economical organisation can only then be judged for its power of increasing wealth if one employs a method of calculating wealth which is independent of money.'[133] Neurath argued that references to biologically determined inclinations cannot decide how wealth is to be conceptualised in economics. His worry was that the biological approach would support those who wanted to limit economics to the theory of prices and objected to developing a theory of real income. Still more generally he observed in 1913:

> Some are of the opinion that to start with, one could reflect, and then when reflection fails, turn to instinct: this view misuses instinct by consciously introducing it as a mere stop-gap . . . Precisely if one values the significance of instinctive action so highly one should not misuse it like that. One should clearly realise that instinct must fail with respect to the complex rational relationships created by the consciously shaped institutions of the social order and modern technology.[134]

Thought, which according to Mach emerged out of instinctual life, cannot hand back the reins whence it once wrested them. At best it can let events and inclinations take their course. Much less, of course, can scientific or social conventions be defended by reference to instinct as rationally binding.

Having rejected ideal-type methodology and the strategy of grounding conventionalism by evolutionary biology, Neurath finally embraced metatheoretical anti-foundationalism in no uncertain terms in 'Problems of War Economics' in the passages leading up to his first Boat:

> Since we can only very roughly delimit our field of investigation, we shall have to include some [phenomena] which a closer analysis would eliminate; the attempt to create a system will also be tainted from the start with considerable faults. It is good to state this right away so as not to be tempted to deliver a complete system come what may. The system builder is a born liar . . . The complete system remains an eternal goal which we can only seemingly anticipate.[135]

---

[132] Neurath 1906b, Neurath and Schapire-Neurath 1910b.  [133] Neurath 1911, pp. 89–90.
[134] Neurath 1913b [1983], p. 5.  [135] Neurath 1913a, pp. 455–6.

## From the Duhem thesis to the Neurath principle

The ideal of knowledge as a complete system of propositions covering all and only the relevant phenomena must be abandoned from the start.

As Neurath's remarks make clear, the principal philosophical system builders are Descartes and Kant. But these were not the only system builders Neurath meant to oppose. Since he was concerned with empirical disciplines, he attacked not only the complete philosophical system, but complete scientific theories as well. It was popular in his day to believe that one might – as Du Bois-Reymond put it – represent 'the process of spatio-temporal events by one immense system of simultaneous differential equations, from which could be deduced, for each moment, the position, direction and velocity of every atom in the world'.[136] As we saw in the Introduction, Neurath denied the possibility of such a Laplacean formula representing the laws that govern all of nature. Incompleteness is a condition of all human knowledge, and the only truly rational stance is one that admits this. Both philosophical and scientific system builders make a similar mistake. While the philosophical system builder aims to provide secure foundations for knowledge, the scientific system builder aims for completeness of explanatory models – the empirical counterpart to the philosopher's a priori schemes. Both deny what Neurath took to be a basic fact: The knowledge possessed by us is but a partial and perspectival picture of reality. Both were 'pseudo-rationalist'. 'Pseudo-rationalism' is a crucial concept for Neurath. His arguments against the intellectual attitude it denotes in the later Vienna Circle will be discussed in section 3.3; here we explore its early roots.

Not only system building, also the all too widespread demand that every act be decided by reason is 'pseudo-rationalist':

> Once reason has gained a certain influence, people generally show a tendency to regard all their actions as reasonable. Ways of action which depend on dark instincts receive reinterpretation or obfuscation.[137]

The Enlightenment tendency to rely upon reason and rationality here outdoes itself.

> Rationalism sees its chief triumph in the clear recognition of the limits of actual insight. I tend to derive the widespread tendency towards pseudo-rationalism from the same unconscious endeavours as the tendency towards superstition. With the progress of the Enlightenment men were more and more deprived of the traditional means which were suited to making unambiguous decisions possible. Therefore one turned to insight in order to squeeze an adequate substitute

---

[136] Du Bois-Reymond 1872 [1884], p. 13.   [137] Neurath 1913a, p. 441.

out of it with all possible force. In this sense pseudo-rationalism, a belief in powers that regulate existence and foretell the future, as well as reliance on omens, have a common root.[138]

This attitude of not facing up to the limits of rationality represents retarded Enlightenment impulses. Neurath designated it 'pseudo-rationalism.'

Against the pseudo-rationalists Neurath pitted his own image of knowledge. We return to the passage leading to the first Boat:

Of course, someone could espouse the idea to suspend all political economists from their business, as was demanded by Kant for the field of philosophy in his *Prolegomena*. But this is not the way of science, much less, by the way, that of philosophy. It is always a question of the totality of problems with which we struggle throughout the centuries.[139]

Even if the 'scandal' of foundationless natural science, for instance, could not be respectably resolved, it would still not be possible to suspend its pursuit (and economics was no different). The reason was Duhemian holism and the historical dependency of thought. Neurath continued:

We are never in the position to place certain indisputable sentences at the very top and then clearly and accurately display the whole chain of ideas, be it in logic or in physics, in biology or in philosophy. That which is unsatisfactory seeps through the whole of the realm of ideas, it is detectable in the first premises as in the later ones. It is of no use to be careful and supposedly renounce knowledge already gained in order to proceed from a *tabula rasa* and improve things henceforth, as Descartes had the audacity to try. Such attempts only end with rough masquerades of insight which tend to be worse than all that preceded them. We cannot but declare truthfully that the current state of knowledge has been presupposed and that we shall try to improve matters by making changes here and there.[140]

Instead of relying on indisputable facts, a theorist must integrate his results into the totality of beliefs about the world. Fully aware of the socio-cultural dimension of his view of scientific concept formation, Neurath continued:

Our thinking is of necessity full of tradition, we are children of our time, even if we fight against it as we may; there are only ages which recognise this more clearly than others. What good did it do for Kant to try to tear himself away? Despite his eminent genius we epigones are often able to show how some of his trains of thought can only be explained by reference to the thinking of his contemporaries and elders, but impossibly so by reference to an unprejudiced view of the world. We are like sailors who are forced to reconstruct totally their boat

---

[138] Neurath 1913b [1983], p. 8.   [139] Neurath 1913a, p. 456.   [140] Ibid.

on the open sea with beams they carry along, by replacing beam for beam and thus changing the form of the whole. Since they cannot land they are never able to pull apart the ship entirely in order to build it anew. The new ship emerges from the old through a process of continuous transformation.[141]

Without descriptive, normative or metatheoretical foundations, there are no indisputable sentences. The current state of knowledge has to be presupposed, and the prejudices of the age can only be partly overcome.

To recapitulate: Neurath's unpacking of the entanglements of scientific reason extended and radicalised Duhem's result. First, he extended it horizontally to all sciences and, second, he showed it to be utterly inescapable by also extending it vertically. Not only Duhem's holism was in play, but also the historical conditioning of concept formation. Once the possibility of foundations for everyday testimony is denied and yet ordinary natural language is held indispensable nevertheless, the possibility of wiping clean the slate and making a fresh unprejudiced beginning is undermined. Every theory cannot but exhibit in some part an unquestioned reliance on the intellectual tradition to which it belongs, a reliance which can be, but has not yet been, called into question.

## 2.4 RATIONALITY WITHOUT FOUNDATIONS

Neurath's rejection of Descartes and Kant – and their foundations – had momentous consequences. Together with his earlier rejection of Wundt's (and Tönnies') distinction between theoretical and practical science it also overturned the Aristotelian distinction between *theoria* with its apodictic certainty and *praxis* with its situational understanding. All science must be understood as practical knowledge, infected with uncertainty and reliant upon educated guesswork. All science has to be viewed as the practice of human agents and cannot be fully described in abstraction from this context. Neurath offered a positive proposal: only by repositioning thought in relation to the will is it possible to provide parameters for rationality.

### 2.4.1 The primacy of practical reason

Given this radical anti-foundationalism, how could Neurath retain the Enlightenment idea of controllable rationality for science? Neurath's

[141] Ibid., pp. 456–7.

Boat teaches that rationality need not be lost just because foundations are missing. There remains the rationality of permanent reconstruction. The upshot is a radical conventionalism. The rationality of science can be determined, but only in full awareness of the relation between theory and pre-theoretical goals. By bringing instrumental considerations into play, Neurath widened the sphere of what scientific metatheory has to consider: Not only are the means and objects of cognition essential, but also conation. The third 'holism' we must consider is that of practical reason.

Neurath suggested that theoretical rationality requires appeal to external criteria, as does practical rationality. Against Descartes Neurath held that both types of reason are characterised by provisional rules:

> It was a fundamental error of Descartes that he believed that only in the practical field could he not dispense with provisional rules. Thinking, too, needs provisional rules in more than one respect. The limited span of life already urges us ahead. The wish that in a forseeable time the picture of the world could be rounded off makes practical rules a necessity.[142]

Neurath stressed that in pure theory, one 'very often finds oneself in the position of having to choose one of several hypotheses of equal probability'.[143] (The allusion to the Conventionalists' problem of motivating the choice of hypotheses is plain.) It is also no accident that, for Neurath, on the scale of rationality the system builder ranks with the superstitious. Neurath's criticism is eminently practical. The conduct of science suffers from illusions of approximations to complete knowledge. It is barred from achieving what still can be achieved in the absence of such pseudo-rationalist pretensions. Thus he concludes that 'the complete system' represents both an objectively mistaken aim – we can never reach it – and a dangerously misleading strategy – we can only seemingly anticipate it: 'The world of insight is only a small island in the sea of the unknown.'[144] The thinker is in no better position than the agent.

Neurath's conventionalism underlines the importance of decisions. Cognition works out the means and determines the instrumental aims. What sets unconditional values and fixes the final ends of human action is will – conation, not cognition. In this the will can at best be informed, its sympathies schooled. ('But this is very much', as Neurath put matters in another context.[145] On this rests the possibility of all Enlightenment efforts.) The unconditional values that can provide parameters for

---

[142] Neurath 1913b [1983], p. 3, translation altered.   [143] Ibid.
[144] Neurath 1913a, pp. 441–2.   [145] Neurath 1934 [1983], p. 109.

assessing the instrumental values studied by science lie outside science itself. Should science aim to provide control of the environment, or pleasure to the inquirer, or should it help to emancipate the toiling masses? This decision is not predetermined. The 'social will' of those concerned must turn hypothetical aims and norms into unconditional ones that guide the conduct of science. (As elsewhere, so here, '"non-action" is also an action'.[146])

The continuity of science with everyday reasoning (that Mach had already insisted on) does not deny the differences between them. Science is a more deliberative, reflective activity, and in this lies its strength. Moreover, the cognitions and conations constituting the planks of Neurath's Boat are socially mediated ones. The cognitions at issue are shared beliefs; the instrumental parameters governing the advisability of their adoption are set in turn by shared desires and aspirations. But in science, unlike in most of everyday life, it need not be common fate or the vicissitudes of individual life histories that sets these ends. They can be agreed together.

### 2.4.2 Determining the conventions of science

Neurath's Boat expresses the general claim that one must presuppose some knowledge – however vague – in order to gain more. We can see how knowledge is gained and sciences are built up in his 1911 paper, which presented details of his earlier proposals for a reconceptualisation of economics. Seeking to delimit the area of economic investigations and to establish the place of value-theoretical considerations amongst them, Neurath writes:

Admittedly, such a delimitation of the concepts [of an individual science] is of necessity of a conventional nature, but such a convention is bound to the peculiarity of the objects since the commonality of certain marks is not itself the result of a convention.[147]

For Neurath the delimitation of scientific rationality was conventional first in this respect: once we fix by convention what types of objects a discipline will study, its results will be empirical, for whether the objects picked out do or do not share certain properties, and which of the objects do so and which do not, cannot be determined conventionally. The objects of an inquiry are not simply given to us, but have to be

---

[146] Neurath 1913b [1983], p. 2.   [147] Neurath 1911, pp. 52–3; cf. 1931c, [1973], p. 388.

constituted by us. Reality does not determine what there is to be known in this sense. The empirical ground is provided by the relations of similarity and dissimilarity between conventionally determined types of objects. Empiricism demands the empirical disconfirmability of claims attributing given properties to given objects; it does not demand the intrinsic givenness of these properties.

Controllable rationality in science is thus retained at the level of the direct description of the objects of inquiry. But what about the 'control' of high-level principles and conventions? Their control becomes possible once the practical context of all theory is taken account of. Neurath held that theoretical thought, like practical thought, depends on provisional rules. In particular it requires an 'auxiliary motive':

> We have seen that in many cases, by considering different possibilities of action, a man cannot reach a result. If he nevertheless singles out one of them to put it into operation, and in so doing makes use of a principle of a more general kind, we want to call the motive thus created, which has nothing to do with the concrete aims in question, the auxiliary motive, because it is an aid to the vacillating, so to speak.[148]

Auxiliary motives are devices to deal with decision under uncertainty, e.g., the drawing of lots, leaving action to be decided by further events, reliance on oracles, omens and prophecies, the advice of people in authority, even the principle of majority. It is important that Neurath did not speak of auxiliary hypotheses. His talk of auxiliary motives renders plain that his concern was with aids for conation, not cognition. Importantly too, auxiliary motives represent strategies that must be adopted consciously for their force depends on the decision by which they are adopted and sustained. Reliance on instinct does not count as an auxiliary motive. Auxiliary motives are especially important for cooperative ventures, yet their application 'needs a prior high degree of organisation'; moreover, abiding by the auxiliary motive chosen requires 'character' of the participants.[149]

The recognition and systematic use of the auxiliary motive represents 'the culmination of rationalism'.[150] How then is true rationalism to be saved from conceptual caprice? The rational employment of auxiliary motives requires different standards than those derived from the foundational ideal. The question of their truth does not arise, only of their functionality. It is clear from Neurath's point of view that the adoption of an

---

[148] Neurath 1913b [1983], p. 4.   [149] Ibid., p. 10.   [150] Ibid., p. 11.

auxiliary motive is rationally defensible if it helps to select and sustain the selection of a desirable course of action.

Neurath's pragmatism and the notion of the auxiliary motive provide the key to conventional theory choice. Scientific knowledge is an instrument whose use can only be evaluated relative to the ends pursued. Unlike questions of the good life, the goal of the activity is much more clearly defined, namely, science aims to provide control of the environment. Besides this broadly Machian, even Baconian perspective, we might add intellectual pleasure as an aim of science or, for social science, the emancipation of the working class. Still science's instrumental nature would not be disputed. There may be room for disagreement about the tasks relative to which the criterion of utility is to be applied, but the pragmatic nature of the criterion is not in question.

So what was Neurath's answer to the questions about Mach's principles of economy or simplicity and the unity of science? Given that 'the differences between thinking and action are only of degree' and that 'thinking too needs preliminary rules in more than one respect',[151] the role of the auxiliary motive is plain. Rather than view simplicity and unity as absolute norms to which all theorising must adhere, Neurath chose to view these metatheoretical concepts in the wider practical context of all theorising. As a pragmatist, he answered that simple theories are generally most economical in terms of the intellectual effort required to comprehend their applications. It does not follow that simple theories must always be preferred. On occasion a more complicated theory might be adopted, either because it allows for greater simplification of our overall theory, or because it opens promising new fields of inquiry hitherto unknown. The aim of simplicity is an auxiliary *motive* of scientific theorising. Its rational legitimacy derives from the utility of the practical decision to adopt it. The question 'Why unity?' is answered similarly. The aim of unity is an auxiliary *motive* of scientific theorising, whose rational legitimacy derives from the utility of the practical decision to adopt it. Thus Neurath's answer to the problem of conventionalist metatheory. By attending to the relation of theory and practice, scientific metatheory restores controllable rationality within its own domain.

Finally, what of the legitimacy of critically engaged social science? Theoretical rationality requires appeal to external criteria as does practical rationality. The aims and goals of the new discipline of war

---

[151] Ibid., pp. 2–3.

economics can thus be rendered controllably rational as well. Neurath's economics aimed to provide the conceptual means for rethinking the economic organisation of society. It did not require justification over and above that claimed by any other economic theory, for instance, by neo-classical economics. Whatever kind of justification neo-classical economics might derive from its claimed utility for dealing with capitalist economies, Neurath's economics could derive the same standing from its utility for reform or revolution. In principle it is up to the practitioners to determine the context for applying the instrument of science. (How the 'social will' involved here is determined is the subject of the sociology of science.)

Neurath began, we have seen, with his father's hopes to advance the 'rational' reorganisation of economic life but resisted the intrusion of ideal values into science. We have seen how he marshalled will, decision and utility to serve instead. Fully developed, we suggested that Neurath's theory of science represents the non-reductionist *Aufhebung* of Mach's naturalism. How far had Neurath advanced towards this end by 1913? It is clear that the anti-foundationalism of the first Boat already spelt the rejection of autonomous epistemology: 'We cannot but declare truthfully that the current state of knowledge has been presupposed and that we shall try to improve matters by making changes here and there.'[152] It is also clear that Neurath envisaged that action should play the central role in the enterprise of science. The conventions of science can be understood by realising that they involve decisions taken or acceded to by the participants. Here the distinctive potential of Neurath's image of knowledge began to be realised. Clearly, however, much remains to be explained and much to be explored.

### 2.4.3 *The second Boat: one world*

Neurath's first Boat argued against foundationalist theorists of knowledge. Neurath's second Boat was equally directed against 'pseudo-rationalists' but of a different kind. Here the enemy was the unbridled relativism supposedly encouraged by the absence of foundations. The second Boat complements the first and provides a thematic link to the third. It featured in his extensive critique of Spengler's *The Decline of the West*, written, as we learnt in section 1.3, during his imprisonment in 1919 and published in 1921. Spengler's book pretended to be a world-

---

[152] Neurath 1913a, p. 456.

historical analysis of cultural trends issuing in the pronouncement of the impotence of rational inquiry, the prediction of the inevitable decay of Western democracies and the rise of a new martial age. Neurath's critique sought to expose it as 'a treasure chest for anyone who seeks excuses for unscientific behaviour'[153] and to clarify the proper nature of rational, scientific inquiry. In so doing he hoped to counteract the disastrous effects Spengler's doctrines threatened to have on the fledgling democratic consciousnesses of the first German and Austrian republics.

Neurath's was a metatheoretical reflection in an eminently practical, self-consciously educational context.[154] His pronouncements ring familiar, yet they are put forward with a new urgency. The introductory paragraphs restate his critique of pseudo-rationalists with rarely paralleled vigour: 'The necessary unity in action is undermined if insight by itself is to bring the final decision.'[155] The arguments which the Boat serves in Neurath's attack on Spengler make clear again that his rejection of philosophy was not motivated by a reductionist scientism. In *Anti-Spengler* Neurath took a respectful attitude towards existential quandaries, which 'scientific philosophy' typically marks as 'problems of life' that it cannot address.[156] Indeed, the context of Neurath's second Boat brings out an important presupposition of his conception of science.

Spengler's philosophy proclaims an insuperable relativism. Not only are the emotional states of members of foreign cultures ineluctably alien and unfathomable, so too are their cognitions. Neurath's criticism focuses on Spengler's view that 'the statements of one culture about any facts cannot be judged by another . . . "Truths exist only for a certain kind of men".'[157] To rebut Spengler's claims Neurath is forced to address worries that arose from his own anti-foundationalism. (Consider the remark: 'I always speak of "systems of hypotheses" without wanting to delimit exactly what is denoted as a hypothesis, what as reality.'[158]) Neurath first presents principled considerations about the impossibility of entertaining beliefs and concepts radically disjoint from those of the people around us. Second, he considers the mechanism by which people interpret each other and explores the nature of the commonalities presupposed in this process. Third, he shows that these commonalities extend to communities that Spengler claimed to be beyond the possibility of communication. Finally, he suggests that the commonalities

---

[153] Neurath 1921a [1973], p. 206.   [154] See esp. the final paragraph, ibid., p. 213.
[155] Ibid., pp. 158–9.   [156] Cf. Carnap 1928, §183.   [157] Neurath 1921a [1973], p. 197.
[158] Neurath 1916 [1983], p. 23.

presupposed for communication provide a sufficient basis for the objectivity of science.

Neurath begins his argument with a consideration of the 'basic principles' involved in considering 'world-views'.[159]

> We cannot start from a *tabula rasa* as Descartes thought we could. We have to make do with words and concepts that we find when our reflections begin. Indeed, all changes of concepts and names again require the help of concepts, names, definitions and connections that determine our thinking. When we progress in our thinking, making new concepts and connections our own, the entire structure of concepts is shifted in its relations and in its center of gravity, and each concept takes a smaller or greater part in this change. Since in the field of world-views our language, our writing, our thinking are, at least until now, constituted in two dimensions, one should after every advance of thought really repeat in the new sense what has hitherto been said – and this holds likewise for every account in books.[160]

We recall these points as part of the negative argument of the first Boat. Next Neurath considers a way in which it might be thought possible to avoid the anti-foundationalist conclusion. He envisages an enterprise inspired by Mach's old project of a 'universal script': did it possess a superior epistemological basis? Neurath continued:

> One cannot complete a piece once and for all and then go on to the next. This kind of relationship is coped with in mathematics by indicating functional connections. Will we someday learn to present philosophical structure in this way too? There we are not dealing with clearly outlined concepts as in mathematics, they are barely defined in their internal parts; hazy edges are essential to them; these concepts are partially to clear up an indefinite confusion which is tangled up in the most varied ways.

What frustrated the Machian dream is that concepts in 'philosophic structures' are also afflicted by what in 1915 Neurath had called the 'hazy edges', or 'blurred margins', that afflict the concepts of scientific theory.[161] As illustration he offered another simile:

> Not infrequently our experience is like that of a miner who at some spot of the mine raises his lamp and spreads light, while the rest lies in total darkness. If an adjacent part is illuminated, those parts vanish in the distance which were lit only just now. Just as the miner tries to grasp this plenitude in a more concise fashion by plans, sketches, and similar auxiliaries, so we endeavour to gain some yield from immediate observation by means of conceptually shaped results and link it up with other yields. What we set down as conceptual rela-

---

[159] Neurath 1921a [1973], p. 198.   [160] Ibid., p. 198, translation altered.
[161] Neurath 1915, 1916.

## Rationality without foundations

tions is, however, not only a means for understanding, as Mach holds, but also itself cognition as such. A god would 'look at' the logical connections even if, unlike humans, he had no need of them in order to cope with the remaining plenitude.[162]

Once again Neurath gives expression to what earlier we called 'cognitive voluntarism'. The miner simile suggests that the concepts of science constitute cognitive results in their own right. They create structure and so make it possible to comprehend experience. Neurath continued:

> That we always have to do with a whole network of concepts and not with concepts that can be isolated, puts any thinker into the difficult position of having unceasing regard for the whole mass of concepts that he cannot even survey at once, and of letting the new grow out of the old. Duhem has shown with special emphasis that every statement about any happening is saturated with hypotheses of all sorts and that these in the end are derived from our whole world-view. We are like sailors who must reconstruct their ship on the open sea but are never able to start afresh from the bottom. Where a beam is taken away a new one must at once be put there, and for this the rest of the ship is used as support. In this way, by using old beams and driftwood, the ship can be shaped entirely anew, but only by gradual reconstruction.[163]

So much for the 'basic principles'. More clearly than before Neurath's stress on the tangle of concepts dismisses the thought that it is possible to overcome indetermination, namely by a kind of ideal language ('philosophical structure') untainted by the foundationlessness of ordinary language. Even more radical problems for scientific method not yet noticed will arise from the lack of clear definitions and hazy edges. The discussions marked by his third Boat will make this clear.

What makes the second Boat relevant to the question of cultural relativism? Neurath drew attention to the socially shared content of our cognitions, our beliefs: 'Let us start, with Spengler, from the fact that there are many people who communicate with each other.'[164] We cannot repudiate in its entirety the network of beliefs we share with other speakers of our language. But how widely would our beliefs be shared across humanity? Neurath rejects Spengler's claim that truth exists only relative to certain types of humans. Spengler individuated these types by reference to the 'arch-symbols' of their cultures, but, Neurath argues, no good reason is provided for doing it his way. 'Indeed, why does Spengler not say even more generally: "truths exist only relative to one definite

---

[162] Neurath 1921a [1973], p. 198; translation altered.  [163] Ibid., pp. 198–9; translation altered.
[164] Ibid., p. 199.

individual" or more generally still: "relative to a moment"?'[165] Against cultural relativism Neurath sets a rudimentary sketch of the hermeneutics of communication that are presupposed even by science – and that support its claim to objectivity.

Communication with members of our own communities – with whom we share many beliefs – does not differ in principle from communication with members of alien cultures. 'Putting oneself into other people's frames of mind always starts from the existence of some common features to which the rest is tied.'[166] What poets do – 'take details familiar from life, intensifying, diminishing and mixing them in all sorts of ways' – is only the conscious use of a principle employed in ordinary communication. Other minds are known by mostly unconscious inferences that exploit the commonalities between people.

> All [human] expressions are woven into structures ... Emotions and views, whose expressions they are, are experienced in our own case immediately; in the case of others we infer them. A foreign language is understood in so far as we grasp what the words mean, with what emotions, what trains of thought, what views of the world they are commonly connected.[167]

The inferences exploiting the commonalities with others are drawn in a process of interpretation. But can enough such commonalities always be established to allow for the possibility of communication?

> In that we do recognise [others] as people there already lies the assumption that they have something in common with us ... In order that two people might talk to each other at all they require certain things in common. If these were lacking, the two 'humans' would confront each other as two quite alien creatures; words and gestures would not even be 'meaningless signs', they would be mere changes.[168]

The very presuppositions of attempts to interpret another organism as a fellow human being, Neurath held, belie relativism. Not only do we presuppose a similarly intentional constitution of mind in others, a mind that possesses beliefs and desires, but even broadly similar types of contents for beliefs and attitudes are presupposed.

Note then that Neurath does not deride empathy or 'understanding' (*Verstehen*). Just such intuitive fitting of phenomena into a 'form of life' – a term which, like 'language game' predates Wittgenstein[169] – was charac-

---

[165] Ibid. [166] Ibid. [167] Ibid. [1981], p. 145 (this passage dropped in translation in 1973).
[168] Ibid. [1973], pp. 199–200.
[169] Toulmin 1970 (quoted in Eschbach 1988) traced 'forms of life' to Eduard Spanger's *Lebensformen* (1914), Weiler 1970 traced 'language games' to Mauthner 1901.

teristic of the understanding of social life in Tönnies' 'community' (as opposed to 'society').[170] Of course, unless it is further constrained, it is not the stuff that science is made of. Controllable rationality imposes further criteria of confirmability on such empathetic interpretation. What Neurath stresses is that empathy has to proceed from the same behavioural and environmental data as conscious inferential reasoning about other people. All interpretation is grounded in a common world. Thus he notes that:

> [E]very cosmos comprises facts that are quite unambiguously connected with the facts of other cosmoi, and alongside the changing aspects there arises something lasting and common. The facts that the seasons follow each other, that fire burns and wine makes you drunk, are common to all world views. To determine the upper limit of what is in common is a task that many assign to science.[171]

Broad structural similarities of environment and human constitution and similarities of interactions between humans and their environment account for the shared aspects of all human experience. Only on this basis is interpretation possible. But this basis also provides the intersubjectively accessible ground of science. In order to defend the objectivity of science against the relativist challenge, Neurath appeals to the presuppositions of any and all communication. Spengler had allowed for some communication to start with. Neurath argues that, if that is granted, then communication can occur between any two humans and that already provides enough common ground to pursue science. This common ground is the one world in which we act.

The context of Neurath's second Boat thus brings out not only that an intersubjectively discernible world provides the natural context of scientific thought. It also reveals another important naturalist presupposition in Neurath's thought: processes like intersubjective communication and interpretation are part of the world that we must consider 'natural' for humans. Note, however, that his project remains incomplete. Clearly, unifying world-views were regarded as belief-systems, not as systems of ownerless propositions or sentences. How did his radical conventionalism sit with this psychologistic naturalism? Do the conventions of science reduce to decisions to believe? At this point Neurath's alternative conception of knowledge did not advance the action-theoretic approach indicated by the first Boat. Only in the Vienna Circle proper did Neurath take the final step in the development of his scientific

---

[170] Tönnies 1887, Bk. 1, §10.   [171] Neurath 1921a [1973], p. 206.

metatheory, by completing the linguistic turn in his own characteristic way. Then at last he moved from a rough conception of scientific belief-systems to a more detailed theory-sketch of science as a distinctive form of public discourse.

## 2.5 A THEORY OF SCIENTIFIC DISCOURSE

The first two Boats leave us at sea, having to presuppose what we think we know. There is no way to find an Archimedean point untainted by the historical contingencies of concept formation, the facticity of thought. As Neurath put the matter in the days of the Vienna Circle proper: '[I]t is impossible to go back behind or before language.'[172] With the Circle, of course, Neurath took the linguistic turn. Philosophy concerned not the objects of scientific inquiry but rather its linguistic representations. But in a way the resultant view was not unfamiliar for him. 'As makers of statements, we cannot, so to speak, take up a position outside the making of the statements and then be prosecutor, defendant and judge at the same time.'[173] What is new in such statements is the emphasis on statements. The self-conscious exercise of the linguistic turn had one all-important consequence. It led Neurath to confront the infirmity of his anti-foundationalism as expressed in the Boat: first rushing towards an action-theoretic framework for the theory of science, then seemingly falling back into world-view psychologism. In the Circle, Neurath focused on the making of statements and of proposals for how to make statements. The action-theoretic approach was applied to the domain of linguistic representations. Personal beliefs became secondary to public discourse.

The third Boat marks the articulation of Neurath's theory of science as a theory of scientific discourse. What prompted this articulation was, on the one hand, the challenge from Carnap's 'rational reconstruction' of scientific knowledge within the framework of 'methodological solipsism'. Carnap intended to circumvent the social basis of science by the reconstruction of a clean language of experience. The Vienna Circle's protocol sentence debate was the forum for Neurath's response to this challenge.[174] Neurath had to argue that the social character of science cannot be circumvented and to show how the empirical base of science is to be conceived and empiricism upheld.

---

[172] Neurath 1931b [1983], p. 54.   [173] Neurath 1932a [1983], p. 61.
[174] For a detailed reconstruction and analysis of the debate from 1928 to 1935 see Uebel 1992a, chs. 2–9.

The articulation of Neurath's conception of science as a distinctive form of public discourse was also prompted by another source. Neurath had become a card-carrying 'socialist' in Saxony in 1918, but he became a 'Marxist' only back in Vienna. In *Anti-Spengler*, Neurath's Mach had not yet properly confronted his Marx. Before turning to the philosophical debate about what followed from the linguistic turn with, amongst others, Carnap, we must first consider Neurath's turn to what Carnap described as his undogmatic Marxism.[175] The roughest of contours of the context must do here to provide the required backdrop for the emergence of Neurath's theory of scientific discourse. Part 3 will further explore details of the interaction between the 'political' and the 'philosophical' discussions in which Neurath was involved.

### 2.5.1 Anti-philosophy, Marxism and radical physicalism

In the Vienna Circle Neurath displayed little patience with philosophy. 'The last consequence of empiricism: science *without philosophy*!'[176] The contrast to earlier expressions of tolerance is striking.[177] Of course it takes a philosopher to declare the end of philosophy, but is Neurath's talk of 'anti-philosophy' merely an artificial limitation of traditional philosophy, an instance of ultimately intra-academic polemic (as much of the Circle's rhetoric is nowadays understood)?

Among the reasons for not thinking so is, first, Neurath's ever-ready vigilance against new 'metaphysics' not only in Wittgenstein's but also the Circle's philosophies, an attitude that finally unnerved even Carnap. Nor must it be forgotten that in decrying 'school philosophy' Neurath felt a happy concordance with Marx, whom he quoted on this point:

Where speculation ends – in real life – there real, positive science begins. Empty talk about consciousness ceases. When reality is depicted, philosophy as an independent branch of knowledge loses its medium of existence.[178]

Neurath once chose to tell readers of *Der Kampf* that 'what is correct does not become more correct by having also been taught by Marx and does not become false because Marx is of a different opinion.'[179] The quotation above therefore indicates a well-considered agreement with Marx.

That 'outside of science there are no meaningful sentences expressing

---

[175] Carnap 1963, pp. 23–4.   [176] Neurath 1932e, p. 311, original italics.
[177] Consider the tolerance of 'philosophical ideas' (1910a), and 'philosophic structures' (1921a).
[178] Marx and Engels 1846 as cited in Neurath 1931c [1973], p. 350; Neurath 1931d [1981], p. 411 quotes only the last sentence.   [179] Neurath 1923b, p. 288.

philosophical systems'[180] was by 1931 a well-established doctrine of the Vienna Circle, generally traced back via Schlick to Wittgenstein. Neurath's alternative to philosophy was different from Schlick's and Wittgenstein's, however. He was happier to associate himself with whatever Marx seemed to be saying. Recognition of this fact has tempted some to attribute his dissension from the Circle's orthodoxy (and his wholesale rejection of 'philosophy') to his Marxist convictions and to discount the possibility that Neurath had an argument, perhaps even a 'Marxist' one, that told against 'philosophy'. We shall point out below that Neurath did have such an argument and moreover, that it was an argument that brought to the fore a train of thought already operative in the first Boat.

The precise shade of Neurath's Marxism is still unclear.[181] It would appear that this dark horse was recognised as such already in his time. As we saw in 1.3 Neurath's Munich engagement was considered adventurist and un-Marxist, because he thought socialisation programmes could be implemented independently of how issues of political power were resolved. We may add here that when Neurath styled himself a 'Marxist' throughout his Vienna period after the First World War, he was again attacked for several heresies.[182] For his part Neurath opposed the Kantianisation of Marx proposed by such party luminaries as Max Adler, who, in his turn, once left a lecture of the Verein Ernst Mach deeply shocked by Neurath's 'undisciplined' thought.[183] Neurath's peculiar mixture of 'scientism', 'socialism' and 'utopianism' always threatened to offend. Yet he also knew how to be diplomatic. Once Reichenbach, in his capacity as radio-producer in Berlin, confirmed a planned broadcast by Neurath entitled 'Of White, Red, Brown and Black People' for 17 June 1931, 5.45–6.05 p.m., and asked him especially to avoid offending the churches. Neurath responded: 'I know how to handle this since I spend my life on the most different levels of neutrality at the same time.'[184] Given the delicate task of delimiting himself against numerous, sometimes incompatible partylines – as we saw in section

---

[180] Neurath 1931c [1973], p. 407; translation altered.
[181] Cf. Glaser 1976, Nemeth 1981, 1990, Müller 1982, Cat, Cartwright and Chang 1991, Hegselmann 1979, W.R. Beyer 1982 and Paul Neurath 1994; see also the following note.
[182] E.g. Pick 1920, Leichter 1923, H. Bauer 1923. For more recent Marxist criticisms see esp. Mohn 1978 and Freudenthal 1989.
[183] See Neurath 1931c [1973], p. 353 and 1931d; for the incident see Neurath to Carnap, 20 May 1932, RC 029-12-49 ASP.
[184] Reichenbach to Neurath 28 May 1931, HR 013-41-56 ASP, Neurath to Reichenbach, 30 May 1931, HR 013-41-55 ASP.

1.4.3, not only in Red Vienna but also in Moscow – Neurath's failure to identify more clearly his own brand of Marxism may well have been intentional. In any case there was too much practical, educational work to do instead: the exhibitions and publications of his museum, the lectures of the Verein Ernst Mach, and seminars for trade unionists, adult-education workers and party speakers.[185]

None of the above excused Neurath, especially after his return to Vienna, from having a position on numerous issues dividing the left at the time in Europe. Two examples assure us that his positions were finely nuanced. While he was obviously prepared to assist the Bolsheviks with building socialism in their country with his museological expertise (as we saw in section 1.4.3), Neurath did not cede to the Third (Communist) International the leadership of international socialism. And while he believed that the capitalist mode of production provided the very organisational elements that a revolutionary socialist could build upon (cartels, etc.), he did not believe in the gradual change to socialism but insisted on the radical break of 'total socialisation'! We need not follow Neurath through these issues here. But we must note one aspect of his own late socialisation into the Austrian socialist tradition, namely, how he came to perceive the epistemological challenge to any renewal of Machianism issuing from the Marxist tradition. This topic proved of decisive importance for the further development of his theory of scientific discourse. Otto Bauer, whose socialisation plans we compared with Neurath's in section 1.2 and who, as we saw in section 1.3, saved Neurath from the full length of his prison sentence in Bavaria, is most likely to have played yet another, unsuspected and ultimately larger, role in this respect.

Neurath was by no means the only left Viennese intellectual who returned at the end of or after the First World War with a manuscript written during periods of incarceration. What to Neurath was *Anti-Spengler*, was to Bauer 'The World-View of Capitalism' and to Friedrich Adler (son of Victor and translator of Duhem) *Ernst Mach's Overcoming of Mechanical Materialism*.[186] Both dealt with the question: Mach or Marx?[187] Ever since 1904, when a young Max Adler (no relation to Friedrich and Victor) took hold of a rather contingent feature of the German revisionism debate – Kautsky's Machist and Bernstein's Kantian tendencies –

---

[185] Cf. Dvorak 1982, Stadler 1989, and various memoirs from the period in Neurath 1973 and Stadler 1982a.   [186] O. Bauer 1924, F. Adler 1918.
[187] For overviews of this discussion in Austria see Stadler 1982b, pp. 81–110, in Russia see Cohen 1968.

and set to providing a synthesis via a 'social a priori', this question found an audience, indeed, a much increased one after Lenin's *Materialism and Empirio-Criticism* of 1908.[188] Unlike Max Adler, who hoped to borrow from Kant the normative philosophical foundations which he found lacking in Marxism, Friedrich Adler declared Mach's epistemology more or less satisfactory for Marxism as he understood it: social science with a large-scale developmental hypothesis.

Otto Bauer had a more sophisticated, open-ended idea about the relation between Mach and Marx. His 'World-View of Capitalism' was a study of the relation between the modern philosophies and the social and political conditions under which they were conceived. It extended Engels' classical analysis in *Ludwig Feuerbach and the End of Classical German Philosophy* into the immediate pre-war period. The study was written in a Tsarist POW camp and published only in the *Festschrift* for Kautsky of 1924, but Neurath is bound to have come across its theses in discussions with Bauer soon after his return to Vienna. Bauer's ideas could not but appeal to Neurath who had long been in the business of overcoming Machian reductionism. Summing up the pre-war development Bauer wrote:

Thus the whole mechanistic conception of nature, and all the philosophical systems based upon it, has been dissolved in modern positivism and relativism. But the self-dissolution of the classical world view of capitalism has been completed only within the limits of bourgeois thought. The task of freeing modern epistemology from these limitations has still to be accomplished.[189]

Bauer himself did not advance that task, but, we claim, Neurath did, prompted by his suggestions.[190] Bauer elaborated:

Once materialism, together with the mechanistic conception of nature, lost its lustre, younger Marxists tried to connect Marx's conception of history with the more recent epistemologies, with neo-Kantianism here and Mach's positivism there . . . Only the self-dissolution of the world-view of the capitalist period of history breaks its power over us; it alone gives us the courage to refuse to connect the Marxist conception of history with the systems of bourgeois philosophy, and instead to try to comprehend these systems themselves through the Marxian

---

[188] M. Adler 1904, Lenin 1908. After the Bolshevik revolution and the resultant schism of the European left, the question 'Mach or Marx?' assumed even greater importance.

[189] O. Bauer 1924 [1978], p. 217.

[190] Note also that O. Bauer 1924 is listed in the bibliography of the 1929 collaborative pamphlet (Carnap, Hahn and Neurath 1929 [1981], p. 316), presumably at the suggestion of Neurath, under the category 'Sociological Foundations of the Development of World Conceptions', alongside only his own 1928a and Zilsel 1929. Later, Neurath referred to Bauer's *Weltbild* in his 1931c [1981], p. 495.

conception of history in their historical dependency, in their temporal determination, and thus to free ourselves from their spell . . . The dissolution of the mechanistic conception of nature is completed by the epistemologies of the moderns, especially Mach's. But inasmuch as this dissolution has led us to place the systems based upon a mechanistical conception of nature in their social and historical context, it has led us beyond the moderns, even beyond Mach.[191]

Scientists too, in short, make their statements sometimes in conditions 'not of their own choosing.' A theory of knowledge which leaves the socio-political dimension of knowledge production out of consideration will not do. Bauer concluded:

Marxism requires a theory of knowledge which Mach and Avenarius, Poincaré and James cannot provide. Such a theory would demonstrate in detail the procedures, the mental process, by means of which human beings create their world-view according to the example of their own labour, as a reflection of the social order in which they live, or according to the social order for which they fight, according to the requirements of their economic and social, political and national struggles.[192]

Beyond occasional caustic comments, Neurath never expressed a view about how to understand Engels' notorious 'in the last instance' formulation about the determination of the superstructure by the base, by the mode of production. All the same his own discourse theory, we shall see, provides a model for the practice of science. The gradual articulation of that model may be viewed as marking Neurath's gradual realisation of concrete ways of rationalising the processes of knowledge production by intervention. After his experiences as a 'social engineer' in Munich, Neurath was concerned to integrate the reflexivity of an investigator who realises that he is part of the scene investigated. Bauer's remarks pointed the way for Neurath's own development.

Neurath favoured a scientific interpretation of Marxism: party programmes aside, Marxism is social science.[193] With this choice he placed himself in the Viennese tradition of Karl Grünberg, the 'father' of the Austromarxists and the first Marxist Ordinarius appointed at a German-speaking university.[194] Still this self-placement marks at best a minimal agreement with orthodox Marxism. If Neurath fits the description of a 'scientific' (or 'metascientific') Marxist in the early 1930s, then he does so not because he thought historical materialism compatible with a mechanistic universe, or an indeterminist one, nor because he saw the dialectics

---

[191] Bauer 1924, pp. 462f.   [192] Ibid., p. 464.   [193] Neurath 1931c, 1932a.
[194] Bottomore 1978. Viennese intrigues (Stadler 1979) made Grünberg accept a call to become the first director of the Frankfurt Institut für Sozialforschung in 1924 (Jay 1973, pp. 9–12).

of nature revealed, nor because he had located another *synthetic a priori*. Rather it was because he detected in Marx not only general social scientific merits, but also the same naturalistic rejection of transempirical claims that provoked him to reject 'philosophy'.

Neurath twice quoted the following passage from Marx's *German Ideology* (first partly published only in 1927):

Only now, after we have considered the primary historical relationships, do we find that man also possesses 'consciousness'; but, even so, not inherent, not 'pure' consciousness. From the start the 'spirit' is afflicted with the curse of being 'burdened' with matter, which here makes its appearance in the form of agitated layers of air, sounds, in short, of language. Language is as old as consciousness, *language is practical consciousness that exists also for other men and for that reason alone it really exists for me personally as well*; language, like consciousness, only arises from the need, the necessity of intercourse with other men.[195]

It has been remarked of this passage that it 'vigorously attacked' the 'idea of a private language'.[196] Certainly Marx and Engels leave us only with the common intersubjective language.

What is interesting for present purposes is that these quotations date from the year 1931. The importance in 1931 of a new juncture of Neurath's Marxism and his empiricism is one of the central insights unearthed in the story we shall tell in part 3. Our concern here is to point to one particular effect of this juncture on his stance in the Vienna Circle's protocol sentence debate. That year Neurath began to inveigh publicly against Carnap's methodological solipsism. Neurath's battle-cry was 'physicalism'. His understanding of this term differed from Carnap's, as we shall see presently. What should be noted in the present context is that from 1931 onwards Neurath used a kind of private language argument to defend his physicalism against Carnap.

### 2.5.2 The forward defence of naturalism

What robs us – according to Neurath's first two Boats – of the possibility of starting from scratch is the historical conditioning of thought, its contingent determination by historically formed concepts. With his third Boat Neurath insisted again on the impossibility of avoiding the

---

[195] Marx and Engels 1846 as cited in Neurath 1931c [1973], p. 351, our italics; 1931d [1981], p. 411 quotes all but the first of these sentences.
[196] Rossi-Landi 1968, p. 27, musing on the specificity of Sraffa's influence on Wittgenstein. Earlier, the parallel between 'Marx's aperçu: language is practical consciousness' and Wittgenstein's 'to imagine a language is to imagine a way of life' was noted by Cohen 1963, pp. 149–51.

## A theory of scientific discourse

historically grown common language. In 1913 all we could do was to 'try to improve matters by making changes here and there'.[197] In 1932 this meant 'purifying this ordinary language of metaphysical components' as far as possible, creating a 'universal slang' that still in principle 'is the same for the child and for the adult' and 'the same for a Robinson Crusoe as for a human society'.[198] In the protocol sentence debate Neurath was called upon to defend his long-standing conception of scientific knowledge. Yet he was less challenged to defend his anti-foundationalism than his conviction that even abstract theory must take its start from empirically discernible elements (section 2.3.3). The idealisation of knowledge that Carnap offered seemed to relieve us of the need to deal with the tangle of cognitive reality.

To appreciate Neurath's response let us first consider Carnap's stance in the protocol sentence debate up to 1932. A 'rational reconstruction' makes no claim to reflect either the way knowledge is arrived at, nor how it is justified in practice. Rather it presents how scientific claims can be understood and justified 'in principle'. Carnap held that all scientific statements can be reconstructed either in the language of physics or in a phenomenalist language. Carnap postulated a certain order of epistemic priority: first 'auto-psychological' objects are cognised, then physical objects, then other minds, and finally 'cultural' objects. In order to reflect this, Carnap adopted a language with an auto-psychological base and adopted the position of 'methodological solipsism' – the doctrine that empirical knowledge is built up from an individual's phenomenal experiences.[199] Carnap's use of rational reconstruction held out the promise that we can do without historically grown concepts and replace them instead with mathematised 'philosophic structures'[200] without losing sight of science itself.

Carnap's *Aufbau* attempted a rational reconstruction of empirical knowledge of a very radical sort. Every statement of science was to be translated – via explicit definition – into a statement about the experiential given of an individual subject. Moreover every statement was to speak only of the structure, not the content of experience. (Carnap tried to reconstruct even sense-data out of undifferentiated whole experiences and to give a structural description of them.) By combining reductionism with this 'structuralism', Carnap sought to save empiricism from the challenge of sceptical doubt about 'intuition' that Schlick had raised in

---

[197] Neurath 1913a, p. 456.  [198] Neurath 1932d [1983], pp. 91, 96.
[199] Carnap 1928, §§ 54–60, 64.  [200] Neurath 1921a [1973], p. 198.

1918: empiricism was in trouble if it had to rely on the experiential, intuitive content of experience, for intuition was inevitably vague and private.[201] Carnap aimed to render appeal to intuition unnecessary. By 'structuralising' experience Carnap sought, first, to rid the evidential ground of science of the vagueness Schlick had complained about. Second, Carnap sought to render our evidence for scientific statements public in principle. Structures, unlike private experiential contents, can be compared intersubjectively.[202] In addition the reduction of all scientific statements to the auto-psychological base amounted to the defence of the unity of science: there is only one object domain and there is only one kind of access to it.

Recent scholarship argues that Carnap's interest was not in the phenomenalist reduction of all physical object talk.[203] Nor did his epistemology aim for the irrefutable justification of knowledge claims. His interest lay rather in establishing science's claim to objectivity by means of the structural explication of its concepts. Given his assumption of the epistemic priority of auto-psychological objects, however, we must note that Carnap could not have succeeded in explicating the objectivity of science as the structural determination of its concepts unless he also supplied a phenomenalist reduction. If objectivity rested in the structure of representation and if human cognition was reconstructed as proceeding from the immediately given, then the physicalist language would remain unintelligible without its reduction to the language of immediate experience. Given his assumption of epistemic priority, phenomenalism was not dispensable.

Carnap's *Unity of Science* of 1932 repudiated some but by no means all of the *Aufbau*'s underlying conception. For Carnap the 'thesis of physicalism' amounted to the thesis of the universality of the physical language and that means that all scientific statements are translatable into the physical language.[204] To prove this universality would also be to prove the thesis of the unity of science[205] – albeit in the opposite direction from that of the *Aufbau*. In addition to this difference Carnap's 1932 proof proceeded in the explicitly 'metalogical' vein. Against Wittgenstein Carnap held that metalinguistic discourse about the logical form of linguistic expressions is possible without paradox. According to the 'thesis of metalogic' (or 'thesis of syntax'), 'all propositions of philos-

---

[201] Schlick 1918/25 [1974], pp. 29, 38.   [202] Carnap 1928, §66.
[203] E.g. Coffa 1985, 1991; Friedman 1987, 1992; Proust 1986; Richardson 1990; Creath 1990; Oberdan 1990; Ryckman 1991.
[204] Carnap 1932a [1934], p. 42.   [205] Ibid., p. 96.

ophy which are not nonsense are syntactical propositions, and therefore deal with linguistic forms.'[206] Carnap accordingly drew the distinction between talk in the correct 'formal mode' and in the potentially misleading 'material mode' of speech. The former is the proper language for philosophical inquiry. It speaks only of intra- or inter-linguistic matters ('syntax'), whereas the latter illegitimately topicalises word–world relations.[207] The thesis of physicalism represents the proper formal-mode expression of the unity of science thesis, which in the material mode appears as a claim about the unity of the objects of science.[208]

Carnap defines science as 'the system of intersubjectively valid statements.'[209] Only those statements belong to science which different people can verify by reference to their own 'direct' experience. This direct experience finds expression in evidence, or 'protocol', statements, which are formulated in their own so-called 'protocol languages'.[210] On this basis Carnap proved the universality of the physical language for science by showing that the protocol language is but a sub-language of the language of physics. Physicalism demands that

| [formal mode:] statements in protocol language, e.g. statements of the basic protocol, can be translated into physical language. | [material mode:] given, direct experiences are physical facts, i.e. spatio-temporal events.[211] |

Carnap's argument proceeded in the formal mode which permitted the discussion of translation relations between different languages, but for ease of exposition he sometimes slipped into the material mode. First, Carnap assumed the standpoint of the objector. Given his verificationist criterion of meaning, he concludes that, if physicalism were not true, 'every protocol language could therefore be applied only solipsistically; there would be no intersubjective protocol language'.[212] Then he assumed the physicalist standpoint. Carnap finds that in this case 'every statement in the protocol language of a subject $S$ can be translated into a physical statement and indeed into one which describes the physical state of $S$'s body'.[213]

Carnap's argument was directed against the thesis that methodologically solipsist protocol languages, which do not presuppose the existence of physical objects, are not translatable into the physicalist

---

[206] Ibid., p. 38 fn.
[207] Ibid., pp. 38–9. This 'syntactic' restriction was dropped in Carnap's later semantic theorising (cf. p. 183 below, note 48). [208] Ibid., p. 37. [209] Ibid., p. 66. [210] Ibid., pp. 50–2.
[211] Ibid., p. 76. [212] Ibid., pp. 80, 81. [213] Ibid., pp. 87–8; cf. pp. 60–6.

language. It is important to note that Carnap maintained both a methodologically solipsistic protocol language and a physical language.[214] The very point of Carnap's paper was, after all, to show that this protocol language was fully translatable into the physical language. Otherwise the primitive protocols would stand wholly outside of the realm of intersubjectivity and hence of science. It might be thought that the distinct nature of the protocol language had thus become unimportant for Carnap, especially since he also admitted the possibility of non-phenomenal protocols. But in the case of a conflict of a person's protocol with 'the facts', we can only change the translation of this protocol into the physical language, not the original protocol itself: '[A] protocol sentence, being an epistemological point of departure, cannot be rejected.'[215] Protocols become revisable and need support only after they have been translated into the language of science; in their original form they are incorrigible.

It was this point that Neurath attacked. Behind the appeal of a methodologically solipsist language lies the assumption that this language is capable of being known and used independently of scientific fact, or independently of facts about the physical realm and other minds. Neurath argued against this assumption. But Neurath was not concerned about whatever may have been Carnap's own point in retaining this kind of protocol language. For him this language represented an opening for metaphysical infiltration that had to be closed.[216] The mirage of a private protocol language threatened to promise 'foundations'. (As if to prove the point, subsequent analytic philosophy – led by Quine – took Carnap to be providing precisely this.[217]) Since Neurath never objected to any technical shortcomings in Carnap's logical edifices, however, we must ask how he could have argued against Carnap. For instance, how could he dispense with the argument, later advanced by Quine, that Carnap's phenomenalistic reduction in the *Aufbau* failed?

Having already cast aspersions on Carnap's concern with the 'ideal language' in his 1928 review of the *Aufbau* and even in preceding discussions, Neurath had come to adopt an argument of Heinrich Neider, then a student member of the Circle. According to this argument protocols as reconstructed in the *Aufbau* are not intersubjectively testable.[218] For Neurath this meant that the physicalist language must be primary.

---

[214] Ibid., pp. 44–5; cf. Carnap 1932b [1959a], p. 166.
[215] Carnap 1932b [1959a], p. 191 (cf. p. 170).   [216] Neurath 1932d [1983], p. 93.
[217] Cf. Quine 1951, 1969.
[218] Neurath 1928b; Neider in Haller and Rutte 1977; cf. Uebel 1992a, ch. 4.

Carnap, by contrast, responded by drawing a distinction between the language of science in practice, in which the protocols are physicalist, and the methodologically solipsist language, which should be used to explicate scientific knowledge claims.[219] On 26 February 1931 Neurath presented against this position his 'new idea . . . to do everything with the physical language alone, even syntax etc.'[220] 'Recourse to the given is in any sense superfluous.'[221] Neurath insisted that it was impossible to go 'back behind and before' natural language. This argument was first recorded during a private *Besprechung über Physikalismus* on 4 March 1931 with, amongst others, Carnap and Hahn:[222]

> We begin with the historical natural language. Its sentences are *Ballungen*, i.e. mixtures of expressions (precise and imprecise concepts), e.g. 'The screeching saw cuts the blue wooden cube.' In these *Ballungen* there are elements which have to be transformed to different degrees. A human being whose sensory world undergoes change relates to his earlier sentences as to those of another human being. Intersubjectivity not only between different humans, but for the same human in case of change of the sensory complex . . . Once we transform the *Ballungen* (conglomerations) we obtain the standardisations (sentences of the physical language). The physicalist language makes possible the most far-reaching understanding between human beings and that of human beings of themselves.[223]

We may split up Neurath's highly compressed argument into a 'cannot go behind' and a 'cannot go before language' part. It concerns what follows once it is conceded – and why it has to be conceded – that we cannot go 'behind language'. For the rest of this section we will explain why that has to be conceded. In part 3 we shall learn of the momentous consequences that follow.

Let us begin with how one cannot go 'behind' language. Neurath started with an objection to Carnap's assumption that it was possible to use a phenomenological language in epistemological reconstruction. Instead of presenting arguments concerning the logical insufficiencies of the *Aufbau*, Neurath put forward what amounts to a private language argument. In fact the conflict between Carnap and Neurath is usefully conceptualised as one between two kinds of private language arguments of different strength. Carnap's we have seen above. The details of

---

[219] Neurath 1930a [1983], p. 47, Carnap 1930a [1959a], p. 144, 1930b, p. 77.
[220] Carnap's diary, 26 February 1931, RC 027-73-05 ASP.
[221] Circle protocol, 26 February 1931, RC 081-07-11 ASP.
[222] The next day, 5 March, at an 'official' *Sitzung* of the Circle Carnap presented his thesis of physicalism. Cf. Uebel 1992a, chs. 6–7, 1992b.   [223] RC 029-17-03 ASP.

Neurath's private language argument, however, require excavation from a complex context. Neurath did not elaborate it *in extenso* anywhere, though he sketched it in several places.[224]

Neurath argued that for a language to be usable by an individual over time there must be constancy of use. For Neurath a phenomenal language does not even 'come into consideration' for it does not allow for the mechanisms whereby the constancy of an individual's language use can be ensured. These in turn are required for any empirical testing or 'checking' to take place.[225] It is this precondition of the testability of even an individual's 'inner speech' that distinguishes Neurath's physicalism from Carnap's. Neurath's argument rules out Carnap's methodologically solipsist protocol language. His reasoning may be explicated as follows. Once on a solipsistic base there is no escaping solipsism of the moment. If physicalistic statements (about, for example, instrument readings) need be translated into phenomenal terms directly related to a scientist's experience in order to be meaningful, then no touchstone is available by which the constancy of that scientist's language use can be established. For what could such phenomenal terms refer to but my experience now? Yet we rightly assume the general constancy of our language use because we use a language whose constancy is itself continually in check. Given no contrary signals from our speech community we are justified in assuming our use is constant. Clearly if language use is to be controllable in this way, it must be an intersubjective language. If the protocol language is to be a usable language and if it is to provide epistemic justification, then it cannot be a phenomenalist one. The reason lies in the constitutive or capacitating contribution that social interaction makes to an individual's language use.

Neurath's argument does not focus on the private reference of a phenomenal language – which Carnap could have claimed to have shown to be co-extensive with the physical one had he not forsworn the 'material mode of speech'. Nor does he focus, as Carnap had, merely on what was necessary for communication. Rather Neurath focused on what was required for individual agents to justify the way they comprehended and systematised their own experience. As he put it in the second *Erkenntnis* paper dedicated to this matter (the paper that also featured the third Boat):

---

[224] Neurath 1931b [1983], pp. 54/5; 1932a [1983], pp. 62, 63, 65; 1932f, pp. 105–6; 1932d [1983], p. 96; 1934 [1983], p. 110; 1941a [1983], pp. 228/9. For evaluation and comparison of Neurath's argument see Uebel 1992a, chap. 10; 1992c; 1995a.

[225] Neurath 1931b [1983], p. 55.

## A theory of scientific discourse

If Robinson wants to join what is in his protocol of yesterday with what is in his protocol today, that is, if he wants to make use of a language at all, he must make use of the 'inter-subjective' language. The Robinson of yesterday and the Robinson of today stand in precisely the same relation in which Robinson stands to Friday . . . Therefore it does not make sense to speak of monologising languages, as Carnap does, nor of different protocol languages that are later related to each other . . . If, under certain circumstances, one calls Robinson's protocol language of yesterday and today the same language then, under the same conditions, one can call Robinson's and Friday's the same language.[226]

The crucial question is: 'How I can know that my experience now is of the kind that I anticipated five minutes ago?' Anything that solves the problem of reidentifying intrasubjective objects will also solve the problem of reidentifying intersubjective objects, i.e., it will also answer the question: 'How can we know that the objects before each of us are the same?' There is a tempting line of reasoning that goes from the corrigibility of statements about our common experience to the conclusion that the evidential basis for science must be sought in language that speaks only of the experience of an individual speaker. Neurath showed what was wrong with this argument. There is no radical discontinuity between the verification of statements about other minds and one's own. First-person statements are not distinguished in terms of their supposed certainty. A speaker's sense of the meaning of the expressions of his own language is shaped by the processes of communication; it does not precede it. Even the self-understanding of a language user was as problematical for Neurath as was the understanding of other minds. (Neurath here radicalised the problem he discussed in *Anti-Spengler*.) Robinson required the intersubjective language for the coherence of his experience 'even before Friday arrived'.[227]

The conclusion of this argument is plain. It is expressed in Neurath's third Boat:

There is no way to establish fully secured, neat protocol statements as starting points of the sciences. There is no *tabula rasa*. We are like sailors who have to rebuild their ship on the open sea, without ever being able to dismantle it in dry dock and reconstruct it from the best components. Only metaphysics can disappear without trace. Imprecise 'verbal clusters' [*Ballungen*] are somehow always part of the ship. If imprecision is diminished at one place, it may well reappear at another place to a stronger degree.[228]

---

[226] Neurath 1932d [1983], pp. 96–7.    [227] Neurath 1934 [1983], p. 110.
[228] Neurath 1932d [1983], p. 92.

After the linguistic turn the historical conditioning of thought and language become compressed, concretised – as it were: *verdichtet* – into *Ballungen*. Neurath's private language argument backed up what he had assumed with Tönnies' semiotics all along: There is no way to avoid the hazards of the common, physicalistic natural language.

Carnap never accepted Neurath's argument. In response he admitted that the protocol language could be physicalistic from the start. Indeed he agreed that this was preferable. But Carnap did not accept Neurath's naturalistic constraint that even rational reconstructions must not presuppose 'unrealisable'[229] assumptions concerning human cognition. For him the possibility of pursuing the strategy of methodological solipsism is not precluded; it is simply more convenient to think of the protocol language as straightforwardly physicalistic.[230] Carnap continued to assume that we can explicate scientific knowledge claims by means of whatever language form we fancy. That this does 'not make sense', however, was Neurath's central point.[231] Neurath denied not only phenomenalism but also the epistemic self-sufficiency of individual subjects – just what Carnap's rational reconstructionism retained.

But what about Neurath's claim that one cannot go 'before' language? We take this to be a slogan that summarises Neurath's disapproval of traditional epistemology:

within a consistent physicalism there can be no 'theory of knowledge', at least not in the traditional form. It could only consist of defence actions against metaphysics, i.e. unmasking meaningless terms. Some problems of the theory of knowledge will perhaps be transformable into empirical questions so that they can find a place within unified science.[232]

Because of the last sentence it seems appropriate to use a contemporary label and say: the only epistemology Neurath would abide would be a naturalistic epistemology. So we may rephrase our question. Did Neurath's private language argument leave us with only naturalistic epistemology? Even if we cannot assume a phenomenalistic language (nor leave language behind), does this make it impossible to ask the questions of traditional epistemology in scientific language? What if epistemological questions are reserved for *formal* science, namely, the 'logic of science'? The latter kind of rational reconstruction might even begin with physical object discourse, but freely help itself to in principle 'un-

---

[229] Neurath 1933a [1987], p. 3.
[230] Carnap 1932c [1987], pp. 470 and 457, respectively; cf. his 1936/7, p. 9, 1961, pp. v, vii; 1963, pp. 945–6.   [231] Neurath 1932d [1983], p. 97.   [232] Neurath 1932a [1983], p. 67.

# A theory of scientific discourse

realisable' counterfactual reconstructions in its attempt to explicate and formally justify scientific knowledge claims. Invoking the spirit of 'logical tolerance'[233] Carnap continued to evade Neurath's strictures in just this fashion. Clearly against such a version of rational reconstructionist epistemology another argument is needed.

Here we must remember the other part of Neurath's argument – what follows once it is conceded that there is no going 'behind or before' language. What determines the meaning of statements and terms on Neurath's view? Their use in the community. Language is social. For Neurath meanings as fixed by the linguistic customs of the speech community are not the clean-cut thing rational reconstructions build on. Once the primacy of the physicalist language is accepted, it is the job of *Ballungen* to preclude Carnap's recourse to such clean elements.[234]

The concept of *Ballungen*, of the 'compressions' and 'clusters' of everyday language, will occupy us greatly in part 3. For now we need only note that *Ballungen* preclude definite, once-and-for-all fixed meanings. Even Carnap's physicalistic reconstructions of knowledge claims after 1932 run afoul of the fact that our everyday language is full of *Ballungen*. To see this we need only suppose that we call into question not only empirical evidence statements but also meaning statements. Neurath did just that:

> [I]f Schlick thinks that I have only understood a statement when I know whether it is analytic or synthetic, what about the case when I declare a statement analytic today and reach another opinion tomorrow, declaring that I had been wrong and not understood the statement yesterday. I have no means available at all to reach a final verdict about whether a statement was understood by me or not – this is a typical pseudo-formulation.[235]

As has been noted, the analytic/synthetic distinction comes under fire here.[236] But we can also see why Neurath thought it reasonable to extend his fallibilism as far as he did. If the meaning of language is determined ultimately by consensus (as Neurath's argument still allows), then it does not make sense to speak of 'final' verdicts about linguistic understanding. The groundlessness of the question 'What did I really mean?' undermines the supposition that we can work with clean elements whose meaning we can presume to know. Perhaps if we had

---

[233] Carnap 1934a [1937], § 17.
[234] Naturalists have long charged that Carnap's rational reconstructions presuppose knowledge of meanings rather than explain it. This typical failing can be detected not only in his semantic models, but also in his *Aufbau* (the 'foundedness' problem of the basic relation) and *Syntax* (transfinite analyticity). [235] Neurath 1934 [1983], p. 104.
[236] Koppelberg 1987, p. 32, Zolo 1986, pp. 40–1, Uebel 1992a, p. 251.

complete understanding we would get clean elements; but since we don't, we can't.

This argument works against both Schlick and Carnap. Carnap's presupposition that fixed meanings can be unproblematically assumed is evidenced throughout his writings. By contrast, Neurath holds that '"imprecise formulations" cannot be prevented. To think their exclusion possible is metaphysics.'[237] We must not neglect the messiness of ordinary language and proceed with a logical calculus as a model of knowledge. To do so is to turn away from – indeed, underhandedly overturn the moral of – the first and maybe only essential characteristic of meaning: that it is a social affair and never ultimately fixed. We must investigate cognitive practice and forswear aprioricism across the board. It falls to *Ballungen* to carry the argumentative weight of his unorthodox conception of unified science. How they do will be shown in part 3.

### 2.5.3 Science as discourse: the theory of protocols

What shall we make of Neurath's argument against methodological solipsism and his 'defence' of the naturalistic constraint against Carnap? As we saw, Carnap never accepted the former. Yet we may note that just as Carnap never declared impossible or unprofitable the development of empirical theories of science, Neurath never denied the possibility and profitability of Carnap's logic of science.[238] Their conflict lay, first, in the evaluation of the purely logical investigation of the means of scientific representation: Was it the successor discipline to 'epistemology'? For Neurath the logic of science is but a necessary preliminary for a decidedly non-philosophical, empirical theory of science. For his part Carnap held all along that the question between them was one 'not of two mutually inconsistent views, but rather two different methods for structuring the language of science which are both possible and legitimate'.[239] Was their conflict only of a semantic nature, then?

Ultimately, Carnap was right: their disagreement stemmed from what each wanted a theory of science to do. Consider Carnap's project. Even casual readers of his autobiography are struck by what scholars have painstakingly documented in their textual exegeses of his formal enterprises. Carnap's project, doggedly pursued against all objections, was to establish domains of discourse neutral between competing

---

[237] Neurath 1932c [1981], p. 565.  [238] Carnap 1934a [1937], §72.
[239] Carnap 1932c [1987], p. 457.

camps, to isolate a core of cognitive content in our theories of the world indifferent to opposing metaphysics, to make plain a conception of knowledge unafflicted by interests high or low. If only that could be done, one suspects Carnap to have reasoned, the bitter divisions of humankind would themselves lose their incendiary edges and, with tolerance, a better day would dawn. Neurath disagreed with Carnap on one central point of strategy. He did not believe in the possibility of a neutral stand of science between competing metaphysics, nor of a conception of knowledge unafflicted by interests. Knowledge for him is action and so connected to practical aims. No refinement of traditional philosophy will do, not even a minimalism like Carnap's. Traditional philosophy must be rejected altogether. What is needed is a new conception of knowledge that will propel us into change, beyond mere reinterpretation. Carnap's and Neurath's projects differed, for they disagreed on the strategy for a nevertheless common goal – a tolerant and tolerable world.[240]

If this is right, then we must conclude that Neurath owes us a positive account of knowledge, an account of knowledge we can use in changing the world. In fact, though it remained long unrecognised, Neurath did provide the beginnings of such a positive account in the very paper at issue: his theory of protocol statements. Yet first note again Neurath's naturalism: '[T]he work on unified science replaces all former philosophy.'[241] All the concepts used in such work can potentially be scrutinised. Rather than chase after the truth-conditions of statements 'behind or before language', Neurath urged the investigation of the conditions of provisional acceptance for knowledge claims.[242] Neurath's proposal for the form of protocol statements is best read as an analysis of the conditions for making and accepting observation statements, an analysis which makes explicit the conventions governing this particular type of 'move' in scientific discourse.

In effect, protocol statements are *Ballungen*. They represent the *Verdichtung* of sets of conditions whose contribution Neurath's theory aimed to render plain. It is a mistake to seek something behind or at the bottom of the evidence statements of science. Observation reports gain

---

[240] Carnap agreed with Neurath's socialist political opinions (Marie Neurath 1973a), though not with his mixing of politics and philosophy (Carnap 1963, p. 23).
[241] Neurath 1931b [1983], p. 56.
[242] 'We just resign ourselves to a moderate clarification in order to delete or accept statements later.' Neurath 1934 [1983], p. 109.

their distinguished status from the interplay of the different procedures their acceptance is subject to in the discourse of science.

A complete protocol sentence might for example be worded like this: 'Otto's protocol at 3:17 o'clock: [Otto's speech-thinking at 3:16 o'clock was: (at 3:15 o'clock there was a table in the room perceived by Otto)].'[243]

Neurath's proposal represents an attempt to provide a theory of scientific data and, by extension, of theory testing. The core theory is, as it were, formalised by his proposal. It seeks to achieve a particular effect by means of the repeated embedding of an unadorned singular sentence like 'There was a table in the room.' Each of the embeddings specifies a condition which must be fulfilled before we are justified in considering the singular sentence as an acceptable scientific datum. Together they express the combination of conditions whose fulfilment is a necessary but not sufficient condition for the control of laws, generalisations and predictions.

Neurath started from somewhat stylised descriptions of scientific practice; it was public knowledge claims whose acceptance was to be justified. Take Neurath's example, suitably amplified, namely as the claim that there is an oval table in the seminar room of the Institute of Mathematics in the Boltzmanngasse in Vienna on 5 March 1931. When we do what in normal parlance would be called 'accept an observation report', we hold that a number of conditions are fulfilled. We may specifically note among these conditions: (i) the institutional condition, (ii) the doxastic condition, (iii) the stimulation condition, (iv) the 'factual' condition. Condition (i) holds that somebody made the claim that somebody thought that somebody was stimulated as if he observed an oval table in the seminar room, etc.; condition (ii) holds that somebody thought that somebody was stimulated as if he observed an oval table, etc.; condition (iii) holds that somebody was stimulated as if he observed an oval table, etc.; condition (iv) holds that there was an oval table in the seminar room, etc.

Neurath's seemingly baroque proposal can thus be schematised as follows:

protocol (thought [stimulation state {'fact'}])

Thus schematised protocols are 'decomposed' in the following fashion:[244]

---

[243] Neurath 1932d [1983], p. 93; cf. e.g. 1934 [1983], p. 107, 1935a [1983], p. 118, 1935b [1983], p. 129, 1936c 1983], pp. 162–4, 1936e [1983], p. 152, 1941a [1983], p. 220.

[244] Neurath's term is 'deletion of brackets' (1932d [1983], p. 93) For the details of the following exegesis see Uebel 1992a, ch. 11 and 1993a.

(i)   protocol (thought [stimulation state {'fact'}])
(ii)            thought [stimulation state {'fact'}]
(iii)                    stimulation state {'fact'}
(iv)                                        'fact'

Each of these components (i) to (iii) expresses a separate condition that a singular sentence (iv) has to meet if it is to count as an observational datum (or, in the extended theory, that a test sentence (iv) derived from a theory has to meet if the theory is to be counted as confirmed by it). In other words, when unpacked a protocol statement is really a complex of statements, namely, '(i) & (ii) & (iii) & (iv)'. (Each component alone is not truth-functional due to the opaque embeddings, yet as a whole protocol statements are truth-functional, namely when spelled out as a conjunction of the four conditions: (i) & (ii) & (iii) & (iv). Now the usual truth-tables apply.) It is also important that each of the components (i), (ii) and (iii) contains a proper name. This specification of the maker of each substatement – bar (iv) – allows the substatement to be checked and allows the use of the observation of others in one's own protocol statements. Thus we can specify reasons for discounting certain protocols in our theorising. Depending on the case we can designate them, as Neurath put it, as 'lies' or a 'dream' or 'hallucination statements'.

For the acceptance of an observation report in science, these four formal conditions 'stand fast'. It is less clear whether for acceptance in normal parlance an additional condition should be met as well, one which is not 'formalised' in Neurath's proposal. This is the condition, typically associated with an evidence-reporting statement in scientific discourse, that it be taken as a test of a theory in the strictest sense: in cases of conflict between them we retain the protocol and not the theory. Neurath did not 'formalise' this condition because it is pragmatic through and through. As he pointed out to Popper, it is pseudo-rationalism to insist that we always reject a disconfirmed theory.[245] Sometimes we do not do so for very good reasons (see part 3). Protocol acceptance in the sense of resting the change or retention of a theory on them has (i) to (iv) as necessary but not sufficient conditions.

Neurath's proposal for the underlying structure of protocol statements lays out the steps he thinks must be followed for the conventional determination of the admissibility of scientific data claims. It is an account that springs from the study of the contingencies that scientific practice is subject to. For the naturalistic epistemologist Quine, there is a

[245] Neurath 1935b [1983], p. 124.

mutual containment of epistemology and psychology; for Neurath there is a mutual containment of epistemology and sociology – if one still wants to speak of epistemology at all![246] There is nothing 'before or behind' language, nor is scientific cognition all 'in the head'. What matters to Neurath's theory of science is out there in social interaction. Like Carnap, Neurath put forward a proposal concerning language, but his proposal deals with the use, not merely the form of language. Unlike Carnap, who preferred to keep questions of use for the theory of pragmatics (to be dealt with at a later date), Neurath was concerned with what scientists do. Neurath proposed conscious intervention in the discourse of science. He took Conventionalism literally and placed voluntarism at the centre of his theory of science as social practice.

Neurath's proposal is itself to be evaluated pragmatically. It recommends itself for several reasons, amongst them that it provides defensible norms for scientific discourse. Requiring that an acceptable protocol fulfil the indicated conditions renders scientific discourse as controllable as it can be without falling into 'pseudo-rationalism'. It allows for reasoned agreement or disagreement in the light of intersubjective evidence. Even the proposed metatheory can be scrutinized – and revised. Neurath, we begin to see, has the makings of a deconstructivist – albeit one with a difference. The 'dénouement' of epistemic norms as descriptions of social practices is intended constructively, towards a conception of controllable rationality. Neurath's project was to fashion a conception of knowledge of use in Enlightenment. For this project to succeed we need to make available and revisable – even, if need be, by revising – the moves in the language game of science. The guidelines for how we should come to accept the 'masses of statements' that make up science must be conceptualised in a way that displays the actual social mechanisms. This is best done in a discourse theory of science. Neurath calls for a realistic assessment of the practice of science, not counterfactual rational reconstructions. He wants a manual for science that can be of use 'in the struggle'.

Of course Neurath's own thinking about science was no less emboldened by utopian fancy than his thought about matters economical. Thus it may be wondered whether the practitioners really do create the conventions governing their discourse: does the 'republic of scholars' constitute a domain of discourse free from domination? Neurath's sketchy remarks

---

[246] Compare Quine 1969. To be precise, for Neurath the mutual containment was one of physicalism and historical materialism; cf. Uebel 1992a, ch. 10.

about the sociology of science indicate his answer. In principle science can be so conducted. Whether it is, and what prevents it, are empirical matters for the history and sociology of science to investigate.

## 2.6 TOWARDS A THEORY OF PRACTICE

We shall not follow the Boat through its last appearances in detail. The fourth Boat repeats the *Ballungen* point, only to direct it against idealisations not at the base but at the theoretical end of scientific theorising. It is deployed 'against *the system* as the "limit" of scientific research'.[247] (This general point is the topic of section 3.2.) It seems indicative of Neurath's general attitude that he remarked that 'similar views . . . were developed by Carnap, Frank, Morris and others' while advertising his programme of 'encyclopedism' which in fact radically broke with the emerging orthodox view of unified science.[248] Neurath called himself a 'Logical Empiricist' because he associated himself with the Vienna Circle and successor institutions and not because he shared all their doctrines. He did so because it represented a practical engagement (if a somewhat rarefied one) towards the renewal of Enlightenment criticism.

Similarly Neurath called himself a Marxist not because he shared the right doctrines (whatever they may be) nor because his thinking had come up the Marxist path. He did so because his association with the workers' movement (for him most vigorously represented by the Socialist party) secured him access to another part of his intended audience. He could do so without compromise because he picked and developed his own 'Marxism'. After the failed German revolution Neurath chided the organised left for having forgotten 'what Marx taught about the participation in the renewal'.[249] This suggests familiarity with the only major early Marx text then around, the 'Theses on Feuerbach'.[250] Consider the third thesis: 'The coincidence of the changing of circumstances and of human activity can be conceived and rationally understood only as revolutionising practice.'[251] In his own idiosyncratic way Neurath had come to develop a discourse theory of science that did not depict reality as an 'object of contemplation', but as 'human sensuous activity,

---

[247] Neurath 1937a [1983], p. 181, original italics.
[248] Ibid. On the encyclopedic Neurath see also Zolo 1986, Mormann 1991, K.H. Müller 1991a, Uebel 1992a, ch. 12, Nemeth 1994, and Cat, Cartwright and Chang forthcoming.
[249] Neurath 1920a, p. 44; cf. our Conclusion.
[250] The 'Theses' were first published in Engels 1886. The *German Ideology*, whose Part 1 was first published in 1927, did not emerge fully, the *Paris Manuscripts* not at all until the mid-1930s.
[251] Marx 1845, Third Thesis. Compare, e.g., Neurath 1931d [1981], p. 411.

practice'.²⁵² Neurath's theory of science was to be a tool for 'revolutionising practice'. Fittingly Neurath described his and his colleagues' metatheoretical efforts in the Vienna Circle as 'the *superstructural* equivalent of the great revolutionary transformation of our social and economic order which is already under way in several respects'.²⁵³

For Neurath, then, the unlikely combination of 'Logical Empiricism' and 'Marxism' spelt a critical theory of scientific practice. Its partisanship was not party-political, however. While a projected co-operation with the equally exiled Frankfurt School was sabotaged by Horkheimer,²⁵⁴ Neurath was 'very glad' to have won over John Dewey to participate in the *International Encyclopedia* project.²⁵⁵ Dewey's first contribution also touched upon the 'human, cultural meaning of the unity of science'.²⁵⁶ He returned to the matter in his monograph *Theory of Valuation*: 'The practical problem that has to be faced is the establishment of cultural conditions that will support the kinds of behavior in which emotions and ideas, desires and appraisals, are integrated.'²⁵⁷ Neurath found an ally in Dewey for his longstanding battle to bring science and life together. Dewey in turn considered him 'the one pragmatist' in the Logical Positivist movement whose members he regarded as 'all scholastics, with the exception of Neurath'.²⁵⁸

The later Wittgenstein is, of course, also treated as a philosopher of practice.²⁵⁹ Fittingly, Neurath's social conception of language, even his contextualist view of 'justification' agree broadly and independently with Wittgenstein's.²⁶⁰ Yet Neurath was not prepared to leave everything 'as it is'. His anti-philosophy was made for intervention in the struggles of the day. As we have seen in part 1, though not 'revolutionising' enough for some, for others it was already too much so. Neurath's developed views in the Vienna Circle period took up again the centrality of action suggested by the first Boat. Marxist reflection allowed him to comprehend better the wisdom of his earlier methodological decisions. In the end Neurath outlined a theory of science that sought to provide the tools

---

[252] Marx 1845, First Thesis. Note also that Marx's idea that the active role of humans in the knowledge process was explored ('but only abstractly') more in idealism than in empiricism is a mainstay of Neurath's histories of philosophy (e.g., 1930a).
[253] Neurath 1932b [1981], p. 576, our italics. For related remarks in letters to Carnap of 5 and 13 March 1933 (RC 029-11-22, 029-11-20 ASP) see Uebel 1992a, p. 261.   [254] Dahms 1990, 1994.
[255] Neurath to Morris, 12 November 1936 (RC 102-52-05 ASP). For an account of the decisive meeting between Neurath and Dewey see Nagel 1959.   [256] Dewey 1938, p. 32.
[257] Dewey 1939, p. 65.   [258] Farrell 1959.   [259] Kitching 1988.
[260] In addition to his private language argument which places knowledge of the world and knowledge of meaning on a par, see also Neurath's quote from Lichtenberg in note 37 of Neurath 1944 – both views also expressed in Wittgenstein 1969. For further parallels see K.H. Müller 1991b.

to empower its practitioners to intervene in, develop and newly create a social practice even under conditions 'not of their own choosing'.

Neurath's last Boat concludes his own *Encyclopedia* monograph of 1944. It stands at the end of the final section 'Argument, Decision, Action' in the chapter 'Sociology and the Practice of Life'. The context echoes that of the second Boat and consciously places his now articulated theory in the pragmatic perspective already familiar from the first: '[A]ltering our scientific language is cohesive with altering our social and private life.'[261] Again he alludes to a synthesis of Mach and Marx[262] and again he affirms 'scientific utopianism', the need to develop as yet unrealised models for use in dealing with reality. Yet Neurath was also by this time very much impressed with the struggle against totalitarianism right and left – indeed, nearly shattered by the non-aggression pact between Hitler and Stalin. He wanted to preserve 'toleration' and promote a 'pluralist attitude of arguing'.[263] The monograph ends:

A new ship grows out of the old one, step by step – and while they are still building, the sailors may already be thinking of a new structure, and they will not always agree with one another. The whole business will go on in a way we cannot even anticipate today.
That is our fate.[264]

This last one-sentence paragraph recalls Neurath's debt, despite ultimate disagreements, to Simmel and Weber and especially to Tönnies. His experience of classical German social science not only prompted his early metatheoretical reflections but also provided the framework for his long-term programme. The project of a controllable rationality, a tool for emancipation, has to be placed in the intellectual context of the classical theorists of modernity – between the topoi of *fate* and *utopia*.[265] Neurath tried to steer his Boat between the Charybdis of cultural pessimism and the Scylla of idealist teleology.

Neurath believed that for reason to fulfil its Enlightenment promise it had to be reconceptualised. Neurath's image of knowledge, the Boat, marks significant stages of his pursuit of this project: from its inception through its theoretical articulation to its organisational realisation. Did

---

[261] Neurath 1944, p. 46. Controversial already with its 'encyclopedic' theses, Neurath's idiosyncratic English did not help matters.
[262] Ibid., p. 19; cf. p. 22 for a rephrasing of the base-superstructure relation as 'asymmetry of prediction'. [263] Ibid., pp. 19f. and 45f.; cf. Neider in Haller and Rutte 1977 and Galison 1990.
[264] Ibid., p. 47. [265] Liebersohn 1988.

Neurath succeed? Once we allow for the sketchy form in which Neurath left his theory, attempts to answer this question lead us squarely into contemporary discussions. Given his self-understanding, of course, Neurath would have found such critical attention 'stimulating'.[266] As he once he wrote to Carnap: 'Now I only want to know whether we too have our metaphysics, but do not notice it.'[267]

[266] Neurath 1941a [1983], p. 217.
[267] Neurath to Carnap, 2 February 1935, RC 029-09-87 ASP.

PART 3

*Unity on the earthly plane*

3.1 TWO STORIES WITH A COMMON THEME

Neurath founded the Unity of Science movement in 1934.[1] For Neurath – a social scientist – the drive for unity was rooted in the great debates between Carl Menger and his thesis examiner, Gustav Schmoller, about the nature of political economy and in Max Weber's insistence that the social sciences are indeed sciences, although sciences of a different type from physics. In this setting, unity of science necessarily meant unity of the social and the natural sciences. For Neurath it meant both more and less: he did not look for a sweeping philosophical union of two great domains of human thought, but rather for the practical unification of the rich variety of special disciplines in all their detail. Unity for him was not a matter of a single metaphysics for natural and social law as Menger

---

[1] The Unity of Science movement was founded at a time of particular interest in unified science and it was by no means unique. It was established just after the Second International Polar Year, for instance, and three years after the current International Council for Scientific Unions (ICSU) was created out of the League of Nations International Research Council (IRC). Although the roots for the idea of ICSU and IRC can be traced back to the attempts of Eduard Suess, President of the Austrian Academy of Sciences, to set up an association of academies, Austria and Germany were not national members until after the Second World War (although Germany was represented at the 1934 general meeting). The Unity of Science Movement shared many common concerns with the IRC and ICSU, especially on standardisation of units, rationalisation of terminology and the international production of catalogues of scientific literature, on the co-operation of large teams of scientists to gather data world-wide and on the unity of scientists across political divides. Especially in line with Neurath's own interests in unity was their stress on the unification of theoretical and applied science and the fostering of what the outgoing president of the IRC, French mathematician Emile Picard, described at the founding of ICSU as the 'increasing amount of penetration of one kind of science into another, a fertile source of new discoveries' (cited in Greenaway, forthcoming). Frank Greenaway in his long story of the ICSU describes the detailed level at which the interpenetration proceeded thus: 'The chemists in particular were continually finding that the pure chemist and the applied chemist were drawn into useful dialogue. An association of manufacturers of fat and fat products wished to benefit from contact with their theoretician colleagues and found themselves extending their own work into the theoretical domain. An area which was bound to expand was that of the many substances of pharmaceutical interest, natural or synthetic' (Greenaway forthcoming, ch. 3).

advocated nor of one epistemology to mend the rift that Weber imposed; and it was not a unification once and forever. Instead Neurath laboured for different combinations of concrete lessons from different sciences at each and every occasion of rational action. Carnap had proposed an *Aufbau*, one gigantic structure into which every valid science could in principle be fitted. Unity for Neurath was not a matter of abstract philosophical principle nor a programme for constructing a complete world picture. It was a tool for changing the world; in his own vocabulary 'an auxiliary motive'. We aim, he argued 'to create a unified science that can successfully serve all transforming activity.'[2] If we are to effect the changes in society that we want, we need to know in detail what to do. For this something far more concrete and practical than an *Aufbau* is required.

When a contrast is made with the Vienna Circle's views on unity, Max Weber is often cited. For example, Stephen Turner and Regis Factor claim:

The Positivists held that the language of physics constituted a basic language for all the sciences, and that this fact enabled the "unification" of science into a single (albeit) elaborate theory with a *single* object and object-language. In contrast Weber supposed that each science departed from certain presuppositions. For Weber one may have a science of jurisprudence, just as one has a science of physics, by making certain presuppositions.[3]

The need for presuppositions arises because the objects of science are not given to it as such. The impossibility of their being given is defended in Weber's own words 'on the grounds of the claim that "reality" consists of an infinitely manifold stream of events' where 'analysis of infinite reality which the human mind can conduct rests on the tacit assumption that only a finite portion of this reality constitutes the object of scientific investigation'.[4]

Turner and Factor explicitly mention Neurath among the Positivists who oppose Weber. But Neurath held nothing like the 'Positivist' view they describe. First, as we have seen, he did not advocate the language of 'physics' but rather the language of *physicalism*, a language describing events that happen in space and time. That is his attempt to characterise 'positive knowledge', a task others associated with the Vienna Circle would try to achieve by using the concept of verifiability. More importantly for the theses we shall discuss in the remaining pages, he had no interest in 'a single . . . theory with a single object'.

---

[2] Neurath 1930a [1983], p. 42.   [3] Turner and Factor 1984, p. 194.   [4] Quoted ibid.

The usual image associated with the canonical Positivist view described by Turner and Factor is the pyramid structure of Figure 3.1. This is a picture that Neurath rejected. In his view the relations among the sciences are far more like those depicted in Fig. 3.2 (Balloons).[5] The sciences have flexible boundaries with no fixed relations to one another. Yet they are all tied both in application and confirmation to the same material world; their language is a language of space-time events. Like the balloons, the sciences will relate differently to each other from one occasion of use to another depending on the point of application.

In the illustration the balloons are tied to wires and trees and other things that shake in the wind. This is to represent the thesis of the first and second Boats that science is not grounded in secure knowledge. Beginning in 1931 Neurath added to this a more radical thesis: not only are there no secure foundations; there are also no fixed connections between the theory and its foundations. This doctrine launches a far-reaching attack on method. Falsification, confirmation and the hypothetico-deductive method are all faulty for they all presuppose that there are strict logical relations between theory sentences and data reports. Fig. 3.2 (Balloons) is an illustration for 1932 and after, for in it the balloons are tied with strings and not fixed on rods.

Part of the reason that Neurath did not expect a unity of the sciences at the theoretical level is because as we shall see in section 3.4.2 his view was consistent with Weber's that '"reality" consists of an infinitely manifold stream of events.' The doctrine is anti-realist both about scientific concepts and about scientific laws. Science does not work by zeroing in on properties that are really there in the stream of events and that are

---

[5] Donald Gillies also uses the image of the balloon, apparently entirely independently: 'I will conclude this chapter with an analogy which is constructed by, so to speak, reversing the direction of gravity in Popper's weighty edifice of science from sinking into a swamp. Let us instead conceive of scientific theories as hydrogen balloons with a tendency to escape from reality (the earth) into the airy regions of metaphysics. These hydrogen balloons are attached to the earth not by large cables, but by a multitude of fine threads and thin strings, rather like those which held Gulliver captive when he first awoke in Lilliput. Each fine thread is a protocol of the form 'Mr $A$ observed that $O$.' So each thread represents the sensory experiences of a particular individual interpreted in the light of a set of theories. The thin strings are formed by twisting together a few of the threads. They represent impersonal observation statements, $O$, which are formed on the basis of individual protocols such as 'Mr $A$ observed that $O$,' 'a Ms $B$ observed that $O$', and so on, but which are more certain than any other protocols on which they are based, just as the strings are stronger than the threads which compose them. Any thread or string may snap or be cut, but although this may alter the position of the balloon, it will still remain attached to earth because of the multitude of other threads and strings. If we cut *all* the threads and strings, however, our theoretical balloon will float away from the reality of the earth towards the airy regions of metaphysics. Our scientific theory will have become a metaphysical theory.' (Gillies 1993, p. 149.)

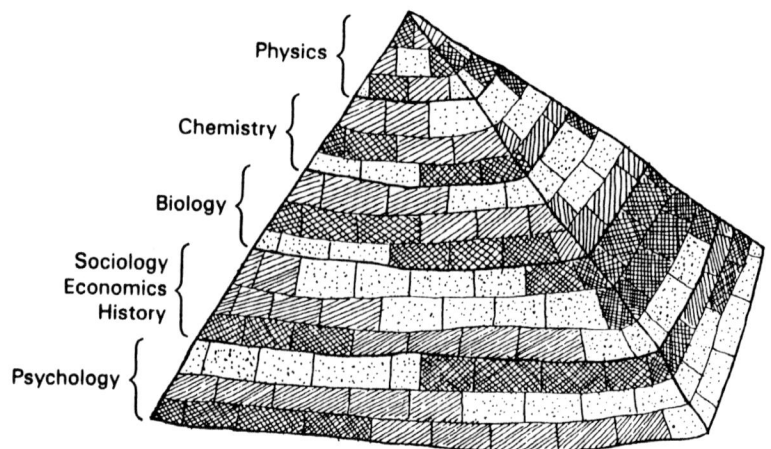

Fig. 3.1 Pyramid

Fig. 3.2 Balloons

causally (or nomologically) responsible for what happens. Rather science works by constructing concepts that we can deploy more or less well to grapple with reality in the ways we need. Different sciences have different concerns and different concepts are already historically given. Trying to fit these concepts into a single theory is a false ideal deriving from the mistaken assumption that they are all small pieces of a single picture. In fact they are not pieces of a picture at all.

Given the linguistic turn of the Vienna Circle and Neurath's rejection of all metaphysical talk, he would not have wished to hold any beliefs about the constitution of reality. Nor do we claim that he did so. It is the appropriate formal-mode version of Weber's thesis that was important for Neurath. In the Vienna Circle period the thesis appeared as a claim not about the world as it is given to us, but instead about the concepts with which we first respond to the world, the concepts of everyday life. These, for Neurath, are vague, complex and congested.

We shall explain what Neurath meant when he said that the concepts of daily life are congestions (*Ballungen*) in section 3.3.1, for this claim is essential to his attack on method, which is our central topic there. Before that we shall describe the evolution of Neurath's views on the unity of science. His implicit agreement with Weber shows that a unified picture is impossible; his belief that full socialisation can be achieved shows that unification at the point of action is necessary. That is our first story. Our second story is that of Neurath's attack on scientific method. The two stories are knitted together by Neurath's conviction that (allowing ourselves to use the more visualisable material mode) '"reality" is an infinitely manifold stream of events.'

## 3.2 SCIENCE: THE STOCK OF INSTRUMENTS

### *3.2.1 From representation to action*

To trace the connection of Neurath's views on the unity of science with his social views and political life, let us return to Neurath's thoughts in 1910. As we saw in section 2.2, by that time he was familiar with Ernst Mach's programme for the unity of science, governed jointly by a principle of 'economy of thought' and by the aim to reduce every concept to the language of experience. Neurath had also been reading the works of August Comte on the relationships among the sciences. He placed Comte in an old tradition that included not only Leibniz but the Catalan Raimon Llull as well. In his 1910 review of Wundt's 'Logik' Neurath observed: 'The history of science shows us that the thought of a universal science comes up again and again.'[6] But this ideal has never been achieved. In the Wundt review Neurath exhibited his lifelong impatience with 'ideal' claims that have no tenancy in reality and with abstract arguments about how matters 'must' proceed or how they could proceed 'in principle'. In the protocol sentence debate he was to make fun of 'the fiction of Laplace's spirit'[7] and of 'the ideal forecast'.[8] In 1910 he had much the same reaction to Comte's hierarchy of the sciences. Comte's characterisation simply fails to get the reality of the sciences right.

Although in 1910 Neurath provided only oblique hints about what a universal science would be like (see section 2.2.3), he had a clear view of what universal science is not. The review provides a chronicle of failed efforts. Neurath remarked that we are often told that science proceeds by 'division of labour (*Arbeitsteilung*)'. He objected that what we actually have would be more accurately described as the 'segregation of labour (*Arbeitstrennung*)'. 'Division of labour' only makes sense when different work groups pursue a single end together. This is decidedly not the case in modern science. We need at least to strengthen the threads that tie the different sciences together.[9] If this can be done, it will allow 'the world to stand before us as a whole again'.[10] It is important to notice for our story that the unification that Neurath envisaged at this time is a unification into a single theory, a single account of the world. It is for good reason that we describe Neurath's agreement with Duhem's underdetermination thesis as a commitment to 'Duhemian holism'. It is, we recall, the 'entire system' that we must decide to keep or to remodel. What we see here is a wistful yearning in Neurath for a unification that he doubted would ever obtain. Even though 'the complete system remains an eternal goal which we can only seemingly anticipate',[11] nevertheless 'It is a great task to comprehend as far as possible the complete order of life.'[12]

In 1916 Neurath again returned to the question of how to systematise a body of scientific theories. 'As we need theories to classify things, so we need theories to classify theories', he urged.[13] He illustrated with examples from the history of optics, using a strategy he would often employ: 'In order to obtain a scientifically satisfactory systematisation, one must first, willy-nilly, try to give a complete survey of combinations of the elementary notions; by the application of certain principles a

---

[6] Neurath 1910a [1981], p. 24. [7] Neurath 1932d [1983], p. 91. [8] Neurath 1931c [1973], p. 404.
[9] Neurath 1910a [1981], pp. 24–5. [10] Ibid., p. 46.
[11] Neurath 1913a, p. 455. (See section 2.3.3.) [12] Neurath 1910a, p. 44. (See section 2.2.3.)
[13] Neurath 1916 [1983], p. 31.

selection from the logically possible combinations could already be created.'[14] Once the field of abstract historical possibilities has been reconstructed, the next task for the historian is to examine the circumstances of the actual emergence of particular theories. The final aim is to obtain a coherent picture of the historical reality of each scientific discipline. In Neurath's own terms, 'After surveying this totality, one could investigate which of these combinations are realised *in "nature"*. As not all chemical compounds are represented in the minerals, so in the world of real combinations of ideas as well, not all possible theories which can be derived from certain elementary notions are *represented*.'[15] Two different attitudes towards a system of hypotheses are possible according to Neurath. First there is realism, in which we suppose that theories reflect assumptions 'about the character of the world'. In this case the unified picture will 'comprise the multiplicity of reality'. Alternatively one could be an instrumentalist like Duhem and suppose that scientific assumptions reflect 'the character of our thinking'. Realist or not, the systematisation of scientific theories should yield a satisfactory explanation 'of the same complex facts'.[16] Again the emphasis in this early paper is on unification at the theoretical level.

During the course of the Bavarian revolution Neurath gave up on this entirely. He lost interest in the single theory of reality and no serious consideration was given to it again. From at least 1919 onwards Neurath's thoughts centred on practice. One remark best captures Neurath's evaluation of the failed plans for socialisation: 'A unified program could have co-ordinated and unified action.'[17] This statement reflects Neurath's new views on the unity of science. For the success of social planning, the disunity among the sciences must be overcome at every decision at the point of action, and this requires that we incorporate a host of different types of factors into our formulations. '[W]hether a forest will burn down at a certain location on earth depends as much on the weather as on whether human intervention takes place or not. The intervention, however, can only be predicted if one knows the laws of human behaviour.'[18] This generates an urgent demand: 'It must be possible to connect all kinds of laws with each other. Therefore, all laws, whether chemical, climatological, or sociological, must be conceived as parts of a system, namely, of unified science.'[19] It is the necessity for predicting complex

---

[14] Ibid., p. 15. Reisch 1994 refers to this strategy as 'scientific planning in hindsight' by analogy with Neurath's methods in economic planning.   [15] Neurath 1916 [1983], p. 15.
[16] All quotations in this paragraph Ibid., pp. 28–9.   [17] Neurath 1920b [1973], p. 19.
[18] Neurath 1932a [1983], p. 59.   [19] Neurath 1932a [1983], p. 59.

events of the world in general and of society in particular that requires that the laws from different sciences be co-ordinated with each other. As Neurath put it: 'All laws of unified science must be capable of being linked with each other if they are to fulfil the task of prediction.'[20]

In Neurath's post-war thought a number of characteristics from his earlier work reappear with a new pragmatic dimension added. One is holism. As we saw in section 1.2.3, Neurath long accepted the holism that supports Duhem's doctrine that empirical facts underdetermine theoretical representation. He knew that our theoretical representation of the future cannot be unique. This is what led him in 1913 to postulate the need for an 'auxiliary motive' behind every rational act. Given the uncertainty and multiplicity of our theoretical representations of the world, we can only make a selection on the basis of extra-logical and extra-empirical factors and these in turn can only be defended by their instrumental utility for the goals we want to achieve. In the context of social planning, predictions are the key step before every truly rational decision. But there is a caution. Our decisions get their rationality only from a 'clear recognition of the limits of actual insights'.[21] The Laplacean ideal is, we know, 'pseudo-rationalistic'. With the war ending, society in crisis and full socialisation a possibility in view, the incomplete knowledge of the present and past raised problems not only for theory choice but more pressingly for the choice of plans of action. The practical implications are clear in the 1919 tract on war economy discussed in section 1.1.2:

If . . . we raise the question of what characteristics a social order must have in order that no one shall suffer hunger, that there shall be no credit crises, no living standards that are determined by chance and privilege, then we may find several equally correct solutions under quite definite presuppositions. If for instance we ask what order of life we may expect in the next decade we must in principle give several answers, since because of our inadequate insight into the presuppositions of events, several possibilities present themselves to us.[22]

Our lack of certainty and the precautions we must take to deal with it became for Neurath not just a question of abstract scientific method but an immediate question of the real plans for life. Neurath's pre-war work on unity of science was not only a descriptive but also a normative inquiry of second order, as argued in part 2. He taught about the proper methods for constructing scientific systems. The methods he proposed also took on a new pragmatic force during and after the war. In 1910

---

[20] Ibid., p. 68.   [21] Neurath 1913b [1983], p. 8.   [22] Neurath 1919b [1973], p. 152.

Neurath had required of scientific systems that they lay out the set of all possibilities; that is also what he demanded from the systematisation of scientific theories in his 1916 paper. The same idea was crucial in his thinking about economies. A general theory of economics must model the whole spectrum of different possible economic systems:

> If in economics we were so far advanced as to study all possible forms of economy in a quite general way, and put to ourselves the question how given forms change if certain rules are permanently observed, and what happens on the other hand if the rules change, then we should need no special theory for war, this already having been allowed for as a special case.[23]

This was in 1913.

In 1919 Neurath was concerned with more practical endeavours. His job was to implement a plan for full socialisation. The surveys of theoretical predictions refer now to social utopias in the sense not only of the ideal but of the possible.[24] In Neurath's socialist framework it is not enough for science to talk about reality; rather it must teach about 'realisations'. But the realisations must be chosen from a full array of possibilities. The trouble with conventional economics is that it generalises too narrowly from current arrangements. If empiricism means just sticking to what has already been observed, it amounts to conservatism: 'Those who stay exclusively with the present will very soon only be able to understand the past.'[25] Just as mechanics gives to the machine builder information about machines that have never been constructed, so too the natural and social sciences will give us information about economic orders that have never been realised. Perhaps it is not possible to construct one single, closed system of unified science. But the sciences must be unified at each point of commonly planned action if we are to build a socialist economy that works. It is in this revolutionary sense that unified science was, for Neurath, 'the great task of consciously cultivating the future and the possible'.[26]

The task of consciously cultivating the future depended far more on the unification of the sciences for Neurath than it did for other kinds of

---

[23] Neurath 1913e [1973], p. 125.
[24] Neurath has in mind the social plans of 'utopians' such as Ballod-Atlanticus, who was appointed by Neurath in March 1919 as head of the Department of Measurement-in-Kind of the Office for Central Planning, and Popper-Lynkeus who was a friend of Neurath's father. In the tradition of the 'ecological economics' proposed by the latter, Neurath's model of a socialist non-monetary economy has been considered the first attempt in that direction from a Marxist perspective (see Martinez-Alier 1987). See part 1.  [25] Quoted Nemeth 1981, p. 61.
[26] Neurath 1919b [1973], p. 155.

socialists. That is because Neurath aimed for full social planning, not just for scattered social improvements. The piecemeal social planners take on easier tasks that require less cross-discipline co-operation. They can arrange improvements in train with advances internal to the separate sciences. If a new medical discovery is made, we set up a new kind of clinic to take advantage; if new emission control devices are developed, we can improve the atmosphere. For piecemeal planning we can, to a considerable extent, choose our problems to match the solutions we have available. This strategy is not possible under full socialisation. There everything – procurement of materials, transportation, production, distribution – must be made to work together.

A social plan that misses the complexity of social life will be of no use:

> A social engineering construction treats our whole society and above all our economy in a way similar to a giant concern. The social engineer who knows his work and wants to provide a construction that will be usable for practical purposes as a first lead, must pay equal attention to the psychological qualities of men, to their love of novelty, their ambitions, attachment to tradition, willfulness, stupidity – in short to everything peculiar to them and definitive of their social action within the framework of the economy, just as the engineer must pay attention to the elasticity of iron, to the breaking point of copper, to the colour of glass and to other similar factors.[27]

The plan stands or falls with the possibility of unification. In 1931 Neurath's message was clear: 'Common planned action is possible only if the participants make common predictions. A common greater error made by a group often yields better results than mutually antagonistic small errors of isolated individualists. Common action presses us toward a unified science.'[28]

We have described a very particular development in Neurath's unity-of-science philosophy. Before the war Neurath had shared the usual theoretical longing for a unified representation of the world. In the end he aimed instead to assemble a kit of standardised tools with which we can co-operate to transform the world. Neurath himself described the encyclopedia project of the Unity of Science movement as a 'kind of instrumentarium, a stock of instruments for science in general.'[29] This shift supported and was supported by the plan for the full socialisation of Bavaria. We claimed in the Introduction that in Neurath intellectual work and political life evolved not separately but inextricably woven

---

[27] Ibid., p. 151, translation altered.    [28] Neurath 1931c [1973], p. 407.
[29] Neurath 1936d [1983], p. 141.

together. They formed a single passion. Here is a striking instance. On the one hand Neurath's project for socialisation demanded that the sciences be unified. On the other his confidence in the possibility of a unified science at the point of action enhanced first his optimism about socialism and in later years his hopes for international co-operation.

Neurath believed full socialisation to be possible because he thought that we had reached a stage of scientific and technological knowledge that would enable us to construct a workable economic plan. Consider his remarks in 1920: 'Socialism would be advanced if the political leaders could draw on socio-technical engineers, who . . . construct the economic organisation that would best realise the socialist economic plan.'[30] Even after the revolution was over Neurath held to this faith: 'The factual relations of development are, in my opinion, ripe for socialism; it is only the organisational moment that comes into question.'[31]

Unification of the sciences and socialisation of the economy remained a single task throughout Neurath's Viennese period. In 1929 on the occasion of a lecture in Prague on the scientific philosophy of the newly formed Vienna Circle, Neurath reminded his audience that 'our thinking is a tool'. He referred to this well-known Marxist dictum again as he emphasised the Circle's goal of a unified science: '[The scientific world-conception is concerned with] the uniform logical treatment of all trains of thought, in order to create a unified science that can successfully serve all transforming activity.'[32] Through Neurath the same pragmatic concerns turned up in the text of the Vienna Circle manifesto. There he urged:

We have to fashion intellectual tools for everyday life, for the daily life of the scholar but also for the daily life of all those who in some way join in working for the conscious re-shaping of life. The vitality that shows in the efforts towards a rational transformation of the social and economic order permeates the movement for a scientific world-conception too.[33]

Neurath's social plans during the Bavarian revolution had philosophical consequences *for* unified science just as his social plans during the Red Vienna period expressed the political role *of* unified science. First, science serves as a weapon against metaphysics. This idea was widely shared among the members of the Vienna Circle. Yet there is a difference. For most of the others the motivation for fighting 'the anarchism of

---

[30] Neurath 1920e, p. 226.   [31] Neurath 1922a, p. 55.   [32] Neurath 1930a [1983], p. 42.
[33] Carnap, Hahn and Neurath 1929 [1973], p. 305.

metaphysics' was chiefly epistemological. They aimed to ensure the purity of our beliefs, to keep us from espousing nonsense. Neurath's motivation was political. In modern history metaphysics had proved a strong negative social force, a force that could be combated by the doctrines of the Vienna Circle. Neurath hoped – indeed believed – that, '[T]he scientific spirit fights off the metaphysical speculation that fosters with its obscurantism all sorts of clerical and political absolutism.'[34] Marxism had the right idea about metaphysics:

> The cultivation of scientific, unmetaphysical thought, its application above all to social occurrences, is quite Marxist. Religious men and nationalists appeal to some feeling, they fight for entities that lie beyond mankind. To them the state is something 'higher', something 'holy', whereas for Marxists everything lies in the same earthly plane. The community of the state is nothing but a kind of large association, whose statutes do not possess special holiness. What may appear in the way of feelings and ideas is regarded as a piece of this order of life and is not put above it.[35]

Though this identification of scientism with Marxism was not uncommon among Marxist intellectuals, it was not a plank in the Vienna Circle's platform. Neurath left no doubt about his views: 'Many who came from the bourgeoisie are worried about whether the proletariat will have some feeling for science; but what does history teach us? It is precisely the proletariat that is the bearer of science without metaphysics.'[36]

Second, Neurath believed that unified science would serve to enhance unity among people. He urged all the sciences to adopt the single language of physicalism in part because he believed that that would promote the communication and solidarity necessary for co-operation. Like many socialist programmes Neurath's project had an international character. In the Vienna Circle's manifesto for instance, he linked the scientific world conception with the drive 'toward the unification of mankind'.[37] In 'Visual Education' he argued for the role of unified language in this project too:

> Variety within a community has to be based on some structure of a society, and therefore we have to look at the social elements which make contacts possible between human beings from group to group throughout the world and a communication of understanding from nation to nation. A certain amount of trust is needed for successful co-operation, and also the possibility of making contact

---

[34] Neurath 1928a [1973], p. 295.   [35] Ibid.   [36] Ibid., p. 297, translation altered.
[37] Carnap, Hahn and Neurath 1929 [1973], p. 305.

through language. If there is not one common language, then languages which are partly translatable into one another are wanted.[38]

When Neurath founded the Unity of Science movement in 1934 the need for unity was once again tragically clear. The movement faced a period of international crisis and devastating war. His idea that unified science can bring different people together received its most succinct expression in his slogan, 'Metaphysical terms divide; scientific terms unite.'[39] The slogan shows the broad promise of unified science. Neurath had long seen unified science as providing the tools for common action; now he stressed its role in building trust among people and among nations. We shall argue in the next sections that Neurath's attack on scientific method was a battle in his war against metaphysics and pseudo-rationalism. Here we have tried to show how for Neurath the project for the unity of science was the Trojan horse of socialism.

### 3.2.2 Unity without the pyramid

From 1934 until his death Neurath vigorously worked for the model of unity of science that remains recognised as most distinctively his own: the 'encyclopedia'. This section is an attempt to answer two connected questions: Why 'encyclopedia'? And why 1934? The account we offer amalgamates the philosophical and the non-philosophical. Its duality tries to capture the basic duality of Neurath's 'encyclopedia', which he conceived as both a theoretical and a practical project. As a model it was to describe how the unity of science could realistically be conceived. As a project it was meant to guide scientists towards a progressive realisation of the model. At both levels the 'encyclopedia' had first and foremost international scope and historical grounding.

The encyclopedia project was launched in 1934 at the Eighth International Congress of Philosophy at the Charles University in Prague. With the help of Carnap and Frank, Neurath had organised a preliminary Conference of the International Congresses for the Unity of Science that took place from 31 August to 2 September. The leading theme of Neurath's address was 'unity of science as a task'. Neurath presented the germ of what he subsequently referred to as the 'encyclopedia-model' and laid out the plans for the publication of a series of scientific works that would cover the ongoing developments in every particular science and try to build connections with the others: the

[38] Neurath 1946b [1973], p. 229.  [39] Neurath 1933a, p. 23.

'Encyclopedia.' The 'encyclopedia-model' and the 'Encyclopedia' would carry out, in so to speak the 'formal' and 'material' modes respectively, the task of unifying science. Neurath placed the emphasis on the historical, the international, the scientific, the antimetaphysical and the co-operative character of his 'encyclopedic synthesis of the sciences':

> Therefore whoever discusses the *unity of science* as a possible task starts from the assumption that co-operative work is increased, that within mankind scientific thinking will win out more and more in all spheres. If he cannot give specific reasons for this but only has mere hopes, he must understand that this is a *historical matter* . . .
>
> Here is a great task of anti-metaphysical empiricism, that above all makes it sharpen the logical instrument in such a way that it can serve science immediately. The task is to repulse traditional metaphysics, especially traditional teleology, traditional anthropomorphism in a new shape in order to create a unity of science that comprises geology as well as ethnology, astronomy as well as sociology, mechanics as well as biology and behaviouristics. And if we are soon to gather at the First International Congress for the Unity of Science we do so not only to advocate scientism, but also to stress our resolution to work together for the logical development of science. Thus the Parisian friends, who have suggested that we come to Paris can welcome us in that old tradition with the call for: *unité et fraternité*.[40]

Both in the 'formal' and the 'material' modes, the idea of 'encyclopedia' appears patterned after the project of the French *Encyclopédie*. That is no coincidence. In the particular case of the 'Encyclopedia' project, the answer to our two initial questions begins to take shape when we consider Neurath's emphases in their historical context.

The forces of Fascism had been rising in Europe since 1922. Benito Mussolini headed the Fascist government of Italy. On 30 January 1933 Adolf Hitler was appointed Chancellor of Germany. During the Bavarian revolution Hitler had been posted to Munich in charge of the propaganda of the German Workers' Party; on his own view Communists and Jews (probably Leviné, Neurath himself and others) were dragging Germany to social, moral and economic annihilation. The following February (1920) Hitler presented in Munich the twenty-five points of the anti-semitic and anti-Marxist programme of his new National Socialist German Workers' Party. On 7 April 1933 Hitler's government promulgated its first anti-Jewish ordinance: the Law for the Restoration of the Professional Civil Service. The new law required all civil servants of non-Aryan descent to retire: 1,600 German scholars

---

[40] Neurath 1935a [1983], p. 119, original italics.

were stripped of their academic positions and most of the best German scientists were forced into exodus. In February 1934 Fascism took over in Austria with Engelbert Dollfuss and the support of the Catholic party. Dollfuss declared liberalism, Judaism and Marxism to be anarchist tendencies and the Ernst Mach Verein was dissolved for political reasons. Carnap and Frank were working and living in Prague, Neurath was consulting in Moscow at the time and, unable to return to Vienna, emigrated to The Hague.

In Neurath's view the forces of irrationality and obscurantism – Nazism and Catholicism – were bringing repression and dispersion. And that is just what the spirit of scientism opposes, fostering rationality, progress and unity. This very spirit had driven the educational task that Neurath so energetically promoted and carried out in Vienna before 1934. An international expansion of that same effort was now in order: the co-operation of the scientists – by way of a 'republic of scientists' – would make available the totality of scientific knowledge and in turn this educational effort would sustain scientific co-operation and progress. Only this could lead to equality, liberty and fraternity in worldwide enlightened peace.

Why an 'Encyclopedia'? As Neurath recalled it in 1937, years earlier in Vienna he had already realised that:

a comprehensive scientific view could serve as an important basis for a sound general education. In this way there occurred to some friends and myself the idea of working up a series of about 100 pamphlets which would deal with all sorts of things, stars, stones, plants, animals and men. A comprehensive alphabetical index was to be included, so that this collection would be used as a dictionary as well.[41]

The French *Encyclopédie* represented a collaborative effort serving just such purposes, as Einstein had pointed out to him at the time: '[Einstein] wrote me that such a well-planned collection would serve the same function for the masses today as did the French *Encyclopédie* for the intellectual groups in France during the eighteenth century.'[42] If the French *Encyclopédie* was a model of co-operation and educational effort that became a liberating *machine de guerre* in the French Revolution, so a Viennese Encyclopedia could have been in Red Vienna, and in 1934 an international Encyclopedia worldwide. Although under more dramatic and pressing circumstances, the international Encyclopedia of Unified Science emerged as a natural extension of Neurath's

[41] Neurath 1937b [1983], p. 179.   [42] Ibid.

educational scientific projects in Vienna, which already enjoyed an international dimension in the exhibition for visual education and the use of ISOTYPE language.

And, in fact, Neurath's plans for the Encyclopedia included the publication of a Visual Thesaurus[43] – for not only does science unite, pictures do as well. The plans were worked out by Neurath and his collaborators at the Mundaneum Institute in The Hague and presented formally at the First International Congress for Unified Science held at the Sorbonne in Paris in September 1935. The followers of this Unity of Science movement voted approval for the project and Carnap and Charles Morris, in Chicago, negotiated its publication with the University of Chicago Press. Neurath closed his address at the Paris Congress, emblematically enough, expressing the 'hope that on the broad basis of scientific empiricism there may develop *unité de la science et fraternité entre les nouveaux encyclopédistes.*[44]

By the end of the Second World War Neurath was referring to the unification of the sciences as 'orchestration'. 'Orchestration' involved again a mobilisation of the scientists from around the world for planning and for peaceful action: 'We intentionally reject the plan of forming anything like a programme, and we stressed the point that actual co-operation in fruitful discussion should demonstrate how much *unity of action* can result, without any kind of authoritative integration.'[45]

Finally, we should note three prime implications of the realisation of the Encyclopedia as a historical task. First, it materialises Neurath's long-held Marxist beliefs that 'our thinking is a tool' and that, in particular, 'scientific attitude and solidarity go together.'[46] Second, it reflects his Marxist-inspired epistemological position that, contra Descartes, science is a historical, public and social enterprise: 'It is not a single individual who can really think new notions through to the end, but only whole groups or generations. Thinking, too, is a collective occurrence.'[47] Third, Neurath also intended the notion of 'encyclopedia' to provide a theoretical model for what unity of science could be like. The way towards this use of the idea had been paved in the philosophical discussions with the Vienna Circle.

Neurath's mature philosophic views on unity are in express contrast to those proposed by Carnap in the *Aufbau* and in the later writings between 1928 and 1931. As we have seen in part 2, Carnap's views on the *Aufbau*

[43] Ibid.  [44] Neurath 1936f [1983], p. 138, original italics.
[45] Neurath 1946a [1983], p. 230, italics added.  [46] Neurath 1928a [1973], p. 252.
[47] Ibid., p. 293.

changed during the protocol sentence debate, in part out of his discussions with Neurath. The changes were primarily concerned with the nature of the starting points for the scientific construction – were these reports of experience or reports about things publicly observable? Little shifted in the nature of the construction itself.[48] Unified science consists of a single logical construction built upwards from a specifiable basis. In this construction precise scientific concepts at each level are connected by logical reduction chains that go from the most complex down to the most simple.

Carnap, as we have seen, believed that two such systems were possible, each grounded on a different basis. By 1930 both systems were adequate and both were indispensable for Carnap. One system instantiates a phenomenalistic reduction. The concepts and high-level statements of science are to be constructed out of the 'bricks' of private experience. In the second system all higher concepts and laws including those of the cultural and psychological sciences are to be reduced to the concepts and laws of physics. Carnap summed up his view in 1931 in this way:

Result in slogans . . . [Methodological solipsism:] . . . The reduction proceeds in the direction of the given of an epistemic subject. Materialistic view: not only the sentences of certain areas of scientific inquiry are reducible to physics, but every sentence of intersubjective science.[49]

Neurath objected to the starting point for both the phenomenalistic reduction and the physicalist. As we saw in part 2 the phenomenalist basis was ruled out on two grounds. First, scientific statements can not be compared to the private content of experience; rather, 'Statements are compared with statements, not with "experiences", not with a world nor with anything else. All these meaningless duplications belong to a more or less refined metaphysics, and must be rejected.'[50] Second, the building of a scientific understanding of the world is a collaborative enterprise. It requires vast amounts of information from different laboratories and from different parts of the globe. Recall the mammoth statistical studies compiled by the Verein für Sozialpolitik that we described in our Introduction. The idea that a body of knowledge could be constructed from individual experience is ludicrous. Science needs an intersubjective basis. Neurath's special form of protocol sentences provides it.

Neurath's opposition to the physicalist reduction is one of the central

---

[48] Carnap also made the significant syntax-semantics shift, but that had little bearing on the unity of science.   [49] Carnap's Munich lecture of 10 Jan. 1931, quoted in Uebel 1992b, p. 114.
[50] Neurath 1932a [1983], p. 66.

themes of his writings in 1931. Here is where Neurath's professional and intellectual background mixes with the philosophical debates to motivate his philosophical position within the Circle. Neurath was not trained, nor did he work professionally as a natural scientist or a mathematician. His interests, we know, lay in the social sciences. Consequently his views on the language, method and structure of science had to take into account the social sciences as genuine empirical sciences on a par with the natural sciences. He rejected the distinction between natural and social sciences as grounded on objectionable metaphysical and epistemological assumptions.[51] But this step towards a broad unity of science does not require either a reduction or an assimilation of one to the other. It is not necessary to be twins in order to be brothers. The scientific aspect that the different disciplines have in common is the emphasis on the importance of producing correlations and of using the empirical language of physicalism. Disciplines such as sociology and political economy will manifest their own distinctive peculiarities and limitations, as Neurath made clear in *Empirical Sociology*.[52] Regarding the unification of the sciences, Neurath insisted that the social sciences neither need to be, nor can be, reduced to physics. Thus he argues:

The development of physicalistic sociology does not mean the transfer of laws of physics to living things and their groups, as some have thought possible. Comprehensive sociological laws can be found, as well as laws for definite narrower social areas, without the need to go back to the microstructure, and thereby to build up these sociological laws from the physical ones.[53]

Neurath's objections to the physicalist reduction went beyond the special issues of the protocol sentence debate. The idea of reduction itself was idle metaphysics, driven by a 'pseudo-rationalist' ideal. For Neurath any view that promises rational justifications based on knowledge that is never available is pseudo-rationalistic. In this particular case the possibility of making predictions from the microstructure to the macrostructure relies on the fiction of Laplace's demon, who guarantees the complete descriptions that are supposed to make the prediction possible:

---

[51] See his 1931d [1983]. [52] Neurath 1931c [1973].
[53] Neurath 1932a [1983], p. 75. He also remarked that 'the danger is that the logical clarity of formulas in physics may be taken as a paradigm for all the sciences' (1936a, quoted in Zolo 1986 [1989], p. 94). Note that at the time Neurath completed the essay 'Empirical Sociology' in 1930 his position was still Carnapian: 'We must create a system of concepts which provides the possiblility of deriving concepts from each other, like a pyramid (cf. Carnap, *Konstitutionssystem*)' (Neurath 1931c [1973], p. 390). 1931 is the turning point in the evolution of Neurath's thinking within the Vienna circle.

Many proceed from an 'ideal forecast'; from the Laplacean mind that knows all initial conditions and all formulas and thus can predict everything. Such a fiction is already metaphysics. For evidently here there are assumptions that are in principle not subject to any empirical test. In reality we are dealing now with more, now with less predictable partial connections, and in some cases we can say nothing about individuals but we can say something about groups of individuals.[54]

But must we take Laplace's demon so literally? Is it not just a metaphor that points to the existence of deterministic laws that operate on finite sets of initial conditions? Certainly not, for any member of the Vienna Circle. The Circle did not talk about laws and facts and one thing making another happen. Its members talked about statements and what can be predicted from them. And that was the extent of what they were willing – and entitled – to talk about. There is a tendency nowadays not to take the linguistic turn of the Circle seriously enough. We are apt to suppose that they assumed a second structure – 'the real world' – running in parallel to the linguistic structure. The two are alike except that the linguistic structure is sometimes faulty and may not provide an accurate copy of the original. From this point of view it is not too worrying if exact predictability is missing from the linguistic structure; the deterministic laws of the original structure are there in the wings to back up our claims about reductionism. That is clearly not the right picture. According to the Vienna Circle the relations under discussion are relations between statements and when these relations are not there, they are not anywhere else either.

Neurath was firm about this matter, not only in general, but specifically about laws of nature. He criticised Popper for speaking of 'the belief that there are regularities that we can unveil, discover', for this is 'gliding into the metaphysical'.[55] He was sympathetic with Schlick: 'Laws are not statements; they are directions for obtaining *predictions* from observation statements.'[56] Natural laws can fare no better than moral laws, which have the same metaphysical standing as God:

But how should we demarcate a discipline as 'ethics' if God is eliminated? Can we make a meaningful transition to a 'command in itself', to the 'categorical imperative'? We could just as well introduce a 'neighbour-in-himself without a neighbour', a 'son-in-himself who has never had father or mother'.[57]

---

[54] Neurath 1931c [1973], p. 404.  [55] Neurath 1935b [1983], p. 130.
[56] Neurath 1932a [1983], p. 62.  [57] Ibid., p. 79, cf. 1931c [1973], pp. 324–5.

Against both Carnap's phenomenalistic and physicalistic reductionism Neurath maintained that the 'control' points for scientific claims were neither constructs from private experience, which is itself merely a philosophical fantasy, nor observations from physics experiments, which have no clear relevance outside the immediate hypotheses of physics that they test. They are, as we have seen, real data reports couched in the different historically determined languages of the different sciences.

But the real data-reporting statements of science use a mix of technical language and the language of everyday life:

> If we want to embrace the entire unified science of our age, we must combine terms of ordinary and advanced scientific languages, since, in practice, the terms of both languages overlap. There are certain terms that are used only in ordinary language, others that occur only in scientific language, and finally, terms that appear in both. In a scientific treatise that touches upon the whole range of unified science, therefore, only a 'jargon' that contains terms of both languages will do.[58]

In section 3.3 we shall show how this claim set off Neurath's attack on method. It also has consequences for the unity of science. Although Neurath rejected the strong programme of reduction, in 1932 – before he had seen the full implications of his doctrine that the terms of everyday life are congestions (*Ballungen*) – he had been willing to admit a weaker requirement – consistency: 'In unified science we try... to create a *consistent system* of protocol statements and non-protocol statements (including laws).'[59] At this stage it seems that a compromise was still possible with the view that unified science aims to construct a single theoretical account of a single world. It is a central thesis of our discussion of Neurath's changing views on unity that he became uninterested in this traditional unified picture of the world. But was it positively ruled out? We know that he made much of the fact that the different sciences are about radically different things – forces, bacteria, aversions, atomic shells, space-time curvature, marginal utilities and so forth. He also rejected the illusionary ideal of an ultimate reduction, either in God's great book, or in some imaginary limit of a better or more complete science. How should we know what will happen as science progresses? We must base our account on how the sciences are in our experience, and on our best empirical predictions for how they might change.

In that case the ideal of a systematic fit within the rich variety of scientific concepts must seem improbable. Nevertheless it appears that a kind

---

[58] Neurath 1932d [1983], p. 92.   [59] Ibid., p. 94.

of single theoretical picture may still be possible – a patch-work with great gaps and large regions of overlap. The consistency requirement allows the assumption that the different sciences together provide an accurate picture of the world: although the vocabularies are different, the world pictured is unique. But this realism does not fit readily with the image of reality as an infinite continuous manifold, an image that, we shall argue in section 3.4.3, Neurath had been sympathetic to since the end of the Bavarian revolution. With the introduction of the idea of *Ballungen* and the beginnings of the attack on method in 1931 Neurath focused his attention on these kinds of questions. By 1935 Neurath had completely lost faith in the existence of logical connections among the statements of unified science. Rather than pieces that fit together in determinate ways, we have instead 'a mosaic pattern of the sciences'.[60] To Carnap's system-model he opposed his own 'encyclopedia-model'.[61] The encyclopedia is a preliminary assembly of the totality of scientific knowledge available. In this encyclopedia, according to Neurath, 'The statements are linked to each other, sometimes more closely, sometimes more loosely. The interlocked whole is not transparent, while systematic deductions are attempted at certain places.'[62] The encyclopedia-model recognises the irreducible locality and imprecision of the cross-connections between the sciences and the incompleteness of the assembly of accepted scientific statements always in progress. Neurath continued of course to endorse the use of logical and mathematical tools and he was especially keen on probabilistic techniques. His important philosophical realisation was that logic and mathematics can not achieve unity. Unified science cannot realistically be fitted into the reductionist myth of 'the system': the unique, complete, and deductively closed set of precise statements. '"*The*" *system* is a great scientific lie', Neurath insisted in Prague in 1934 and in Paris in 1935.

The rejection of the reductive and deductive system-model is just the stand D'Alembert adopted in 1763 in the French *Encyclopédie*:

Insofar as one has reflected on the connection that the discoveries have among themselves, it is easy to see that the sciences and the arts give support to each other, and that consequently there is a chain that brings them together. But if it is often difficult to reduce each science and each art taken separately to a small number of rules and general notions, it is not less difficult to contain the

---

[60] Neurath 1937d [1983], p. 204.
[61] This distinction is implicit in his address to the Prague Congress in 1934 (see Neurath 1935a), but it is clearly drawn in his criticism of Popper (see Neurath 1935b) and in Neurath 1936d.
[62] Neurath 1935b [1983], p. 122.

infinitely varied branches of human science [*la science humaine*] in a system that is unitary.⁶³

Although Neurath did not acknowledge the empiricist and anti-metaphysical spirit behind the French precedent, he did repeatedly acknowledge his predecessors' rejection of the *esprit de système*. It was this precedent that inspired Neurath's use of the term 'Encyclopedia':

> I have suggested the term "encyclopedia" primarily in opposition to the term "system" by means of which a kind of total science based on axioms is postulated, [a total science] that has to be discovered, as it were. Such a notion is especially dubious if one starts to give the outlines of such a system – a fact that has already been pointed out by the leader of the French encyclopedists, D'Alembert.⁶⁴

## 3.3 THE ATTACK ON METHOD

### 3.3.1 Boats and Ballungen

Neurath, as we have seen, was both a holist and an anti-foundationalist from the time of the first Vienna Circle. During the debates in the Vienna Circle proper these views developed and became more detailed. In 1931 they took a new turn. The second half of our story in part 3 describes Neurath's attack on method and how it came about. The advance in Neurath's views is reflected in the shifting contexts of the Boat metaphor reviewed in part 2. The change with which we are concerned in this section appears between the 1921 and the 1932 version. Neurath's view before 1931 can be summarised in three theses: (1) Anti-foundationalism: there are no secure and infallible foundations for knowledge. Not only advanced hypotheses but protocol statements too can be rejected. (2) Theoretical pluralism: the same bases work for different theoretical structures equally well depending on what auxiliary hypotheses are assumed. This is a consequence of Duhemian holism. Data are not enough to fix the structure that sits on top of them; auxiliary hypotheses are required as well. Finally (3) Historical conditioning: both the bases and the auxiliary hypotheses are socially and historically conditioned and thus subject to change. The same is inevitably true of the theories built upon them.

These three theses are illustrated in Figs. 3.3 and 3.4 (Birds/People,

---

⁶³ D'Alembert 1929, p. 13.
⁶⁴ Neurath 1936c [1983], p. 168, translation amended. See also Neurath 1936d [1983], p. 158.

# The attack on method

Fig. 3.3 Birds (pre-1931)

Fig. 3.4 People (pre-1931)

pre-1931). None of the structures is built on solid rock; the groundings are wobbly, not firm. Nor do they determine what is built upon them – the same ground can support different structures. A comparison of the two drawings reminds us that both the empirical base and the theoretical structure alike are historically conditioned. Those with different cultural settings will choose different places to start and will take different things for granted in their constructions. This pair of drawings illustrates Neurath's views before 1931.

In 1931 Neurath introduced a new concept into his philosophy – that of *Ballungen*. After that another doctrine made itself felt: (4) There are no logically determinate connections between data and theory. Even supposing that we adopt a set of auxiliary assumptions, the bearing of the data on a hypothesis is underdetermined. This change is evident in the 1932 Boat:

There is no *tabula rasa*. We are like sailors who have to rebuild their ship on the open sea, without ever being able to dismantle it in dry dock and reconstruct it from the best components. Only metaphysics can disappear without trace. Imprecise 'verbal clusters' (*Ballungen*) are somehow always part of the ship. If imprecision is diminished at one place, it may well reappear at another place to a stronger degree.[65]

*Ballungen* are literally congestions. More intuitively, we can think of a modern city depicted on a map. One sees a big mass, dense in the middle and then spread out on the edges, here and there, with its boundaries undefined. This is a *Ballungsgebiet*. The word comes from the word '*ballen*' which means to squeeze together to form a ball. Neurath refers to *Ballungen* terms as both imprecise and complex. But we must beware of the term 'complex'. For *Ballungen* are not made out of parts. They are rather the primitive concepts from which we start and from which other concepts might be constructed or derived. The terms of everyday life are *Ballungen* and, as we have seen in the protocol sentence debate, the terms of everyday life are ineliminably part of our scientific data reports.

With the introduction of *Ballungen* new problems arise for scientific method, problems that Neurath had not noticed before. The very ideal of exact scientific testing becomes pseudo-rational. The terms of advanced science are precise, those of everyday life are congestions; between the two no determinate relations can obtain. The hypothetico-deductive method will not work because statements involving the fuzzy-

---

[65] Neurath 1932d [1983], p. 92.

## The attack on method

edged *Ballungen* can not be deduced from the exact claims of science. Induction can fare no better, nor can confirmation. All three – falsification, induction, confirmation – are equally misleading, because equally formal, concepts. The logical connection between a hypothesis-system and the data is never so strict and automatic as they require. In Neurath's earlier works before *Ballungen* the relation between a fact and a hypothesis-system appeared straightforward. Either they were, or they were not, compatible.[66] If they were not, we had to decide (though of course not arbitrarily) which to keep. Now choice enters at an even more fundamental level. Content and syntax do not fix whether data and hypotheses are compatible or not. That too requires a decision. Neurath proposes giving up the language of induction, falsification, and confirmation. Instead we should talk of 'shaking':[67] Data can not, logically speaking, strictly falsify a system of hypotheses; they can only shake your confidence in it.

It is precisely this new element of conventionalism that marks the difference between Neurath's 1921 Boat and that of 1932. In 1921 he taught:

> We are like sailors who must reconstruct their ship on the open sea but are never able to start afresh from the bottom. Where a beam is taken away a new one must at once be put there, and for this the rest of the ship is used as support. In this way, by using the old beams and driftwood, the ship can be shaped entirely anew, but only by gradual reconstruction.[68]

The sailors on the second ship it seems had solid boards, cleanly sawn. If they had a gap to fill, there was no question whether a given board would fit or not. If it did, they could use it. If not, it could still be built into the ship by casting out or sawing up some of the surrounding boards and thus reshaping the gap. Gaps on the third ship are not so well defined. Chunks of rather different shapes can be forced into the same gap, or, conversely, the edges of the gap can be bent about a bit to make it accommodate a given piece. Fig. 3.5 and 3.6 (Birds/People, post-1931) show what can happen when the connections that join theory and data are no longer rigid but themselves become subjects of choice. So long as the choice is made wisely, splendid serviceable structures never before dreamt of become possible.

In the *Aufbau* Carnap attempted a deductive logical construction of scientific concepts in a closed system that radiates unambiguously

---

[66] Cf. Neurath's call for consistency as late as 1932, quoted in section 3.2.1 above.
[67] Neurath 1935b [1983], p. 123.   [68] Neurath 1921a [1973], p. 199.

Fig. 3.5 Birds (post-1931)

upwards from a basis of precise and atomic empirical statements. Neurath objected to the employment of a phenomenalistic base for these constructions. But still in 1930 when he wrote *Empirical Sociology*, Neurath believed that the idea of a system model of science was a valid project: 'We must create a system of concepts which provides the possibility of deriving concepts from each other, like a pyramid (cf. Carnap, "Konstitutionssystem").'[69] After the introduction of the notion of *Ballungen* such a system became impossible. In 1934 he addressed the question of the unity of science again, but in very different terms: '"*The*" *system is the great scientific lie*... The progress of science consists, as it were, in constantly changing the machine and in advancing on the basis of new decisions. Still the result is far-reaching unity that can *not* be derived logically.'[70]

[69] Neurath 1931c [1973], p. 390.   [70] Neurath 1935a [1983], p. 116.

# The attack on method

Fig. 3.6 People (post-1931)

Neurath explained what goes wrong in his review of Popper the following year. Neurath criticised Popper for 'following Laplace's spirit' in aiming for 'one unique distinguished system of statements as the pattern or paradigm of all the factual sciences'.[71] Popper falls into Laplacean metaphysics so readily because he 'still uses well-defined theories built up of clean statements as models, so to speak, of the factual sciences'.[72] That was also Carnap's ideal. In different ways they both committed the same error. Both failed to grasp that 'complex (many) statements of little cleanliness – *Ballungen* – are the basic material of the sciences'.[73]

### 3.3.2 Protocols, precision and atomicity

Neurath maintained that theoretical statements are precise and data reports imprecise. To understand what kind of imprecision *Ballungen* introduce it will help to consider its opposite. What kind of precision do we find in the terms contained in theoretical statements? Theoretical terms may be precise because they represent quantities, that is, physical magnitudes that can be associated with exact numbers. But quantification is not necessary for precision in the sense Neurath intended. Already in 1910 Neurath insisted: 'You can have exact relations between non-

[71] Neurath 1935b [1983], p. 121.   [72] Ibid.   [73] Ibid., p. 128.

measurable quantities.'[74] Quantitative representations presuppose a unit of measurement. To these he opposed what he called, following Duhem, symbolic representation, teaching that: 'Exact symbolic representation and quantitative representations do not come together.'[75] This doctrine was central to his positive doctrines on economics, as we saw in section 3.3.1.[76] In section 1.1.2 we also saw how, extending the Austrian utopian tradition, Neurath urged the construction of a moneyless economy like the ancient Egyptian giro system. Economics, he argued, can be an exact science without the introduction of fictional quantities like market prices.

The kind of non-quantitative precision that Neurath wanted is the precision bestowed by logical order. In the Vienna Circle Carnap too was concerned with precision in the language of science, for he, like Schlick, was eager to overcome the unreliability which beset terms whose meanings were defined by ostension. Following Hilbert, Carnap believed that terms with axiomatic definitions – that is, implicitly defined by the set of axioms they satisfy – would escape this unreliability. Neurath agreed. Abstract terms of high-level science could be exact if they were introduced by implicit definitions.[77] The terms of everyday life are different.

The lack of precision of everyday concepts was a central topic in the Vienna Circle in early 1931. In February the meetings of the Circle were devoted to discussion of Waismann's *Theses*.[78] Friedrich Waismann gave a talk in the Circle discussing claims of Hahn's that the concepts of daily life are not sharp and that their analysis can not be univocal. In the first session of that month Hahn had criticised Waismann's Wittgensteinian assumption that all meaningful speech is supported underneath by precise atomic sentences. Hahn thought these could at best be isolated conventionally.[79]

The next month the Circle discussed the issue of physicalism. On 4 March members of the Circle met outside of their regular Thursday schedule to listen to a discussion of Neurath on physicalism. It was in this discussion that Neurath introduced the notion of *Ballungen*:

---

[74] Neurath 1910a [1981], p. 34.   [75] Ibid.

[76] It was central also to his doctrine of the common scientific status of the natural and the social sciences. Compare section 2.3.1.

[77] This too is a well-known Austrian idea. Carl Menger insisted that there is no difference between the natural and social sciences. Economics can be every bit as much a deductive science as physics. Yet his own major work in economics, for which he (along with Walras and Jevons) is credited with first introducing the principle of marginal utility, is entirely free of mathematics.

[78] Waismann 1930.   [79] Cf. Uebel 1992b.

## The attack on method

We always start from historical, natural language. Its sentences are *Ballungen*, and that means mixtures of forms of expression (precise and imprecise concepts), for example, 'The shrieking saw cuts through the blue wooden die.'[80]

Imprecise terms enter with protocol statements and they will persist even when we have achieved a pure physicalistic language for our data reports, cleansed of all metaphysical overtones. The language of science will always mix the precise and the imprecise; it will always remain, as Neurath later put it, 'a jargon'.[81] Neurath was particularly concerned with proper names and perception-terms: 'In protocol statements the name of the observer and terms of perception occur that are highly imprecise.'[82] He gave a detailed example of the complex vagueness brought in by proper names:

'Otto' is itself in many respects an imprecise term; the statement, 'Otto observes' can be replaced by the statement 'the man whose carefully taken photo is no. 16 in the file observes'; but the term 'photo no.16 in the file' has not yet been replaced by a system of mathematical formulas that is unambiguously co-ordinated to another system of mathematical formulas that takes the place of 'Otto'.[83]

It is crucial to Neurath's attack on method to realise that the concepts that make up the data reports against which scientific hypotheses are to be judged can never be equated with any precise scientific concept no matter how long the definition.

Neurath thought of the terms of scientific theory not only as precise but as having clearly settled meanings, primarily because they are introduced by definition. This provides a second contrast between *Ballungen* and the terms of science. The latter are univocal; the former we might characterise as 'plurivocal'. Neurath must have had something like this in mind when he called the imprecise terms of the physicalist jargon 'pluri-terms'.[84] He gives as examples: 'pluri-moon' and 'pluri-table'; and he draws the contrast: 'But we may not speak of a "pluri-point", "pluri-sphere" and "pluri-line" in discussing schemes of mathematical physics.'[85] The problem is 'how to correlate the "pluristatements" with the "mono-formulae".'[86] In explaining the problem Neurath again refers to the lack of precision of the plurivocal term: 'It is misleading to

---

[80] 'Besprechung über Physikalismus', 4 March 1931, RC 029-17-03 ASP.
[81] Neurath 1932d [1983], p. 92.   [82] Neurath 1934 [1983], p. 106.
[83] Neurath 1932d [1983], p. 91.   [84] Neurath 1941a [1983], p. 219.   [85] Ibid., p. 215.
[86] Ibid. This problem, as we will show in the next section, motivated Neurath's formulation of a second 'Neurath principle' as well as much of his attack on scientific method in his debate with Popper.

speak of the "exactness" of data . . . because we have no standard of exactness within the "pluri-terms", only within the schemes and their formulae may we speak of exactness.'[87] Besides precision and univocality, atomicity is at stake as well. Recall the metaphysical fiction that Neurath was attacking. He was opposed to taking 'the ideal language composed of neat atomic statements'[88] as the observation language on which scientific knowledge must rest. Neurath argued: 'Quite properly we shall be forced to take more systematic account of these "imprecisions", whose full revelation is just as important as the renunciation of "atomic propositions".'[89]

In focusing on atomicity Neurath took up not only a metaphysical but an epistemological battle as well. Neurath rejected the doctrine that the basic sentences of science are in direct correspondence with ultimate elements, elements out of which either the world or our experience are composed. Such correspondence was assumed to be epistemologically primitive and incorrigible, although it was a subject of dispute just what the ultimate elements were supposed to be. For Russell in *Our Knowledge of the External World* the ultimate elements are the sensations – colours, sounds, etc. – that can be distinguished in our sense perceptions. For Mach the physical and the mental are built from sensory elements. For Wittgenstein in the *Tractatus* elementary propositions picture elementary states of affairs in the same way that names pick out simple objects, though he never made clear what simple objects are nor what makes certain states of affairs elementary.[90] For Waismann in his *Theses*, inspired by Wittgenstein, elementary propositions provide the grounding connection through which language 'touches' reality. The components of elementary propositions – 'primitive' or 'elementary signs' – single out the atomic 'elements' from which states of affairs in the world are configured. These elements constitute what is unalterable and subsistent in the world and cannot be described. For Schlick the certainty and meaning of the elementary propositions are guaranteed by ostension in the case of physical objects and by confrontation with immediate atomic sense

---

[87] Ibid., p. 219.   [88] Neurath 1932d [1983], p. 91.

[89] Neurath 1936a, p. 45, n.290, quoted in Zolo 1986 [1989]. Zolo makes brief reference to the 'plurivocality' of scientific language and, in general, of all synthetic propositions: 'What "fundamentally" (*grundsätzlich*) distinguished an analytical, logical or mathematical proposition from an empirical proposition was, [Neurath] said, its "univocality" (*Eindeutigkeit*), in the sense that a conflict between analytical propositions could be resolved through the application of previously agreed syntactical rules' (p. 40).

[90] For discussion on this subject see Coffa 1991, and more recently Wedin 1992.

data.[91] For Carnap in the *Aufbau* atomic protocol sentences directly express constituted qualities of individual basic experiences. These atomic constituents are the product of the application of 'quasi-analysis', a method of constructing classes of relations of remembered similarity.

We have already seen in part 2 how Neurath argued for his own model for the protocol sentence. His formulation requires the inclusion of the name of the protocolist and a perception term as well. The rejection of their epistemological immediacy follows as a corollary. Neurath presented the result as an argument from incompatibility: 'Carnap tries to introduce a kind of "atomic protocol" by demanding that "a strict distinction be made between the making of a protocol and the processing of the statement in the scientific procedure";[92] according to him this will be achieved by "not adopting any statements gained indirectly into the protocol".'[93] If Neurath is right about the correct form of a protocol statement, then 'processing' must always take place: 'In scientific protocols it may be useful to phrase the expression within the innermost brackets as simply as possible, for example: "At 3 o'clock Otto was seeing red", and a further protocol: "At 3 o'clock Otto was hearing C sharp", etc.; but such a protocol is not "primitive" in Carnap's sense, because one cannot get around the "Otto" and the "perceiving".'[94]

The rejection of the false ideal of epistemological immediacy buys Neurath a gain straightaway. That the data of science are various so-called 'simple' experiential qualities, like colours, is another fiction, like Laplacean completeness. Data reports in science do not look like that, and Neurath does not need to force them to. The protocols he will use to judge real scientific hypotheses will be real reports, whose innermost content will be cast, as they must be, in the language that real scientists use. For Neurath the problem of interpretation, of how we get from the pure uninterpreted world to the world of experience, does not exist. The world comes interpreted in the protocols of scientists. Their way of responding is of course historically conditioned, so that different scientists with different practices or different theories or in different laboratories respond in different vocabularies.

We know that for Neurath in treating a protocol sentence a decision is called for as each outer shell of the imbedding is stripped away. An episode in the history of a well-known physics experiment will help to

---

[91] See Uebel 1992a, chs. 8 and 9, for why even Schlick – not just Carnap (see section 2.5.2) – fails to qualify as an epistemological foundationalist.   [92] Carnap 1932a [1934], p. 42.
[93] Neurath 1932d [1983], p. 96.   [94] Ibid., p. 96.

illustrate the point.[95] Through the 1880s in Cleveland A.A. Michelson and E.W. Morley perfected an interferometer that would detect the displacement of interference fringes predicted as a result of the earth's motion through the hypothesised stationary luminiferous ether. The result of the experiments with the interferometer, published in 1887, were negative. No measurable fraction of the predicted displacement, no measurable ether drift. Einstein in 1905 accepted the result of the Michelson–Morley experiment and adopted it as a cornerstone of his special theory of relativity. Yet one experimenter, D.C. Miller, did not accept Michelson and Morley's results. After working with Morley, he accumulated thousands of observations, recorded first on the flat country at Cleveland and later at Mount Wilson in California. For two decades after Einstein's pronouncement, D.C. Miller published a number of papers claiming a small positive result. A relevant protocol might have been something like this: 'It is written in Miller's lab manual in Cleveland Ohio that on 2 June 1910 D.C. said D.C. observed a small displacement of the interference fringes in his interferometer.' First we decide to accept that it is written in the manual; next that D.C. Miller did believe he saw a small displacement of the interference fringes in his interferometer; lastly, to accept that the small displacement of the interference fringes really did occur.

We will discuss this row of embedded decisions further in the next section. But here what we want to stress is that the final decision, to accept the displacement of interference fringes, will depend in part on whether we are prepared to use the concepts with which the scientist responds to the world. That is often a delicate and involved decision, requiring all the finesse and thought and care and all the best scientific arguments that can be brought to bear. Neurath we know has one great prohibition – accept an inner content only if it is in the physicalist language, and that we maintain has as much a political as a scientific motivation. But we never have to decide how to convert a 'raw' feel into an articulated experience. As with Kant, Neurath deals only with an already experienced world. But unlike Kant, Neurath needs no transcendental argument to get to that conclusion. He arrives there by looking at what can be found in the physical world around him – real data reports in lab manuals or in public lectures – rather than at imaginary philosophical constructions.

Although *Ballungen* do not figure in the secure atoms that stand at the

---

[95] Our thanks to Douglas Webster for suggesting the use of this example.

base of the *Aufbau* of science, they are atomic in a different sense: they are indivisible, having no parts, and they are primitive. Philosophers sometimes claim that one concept is logically prior to another. For Neurath priority must be some real, positive relation, perhaps an historical one. Primitive concepts may be the concepts of early peoples or they may be the concepts of young children. He took the languages of both to be metaphysically purer than our contemporary language which is tainted with religion and idealism. Early magical language was in many ways closer to that of modern science than is much of the language from periods in between. Since Neurath also rejected the idea of a pre-logical mentality, he could write:

> What we have of systematic and orderly action and speech ... seems to go back to primeval systematic orderliness as found in magic. The scientific tendency to link everything with everything else, to regard nothing as indifferent, clearly already belonged to the age of magic. If we teach the dependence of human fate on empirically describable conditions, we are much closer in our way of thinking to the men of the magical times than we are commonly apt to suspect.[96]

The language of children, too, is much closer to pure physicalism than is the language of the adults around them, who may well find themselves concerned with metaphysical fictions like the moral order.

The concepts of everyday life are similarly primitive in another real sense. They are always among the concepts we begin with as we approach the task of scientific theorising. This is in part why Carnap's *Aufbau* is off to a bad start given Neurath's conception. Carnap produced his basic concepts by placing experiences in distinct categories based on their remembered similarity. But according to Neurath our experiences of the world are like a sand heap; they have no sharp edges and do not fall on either one side of a divide or another. We must beware, though, of talking about experiences, for Neurath wished to deal only with concepts. If we think in terms of primitive experience, we may be inclined to suppose that we can arrive at cleaner, more elementary ideas by picking apart the complexes with which we begin. But primitive concepts are not like that. The concepts we begin with are not composed of parts that are even more 'primitive', in any legitimate sense of the word. This will be important to our story in the next sections. In a 1928 letter to Neurath, Carnap conceded that on those premises the only task ahead is to work out a method or a logic of science that incorporates both the exact

---

[96] Neurath 1931c, p. 320.

theoretical concepts and the *Ballungen* concepts of experience, or in Carnap's own terms, the 'crystal and the dirt': 'A logic, a method of concept formation, must be constructed which takes account of the fact that we are always presented with a mixture of crystals and dirt, which tells us therefore what demands can be imposed on scientific concepts and statements as long as the "ideal language" is not available.'[97] Carnap never really got down to this job; it was left for Neurath.

Before carrying on to the discussion of Neurath's views on the possibility of such a logic, let us return briefly to the protocol sentence debate. The peculiarities of form that Neurath ascribed to protocol sentences play a crucial role in his attack on method.[98] So it is worth stressing some points that are especially important. If we think only about the subject matter of data reports, we can map out three different positions: (a) Schlick and early Carnap on one side take data reports to be about individual experience; (b) Popper on the other side takes them to be about public matters of fact; (c) Neurath and Carnap in his later work stand in the middle, each attempting in different ways to combine the virtues of both sides. Two issues are at stake. The first is *the requirement of intersubjectivity*. We have already discussed Neurath's grounds for the claim that data reports must be about public events and how Carnap too came to see its advantages (see section 2.5.2). The second requirement is *the entitlement requirement*. To be a ground for a scientific hypothesis it is not enough for a sentence to be true, nor even true and accessible to us, as Popper required in his demands for observability. It must also be one we have title to. In Popper's scheme that means that the sentence is about a fact that someone reliable has actually observed. Schlick was entirely concerned with the entitlement requirement; Popper, with that of intersubjectivity. Neurath and the later Carnap tried to satisfy both, each in his own way. Carnap's strategy was to urge that the phenomenalist and the physicalist *Aufbau* are both correct and neither is dispensable. Neurath used his peculiar form for protocols. Recall the analysis of his proposal given in section 2.5.3: The conjunction of the formal conditions (1) to (4)

[97] Carnap to Neurath, October 1928, RC 029-16-01 ASP, quoted in Uebel 1992a, p. 76; see also Haller 1995.
[98] If one were not so insistent on formulating everything in terms of the form that sentences take, Neurath's attack on method would not require this awkward detour through protocol sentences. He was really worried about two issues: (1) using only a non-metaphysical concept of 'data'; and (2) stressing the entire series of decisions that have to be taken in admitting the innermost content of a protocol. Given both of these, the test of any hypothesis will involve judgements about a great many matters outside not only the discipline of the hypothesis under test, but also outside the domain of any exact science. This should become apparent in the discussion of the next several paragraphs.

explicates together what it takes to meet both the entitlement requirement as well as the requirement of intersubjectivity.

We have looked in detail in section 2.5 at Neurath's differences from Carnap. Let us now turn briefly to his differences from Popper to see why Neurath's protocols could never take on the simple form of Popper's. The issue is testing. If a sentence $d$ about an observable physical event is deducible from a hypothesis $H$ and $d$ is false, what follows? The appropriate conclusion from these two assumptions is that $H$ is false. But that does not guarantee that $H$ has been falsified. For that we need more. An account of testing is not supposed to tell when $H$ is true or false, but rather how we figure out whether $H$ is true or false. Carnap's later theory of confirmation stands somewhere between the two aims. If $d$ confirms $H$ ($C(H,d)$ is higher than $C(H)$) this does not mean that $H$ is true. Nor does it mean that we are entitled to some degree to think it is. $C(H,d)$ is a purely logical relation between $d$ and $H$, and having a theory of what that relation consists in does not tell us what it is to test $H$. But for both Popper and Carnap the instructions for what to do were fairly immediate. To confirm (falsify) $H$, find some sentence $d$ such that (1) $C(H,d)$ is higher than $C(H)$ ( or for Popper find some sentence $d$ such that $H \rightarrow d$ and $\sim d$); and (2) – What should be put here? What is needed is some constraint on $d$ that ensures that we 'have' $d$ as a starting point. The question is how to formulate (2). A phenomenalist would say '$d$ is a report of a sense-experience'; Popper, simply '$d$ is true.' The first proposal makes science a private enterprise, thus violating the requirement of intersubjectivity; the second fails to meet the entitlement requirement. Neurath answered more carefully. His proposal for (2) begins 'It is publicly on record that it is observed that $d$', and then moves through all the requirements for stripping down to $d$ alone. (That is, the requirements expressed in the conjunction of the formal conditions (1) to (4) from section 2.5.3.) Entitlement and intersubjectivity are secured at once.

This may seem a footling point. The requirement of entitlement is implicit in the enterprise and surely no-one would be led into foolish mistakes by Popper's 'shorthand' version. But let us recall that, independent of his own concerns, Neurath was always alert to the Duhem problem. The sentence $d$ describes phenomena that cannot be directly related to the theory, for instance the shift in the fringes of an interferometer. We know that before we can be confident about the shift, a large number of other conditions must be assured, including not just the assumption that the interferometer is functioning as we expect but also that the laboratory assistants did not mistake what they saw, that the right data report got registered in the right place in the laboratory notebook, and so forth. Neurath aimed to make this process as transparent as possible.

Still, some will feel annoyed by Neurath's fussiness. The aim is to elucidate the abstract concept of testing. For this purpose a pure philosophic concept of data suffices: the shift in the fringes is the datum against which the hypothesis about the earth's motion through the ether is tested. Neurath, one may be tempted to object, obscured the central point in a sea of irrelevant practical concerns. But what is this pure philosophic concept of data? There are real data reports – protocols in laboratory manuals, for instance, which we decide to accept as the starting point for testing a hypothesis. But the other is just a metaphysical construction, like the Laplacean fiction of knowing enough information to predict exactly what will happen.

### 3.3.3 The two Neurath principles

Let us now turn to some of the details of Neurath's attack on scientific method. The attack was set out in an article against Popper in 1935. In this article Neurath defended the Positivists in general and himself in particular against criticisms Popper had launched in *The Logic of Scientific Discovery*.[99] One of the criticisms deals with the differences between Popper and Neurath about the adequate form of protocol statements. Popper clearly understood Neurath's doctrine to be different from Carnap's but dismissed both as 'psychologistic', that is, phenomenalistic. Psychology of experience, Popper protested, does not suit a logic of justification. In his attack on method Neurath came to a similar conclusion although from his own anti-Carnapian non-psychologistic position and with a new and more radical diagnosis: even without psychologism, strict empiricism is incompatible with the general logic of justification that Popper wanted.

Neurath saw a fundamental problem at the core of Popper's programme. 'Whereas Popper does not want to treat "induction," this "unfounded anticipation", logico-systematically, not even its special forms, he tries to characterise falsification logically, as a general method, as strictly as possible – though he must admit that this cannot be done precisely – and to base the whole of scientific research uniformly on it.'[100] The problem arises from Popper's starting point: a model that, like

---

[99] Neurath 1935b, Popper 1935. For more details on the exchange between Neurath and Popper, see J. Cat, forthcoming. For an account of the debate with a different emphasis, see Zolo 1986 [1989], chapter 4.   [100] Neurath 1935b [1983], p. 123.

Carnap's, portrays science as a system of clean statements and unambiguous deductions. But that is mistaken:

> For example, we do not use completely precise terms when we say, 'Man $A$ formulates: in the room was a table perceived by $A$.' But this kind of formulation as known in everyday language is always needed where predictions are empirically checked by confirming predictions with protocol statements.
> 
> I want to call the sphere in which formulations are not rendered completely precise, the sphere of the 'cluster concept' (*Ballung*). It is characteristic of the encyclopedia of unified science that not only are the statements more or less clearly interconnected, but that there also occurs among them those of the cluster type.[101]

Neurath recalled Duhem's thesis of underdetermination – but with this new argument of his own. He concluded that in science there is no room for the idea of an *experimentum crucis* based on an 'unambiguous logical deduction as justification'.[102]

Neurath's attack leaves scientific method logically open ended.[103] No

---

[101] Neurath 1936c [1983], pp. 161–2.   [102] Neurath 1935b [1983], p. 127.

[103] Interestingly, by 1945 Waismann had reached similar conclusions. He had abandoned the Wittgensteinian position of his *Theses* which raised so much opposition from Hahn and Neurath in 1931: that complete definitions guarantee clear verification paths elucidating the meaning of ordinary linguistic propositions; and that the fact that we understand at least the propositions of our ordinary language shows that there exist precise atomic elementary propositions, constituted by primitive signs that deal with atomic experiential reality directly. In 1945 Waismann saw in the language of science a distinction between mathematical (extendable to mathematical physics) and empirical concepts. Whereas the former are precise, the latter now are not. Empirical concepts lack the complete – or 'closed' – definitions that guarantee by construction the possibility of complete or conclusive verification that precision of meaning requires. Waismann speaks of a 'porosity' or 'open texture' of such concepts, which ought to be distinguished from vagueness in their actual use. 'Open texture' is, rather, the 'possibility of vagueness': the definition of open concepts 'are always corrigible or emendable'. In Waismann's terms, '[W]e cannot [a priori] foresee completely all the possible circumstances in which the statement [containing any such concept] is true or in which it is false' (Waismann 1945 [1952], p. 121). Waismann contrasts the concept of a triangle with the concept of gold: 'The notion of gold seems to be defined with absolute precision, say by the spectrum of gold with its characteristic lines. Now, what would you say if a substance was discovered that looked like gold, whilst it emitted a new sort of radiation? "But such things do not happen." Quite so, but they *might* happen, and that is enough to show that we can never exclude altogether the possibility of some unforeseen situation arising in which we shall have to modify our definition. Try as we may, no [empirical] concept is limited in such a way that there is no room for doubt' (Ibid., p. 120). The difficulty is then to state completely what would count as verification in each case. (This adds to Neurath's and Popper's claim that the basic material-object statements are as a consequence unverifiable in yet another way and must be adopted ultimately on non-observational grounds.)

Waismann stressed a familiar consequence for a logic of testing in science – now seen by him as a relation between linguistic propositions: '[N]o statement of mechanics [or any other formalised language] can ever come into sharp logical conflict with a statement of observation, and this implies that between these two kinds of statements there exist no relations of the sort supplied to us by classical logic' (Ibid., p. 128). This is just the new difficulty Neurath pointed to in his review of Popper of 1935 (Cat, forthcoming). Waismann's conclusion is in fact more general: 'All justification can count strictly either for or against a theory. As a consequence, whether or not a data report or a protocol statement is considered compatible with the system of hypotheses (in a broad methodological rather than strict logical sense) will require a decision based on extra-logical factors. That is the new half of the story. The old half comes from Neurath's anti-foundationalism: 'in our view the cancelling of protocol statements is a possibility as well'.[104] So even once we have decided that there is a conflict between a protocol statement and a system of hypotheses we must still decide which to keep.

The two stages of decision suggest that we distinguish two different theses within what we have called 'the Neurath Principle'. As described in section 2.3.2 this term was introduced by Rudolf Haller to refer to the doctrine that in cases of conflict it is always open to decision whether to change a protocol or to change the theory. One statement of Neurath's that clearly implies this view appeared in 1934:

> All content statements of science, and also their protocol statements that are used for verification, are selected on the basis of decisions and can be altered in principle.[105]

We want to call this 1934 version the *General Neurath Principle*. The statement of the principle is very brief and if we knew nothing more about Neurath's work at the time it would be easy to confuse it with a less radical principle that he presented in 1932. In the 1932 version he had already introduced the notion of *Ballungen* and its peculiar features but he had not yet worked out its implications. We call the earlier version the *Special Neurath Principle*:

> Two conflicting protocol statements can *not* be used in the system of unified science. Though we cannot say which of the two statements is to be excluded, or whether both are to be excluded, we can be sure that not both can be 'verified', that is, it is not the case that both statements can be incorporated into the system.
>
> If, in *such* a case, a protocol statement has to be given up, why not also some-

---

this tends to suggest the ideal of a verification path is plagued with problematic "fracture lines" of different strata of language; that the relation between a law of nature and the evidences for it, or between a material object statement and a sense-datum statement, or again between a psychological statement and the evidence concerning a person's behaviour is a looser one than we had hitherto imagined. If that is correct, the application of logic seems to be limited in an important sense. We may say that the known relations of logic can only hold between statements which belong to a homogeneous domain; or that the deductive nexus never extends beyond the limits of such a domain' (p. 128). Waismann's line of argument against the deductive method of testing is not, however, an exact match of Neurath's. Both the historical and philosophical connections between the two need further study.

[104] Neurath 1932d [1983], p. 95.   [105] Neurath 1934 [1983], p. 102.

## The attack on method

times when, only after long chains of logical argumentation, contradictions appear between protocol statements on the one hand and a system of protocol statements and nonprotocol statements (laws, etc.) on the other hand? According to Carnap we could only be forced to change non-protocol statements and not laws. *But in our view the cancelling of protocol statements is a possibility as well.* It is part of the definition of a statement that it requires verification and therefore can be cancelled.[106]

This earlier principle tells us as does the later that both protocol statements and laws are candidates for revision. But it also admits the conventional logical relations among them: '[L]ong chains of logical argumentation' can reveal 'contradictions' between the protocols and laws. We know that this is exactly contrary to Neurath's post-1931/2 views. We have seen that by 1934 Neurath was eager to exploit the consequences of *Ballungen* to attack the pseudo-rationalist assumption that data reports could speak clearly one way or another in favour or against theory. His view by then denied that protocol statements – loaded with imprecise *Ballungen* – stand in any fixed relation (logical comparability) to the theoretical hypotheses. The former, we noted, even when a decision has been taken to accept them, cannot falsify the latter. At most they can 'shake' our confidence in them. It is in the context of his other writings at the time that what we have called the 'General Neurath Principle' must be read; and in that context it is clear that the role of decision is much more far reaching than anything suggested by the Special Principle of two years before.

In *The Logic of Scientific Discovery* Popper had criticised the Special Neurath Principle. He feared that without a rule for the revision of protocols the principle would lead to arbitrary decisions:

Every system becomes defensible if one is allowed (as everybody is, on Neurath's view) simply to 'delete' a protocol statement if it is not convenient... Neurath avoids one form of dogmatism [Carnap's], yet he paves the way for any arbitrary system to set itself up as 'empirical science' ... Neurath fails to give any such [decision] rules [and] thus unwittingly throws empiricism overboard ... Neurath does not try to solve the problem of demarcation.[107]

Contrary to Popper's claim Neurath had indeed offered a general criterion of demarcation for the empirical sciences: scientific statements must be formulated in a physicalist language and any new additions must be acceptable in the light of all the other statements and methods that we accept. But he did not tell us how to decide which changes were

---

[106] Neurath 1932d [1983], p. 95.   [107] Popper 1935 [1959], p. 97.

acceptable.[108] As Popper charged, Neurath did not provide a method of theory testing. Nor did he want to: 'Various factors determine the methodological scientist in his choice of a model. We deny that the encyclopedia preferred by the scientist can be logically selected by using a method that can be only generally attained.'[109]

The General Neurath Principle – no less than the Special Principle – leaves scientific method open, yet Neurath did not agree that either Principle leads to pure arbitrariness. As we pointed out in section 3.3.1 the General Principle is a result of Neurath's belief that scientific method contains at least three degrees of freedom, or sources of indetermination. The first one occurs in the logical connection between the protocol statements and the system of hypotheses (doctrine of *Ballungen*); the second in the multiplicity of different systems compatible with the same protocol statements (theoretical pluralism); and the third, in the in-principle possibility of revising the protocol statements themselves (the anti-foundationalism of the Special Neurath Principle). But at each level the indeterminism can be eliminated on the basis of a decision. That is what the General Neurath Principle affirms: 'All content statements of science, and also those protocol statements that are used for verification, are selected on the basis of decisions and can be altered in principle.'[110] And for Neurath how scientists go about making such decisions is not an arbitrary matter: 'We make a selection on the basis of extra-logical factors.'[111]

'Extra-logical factors' is a sweeping term. We can attempt to give it more content by pointing to two different aspects of Neurath's thought about how science works. First, it is to be determined, or decided, whether the hypothesis and the data-report stand in conflict. Second, with respect to the decision of whether to keep the hypothesis and reject the data-report or the converse once the two are seen as incompatible, it must be remembered that every decision to accept or reject a part of a protocol sentence has its own consequences. Protocols are constructed to make transparent what our options are.

To illustrate let us go back to the case of D.C. Miller's experiments and consider this protocol statement: 'D.C. says that in Cleveland on 2 June 1910 D.C. observed a small displacement of the interference fringes

---

[108] The distinction between theoretical and protocol statements does not depend on the assumption that one makes exclusive use of pure theoretical concepts and the other exclusive use of pure observational concepts. For Neurath, just as protocol statements were plagued with vague *Ballungen* terms, they were also 'rich' in theory. See Neurath 1935a [1983], p. 118.

[109] Neurath 1935b [1983], p. 123.   [110] Neurath 1934 [1983], p. 102.   [111] Ibid., p. 106.

in his interferometer.' In this case we could assume that Miller was truthful, that he observed accurately, that the interferometer was working correctly, and that the interference fringes had undergone actual displacement. If so the displacement had better be consistent with the hypothesised motion of the earth (and apparatus) through the fixed ether. Or one can assume that the interferometer was unreliable. That was in fact Max Born's judgement of the situation:

> When I was in the United States in 1925/26, Miller's measurements were still frequently being discussed. I therefore went to Pasadena to see a demonstration of the apparatus on top of Mt. Wilson. Miller was a modest little man who very readily allowed me to operate the enormous interferometer. I found it very shaky and unreliable, a tiny movement of one's hand or a slight cough made the interference fringes so unstable that no readings were possible. From then on I completely lost faith in Miller's results.[112]

Or we could have assumed that Miller was hallucinating, in which case we should like to find someone who reports: 'I saw Miller taking some suspicious-looking pills.' And so forth.[113] The decision is arbitrary if we plump for a claim without reasons; for instance if we say: 'The interferometer *couldn't* have been working properly' without appropriate protocols to back it up. Neither Neurath nor Popper recommend that. The reasons of course will not dictate the conclusion. Recall Duhem's underdetermination thesis embraced by Neurath: confronted with the theory as a whole, the data alone cannot dictate the distribution of praise and blame over the multiple hypotheses in the theory. In like fashion the application of the Neurath principle to reject a protocol fails to determine the distribution of praise and blame over the components that the protocol incorporates. Popper looks for something to fill the gap. For Neurath there is nothing. There are only more and more scientific reasons.

Third, what happens when we are confronted with different totalities? 'We select one of the systems of statements that are in competition with each other. The system of statements thus selected is not, however, logically distinguished.'[114] If not logically, then how? On what basis are we to decide which to adopt? The situation is more a philosopher's anxiety than a problem for scientific method. We have seen that 'The practice of living reduces the multiplicity quickly.'[115] We need to get on with the job and must choose, not among a potential infinity of alternative theories, but

---

[112] Born and Einstein 1971, p. 74.  [113] Neurath 1932d [1983], p. 94, 1941a [1983], pp. 220 ff.
[114] Neurath 1934 [1983], p. 105.  [115] Ibid.

among what we have and what we can develop. 'An individual hardly has the power to work out *one* system properly, let alone several systems.'[116]

For those who aim to find a system of beliefs that are true, taking what you can get, as Neurath recommended, will hardly seem a reasonable decision procedure. But for Neurath both the pursuit of correspondence truth and the hunt for the foundations of rationality are pseudo-rational ideals. We can only compare our sentences with other sentences and then again with yet other sentences, using the concepts and the methods we have. The comparison will inevitably require decisions, with no ultimate rule to provide us with the comfortable feeling that there is a right decision to be found if only we are diligent enough in pursing it. We can apply Neurath's comments about Marxism (section 3.2.1) and the state to his own views about method: Popper, along with many others, hankered after entities which lie beyond mankind. To them method is something 'higher', something 'holy', whereas for Neurath everything lies on 'the same earthly plane'.

## 3.4 WHERE BALLUNGEN COME FROM

### *3.4.1 Duhem's symbols*

Where do *Ballungen* come from? To answer let us think in more detail about the job that Neurath expected *Ballungen* to do. Neurath, as was typical of his time, followed the Enlightenment idea that the concepts of science must be precise and its laws universal and exact. This is part of what he was claiming for social science when he argued that the task set for it is not different in kind from that of natural science (recall section 2.3.2). Scientific laws must form, in Neurath's own words, model-systems: 'deductive constructions, closed in themselves'.[117] By the mid 1930s Neurath argued that even the concepts of abstract theory will fail in exactness and precision. But in his first attacks on method he still supposed that scientific theories are model systems 'consisting of systematically connected clean and precise statements'. On the other hand data reports use concepts that are imprecise, congested, dense. Neurath claimed that no deductive or exact relations are possible between these two. Why?

A quick though anachronistic way to see the issue is to raise a few philosophic problems:

---

[116] Ibid.  [117] Neurath 1936e [1983], p. 145. See also Neurath 1935b [1983], p. 122.

1. Neurath assumed that data are imprecise but theoretical claims are precise. That does not show that data cannot rule out theory. A very precise theoretical prediction may, for instance, lie entirely outside the possible range of error of the measurement outcome.
2. Neurath stressed that testing starts with the protocol – the laboratory notebook. Each step on the long road back to the theoretical claim needs supporting assumptions. Recall the interference fringes observed in the interferometer. We suppose that Miller has recorded honestly and correctly what was observed, that the observation was accurate, that the interferometer was working properly and so forth. To get a deductive chain from theory to datum we must reverse the direction. We need not only the theory of light and of the motion of the Earth but also the theory of the instrument, the theory of perception, psychology, sociology and so on. Yet this does not show that deductive falsification is in principle impossible, but rather that it requires the joint co-operation of all the sciences – indeed exactly the kind of co-operation that Neurath promoted in the Unity of Science movement.
3. The complexity of what is described by everyday concepts versus the uniform nature of scientific concepts, which seem to pick out a single significant characteristic, will not serve either. Consider the arguments of John Stuart Mill about the events of economic life. On Mill's account it is precisely *because* these events are so complex that political economy must be a deductive science. There are too many different causal factors to expect that they will recur often enough in the same combination to provide a good base for inductive generalisation. The first principles of political economy must come from elsewhere, for Mill from a combination of introspection and good judgement about human nature. Nor is the great variety of factors necessarily a stop to deductive falsification. In the fortunate circumstances of the experimental sciences the complexity can be controlled so that one single, simple factor operates at a time.

The introduction of Mill is a help, for Mill was a realist about the concepts of abstract science; Neurath – when the claim is properly rendered – was an anti-realist. That, we believe, is the clue to his assumptions about *Ballungen*. The concept has its source for Neurath in three strongly anti-realist settings: that of Duhem, that of the 'young historical school', especially Schmoller and later Weber, and that of the Marxist materialist conception of history, most notably the views of Plekhanov and Labriola. In each case there is something Neurath left behind. In Duhem it is the sometime suggestion that the problem arises from the ineliminable imprecisions of measurement. For Weber it is the special role of values and interests in the choice of social science concepts that drive a wedge between the natural and the social sciences. For Plekhanov and Labriola it is the metaphysics of the Hegelian dialectic. The historical school and the Marxists have Hegel as a common predecessor, and for both holism is crucial. The historicist stand against the scientific study of history and society rests on holism; the Marxists cannot omit it if they are to hope to encompass Marx's *Capital*. It is probably not too crude a dichotomy to claim that Neurath took from Duhem the imprecision of *Ballungen*. The density aspect comes from the holism which dominated German and Austrian philosophical thought after Hegel.

We begin with Duhem.[118] We saw in section 2.3.2 that Duhem perceived a logical gulf between the language of theory and the language of observation. Following Duhem we have called the indetermination of theoretical facts by practical facts 'symbolic indetermination'. We will see below why he used this term. The language of observation is available to both the physicist and the layman; it is used to describe the concrete facts we all experience. For Duhem these facts display 'the complexity of concrete reality'[119] where, following Aristotle, he conceived of concrete reality as a wealth of quantitites and qualities.[120] The language of observation typically contains terms that 'are and always will be vague and inexact like the perception which they are to express'.[121] The language of theory is precise and exact and its relation to the complex reality of concrete things is very different from, and very much more indirect than, the language of everyday life in which our scientific observations are couched. The precise terms of scientific theory do not depict the content of our observations; instead they relate to reality merely, as Duhem put it, 'as a sign to the thing signified'. The language of theory 'does not grasp the reality of things' but instead

---

[118] We discuss Neurath's concerns with the underdetermination of theory by evidence because we are interested in showing the role of these concerns in Neurath's developing philosophical, scientific and political point of view. But we do not mean to suggest that these concerns were special to Neurath nor to the First and Second Vienna Circles, any more than they were original to Popper. By the time of the Great Depression worries about the underdetermination of theory by observation and the special role of auxiliary hypotheses were widespread in the social sciences, especially wherever statistics were in use. Econometrics provides a good case. Jan Tinbergen, who did the first major macro-economic econometric models for the League of Nations was airing falsificationist worries in 1927 (see Mary Morgan 1990). Tjalling Koopman's dissertation of 1937 provides another example. This point is noted by Mary Morgan 1990 and again in Hugo Keuzenkamp's dissertation, 1994.   [119] Duhem, 1906 [1962], p. 162.
[120] Ibid., p. 120.   [121] Ibid., p. 143.

## Where 'Ballungen' come from

'is limited to representing observable appearances by signs and symbols'.[122] The problem does not lie in the fact that scientific concepts are general and abstract. That is true of all concepts, even humdrum everyday ones like 'man' and 'mortal'. In contrast to vague, ordinary-life terms like these, symbols are somehow 'too simple to represent reality completely';[123] a symbol is just 'something more or less selected to stand for the reality it represents'.[124] In that sense symbols are neither exactly true nor exactly false of any given situation. Nor, in turn, are the laws that connect them.

Symbolic indetermination arises in short from the fact that 'the abstract symbol cannot be the adequate representation of the concrete fact, the concrete fact cannot be the exact realisation of the abstract symbol'.[125] Duhem illustrated with the example of heat: 'Let us put ourselves in front of a real, concrete gas to which we want to apply Mariotte's law; we shall not be dealing with a certain concrete temperature embodying the general idea of temperature, but with some more or less warm gas.'[126] A 'dictionary' is required if we want to bring the numbers deduced from theory ('theoretical facts') into correspondence with the descriptions of what is observed ('practical facts'). For Duhem that is the work done by measurement. Consider the example of the gas: '[I]n order to assign a definite temperature to this more or less warm gas, we must have recourse to a thermometer.'[127] But this does not completely remove the problem. Disparity still remains 'between the precise theoretical fact and the practical facts with vague and uncertain contours':[128] the same practical fact can correspond to an infinity of logically incompatible theoretical facts (quantities), and vice versa.[129] Through an elaborate experimental process, with the aid of a string of auxiliary hypotheses, a number can be assigned to the warm gas. But the number assigned is never uniquely determined, even given the experimental procedures and the auxiliary assumptions.

Still it seems progress has been made. Symbolic indetermination began as a problem about the relationship between scientific concepts and those describing concrete reality. The precise concepts of science do not designate anything that can be described in the everyday language that depicts the world as we experience it. It has now turned into a problem that arises from the physical limitations on our testing procedures. For Duhem, 'the degree of symbolic indetermination' becomes simply 'the degree of

---

[122] Ibid., p. 115.   [123] Ibid., p. 176.   [124] Ibid., p. 168.   [125] Ibid., p. 151.
[126] Ibid., p. 166.   [127] Ibid., p. 166.   [128] Ibid., p. 152.   [129] Ibid.

approximation of the experiment in question'.[130] In the case of the warm gas he says: 'We cannot assert that this temperature is strictly equal to 10 degrees; we can only assert that the difference between this temperature and 10 degrees does not exceed a certain fraction of a degree depending on the precision of our thermometric methods.'[131]

Duhem's account of scientific testing offers some hope for bridging the logical gap between concrete facts and the deductive system of theoretical hypotheses. With more and more refined techniques we can zero in on the correct theoretical description to associate with an everyday experience (always of course relative to a chosen set of auxiliary hypotheses about the theory of the instrument, its proper operation and so forth). Neurath did not share Duhem's optimism about this bridge. He insisted on the gulf between scientific concepts and the *Ballungen* of natural language despite Duhem's account. Neurath explains:

> Perhaps one should envisage the possibility of a series of intermediary steps that lead from cluster to formula. What we demand of a cluster-concept is that we can somehow make a 'formula' correspond to it, in connection with a theory...
>
> Without doubt it is something of this kind that Duhem has in mind when he speaks of concrete facts (instead of formulas in common language, or clusters as we say) to set them against the abstract symbols of science – a manner of speaking that, in our opinion, is hardly adequate.
>
> It is a characteristic feature of the encyclopedia that it is incessantly aware of the difference between a cluster and a 'scientific formula' whose mutual connections are complex, without yielding to the temptation to use, or at least to conceive, a closed system of formulas as a model for the aggregate of statements.[132]

It is clear that the ideas about approximation developed by Duhem are not to solve the problems that Neurath associated with *Ballungen.* That is the lesson of our three philosophical objections. Duhem's solution supposes that the problematic difference between the language of observation and that of science is that the first never provides the right kind of information to support a unique description in the second. But his story about auxiliary hypotheses, instruments and measurement shows a way to get closer and closer. We may never be in a position to assign a single value of a quantity to the concrete reality that we record in our everyday language.[133] Nevertheless by settling on a set of auxiliary

---

[130] Ibid., p. 169.   [131] Ibid., p. 134.
[132] Neurath 1936e [1983], p. 148.
[133] Indeed, we would argue that Duhem would see any attempt to do so as fundamentally mistaken since, in our view, Duhem took reality to be essentially qualitative; quantities are pure constructions that we use in our theoretical representations.

hypotheses about the use of instruments Duhem's account seems to provide us with almost full access to conventional methodology. Theories can be confirmed (relative to the auxiliaries assumed) when their predictions plus the estimated range of experimental error lie within the extent of our qualitative observations; and they are falsified when their predictions lie sufficiently outside.

Neurath rejected both confirmation and falsification. Even if we are willing to take our auxiliaries as given for the moment, Neurath still denied that there are any fixed relations between scientific concepts and the *Ballungen* of everyday life. To see why, we must go beyond the concerns about precision and imprecision, which he shared with Duhem, to other concerns of Neurath that Duhem's introduction of instruments and approximations does not touch. We know from Neurath's own discussions of *Ballungen* that not only imprecision but density and complexity play a key role as well. What did Neurath have in mind by density and complexity? We need an answer that explains why he found it natural to suppose that what is dense and complex can not be deduced from the claims of exact science.

### 3.4.2 The congestion of events

We must remember that Neurath grew up steeped in the terms of the *Methodenstreit*.[134] The two great adversaries of the battle over the correct methods for studying society were the Austrian economist Carl Menger and Neurath's Berlin advisor Gustav Schmoller. The central topic of the debate was whether political economy can be a science. The dispute hinged on the opposition between law-based knowledge and knowledge of the concrete particular. We begin with a pair of complementary doctrines, an ontology and a philosophy of science, that formed a backdrop to the *Methodenstreit*. The doctrines were articulated by Max Weber, who saw them as the source of the dispute. His own account of the role of value in the selection of concepts in the historical sciences was a solution tailored to the shape he gave the problem. We shall look primarily at Weber's ideas in the period 1902–6 when Neurath was a student in

---

[134] We talk of the problems for scientific theorising presented by the congestion of events in the context of *Methodenstreit* and of Weber's response to it because this is the environment in which Neurath lived and worked. But problems of this general kind were also troubling in the inter-war period outside the German-Austrian context. For instance Keynes' hypothesis of limited independent variety maintains that the objects of inductive inference can not be infinitely complex nor produced by an infinite number of generators; otherwise inductive methods would be impossible (Keynes 1921).

Berlin, especially at his comments on two of the founders of the German historical school, Wilhelm Roscher and Karl Knies,[135] at Weber's criticisms of Neurath's Berlin dissertation advisor, Eduard Meyer[136] and at his famous essay on objectivity in the social sciences.[137]

Weber's ontology went hand-in-hand with a theory about the relation between perceptual concepts and perceptual experience. The key metaphors for understanding Weber's vision of historical events are 'complex' and 'fluid' (*flüssig*). On Weber's account reality as given to us in experience is a continuum capable of being differentiated in an infinite variety of ways. Any set of concepts will leave out some importantly different 'aspects' of reality and even for a given aspect our concepts will always be open to greater and greater refinement. In the essay on Knies, Weber speaks of 'the logical inexhaustibility of the empirically given manifold of experience of which we describe only "partial aspects."'[138] His concern with methodology centred on the question of how the characteristics 'essential to our knowledge' may be 'differentiated from the scientifically homogenous manifold of reality'.[139] In the essay on Roscher he uses almost identical language to describe 'the logical impossibility of an *exhaustive* reproduction of even a limited aspect of reality – due to the (at least intensively) infinite number of qualitative differentiations that can be made'.[140]

Weber's account is self-consciously Kantian. He urges us to follow Kant in the recognition 'that concepts are primarily analytical instruments for the intellectual mastery of empirical data and can be only that'.[141] The view stands in opposition to realist theories of the relation between empirical concepts and perceptual reality, whether they be emanationist theories modelled on Hegel's story that concepts form reality or empiricist accounts that suppose that concepts copy properties that are already differentiated. But equally important with anti-realism is Weber's stress on the variety of sets of concepts we can use to study the same phenomena and the failure of any set to be unique. This is a Kantianism with a twist that suits it for the study of historical economics rather than Newtonian mechanics.

Weber's resolution of the *Methodenstreit* rests on the claim that nomological and historical sciences do not have a different subject matter, as many had claimed, but they do have a different theoretical purpose and correlatively a different methodology. The nomological sciences aim for

---

[135] Weber 1903–6.  [136] Weber 1906.  [137] Weber 1904.
[138] Weber 1903–6 [1975], p. 179.  [139] Ibid., p. 213, n.8.  [140] Ibid., p. 57.
[141] Quoted in Schön 1987, p. 62.

# Where 'Ballungen' come from

abstract relations of universal validity.[142] Historical sciences want 'knowledge of *concrete reality*, knowledge of its invariably qualitative properties, those properties responsible for its peculiarities and its uniqueness'.[143] The difference in aims of the two kinds of science compels them to use very different kinds of concepts. The historical sciences need concepts that are rich in the content of perceptual reality; the nomological sciences want exact concepts with precise deductive connections to one another and this drives them far away from perceptual reality.

Weber's views on the aim of nomological science and his related claims about the kinds of concepts these aims require provide the second essential component in his account. Weber says that he takes his views on the nature of scientific laws from Heinrich Rickert, and indeed their views are similar. They begin with a shared ontology. According to Rickert the individual event is infinitely complex. It can be endlessly broken into smaller and smaller parts, but more importantly, it has an indeterminate number of aspects of which we can never have total awareness. To overcome this infinity we must invent general concepts that do not picture the world piece by piece but rather represent it in some other way. Rickert distinguished between everyday concepts that we just find and the scientific concepts that we devise out of these. Everyday concepts have fuzzy extensions and that is bad for science; they tend to be given by the senses and sensation inevitably 'introduces an infinite manifold into the concept'.[144] This is just the manifold we are trying to overcome.

One might think that the way to proceed would be to break the complexity into parts, to name each part separately and to look for the laws that govern the parts named. That is the method of John Stuart Mill and it supposes a two-fold realism, realism both about concepts and about laws. Whole events in nature are composed of isolable, recurring properties and the behaviour of these properties is regulated by laws of nature. This was not the view of Rickert. Rickert wanted the content of scientific concepts to be separated from sensation as far as possible. For him science aims for completely general abstract concepts which do not name parts but rather represent the infinitely complex whole all at once. It is essential that these abstract representations be capable of figuring in precise and universal scientific laws.

Weber's views on the nomological sciences matched Rickert's. Weber maintained that the standards of precision and the lack of ambiguity

---

[142] Weber 1903–6 [1975], p. 56.   [143] Ibid., p. 57.   [144] Rickert, 1896–1902 [1927], p. 35.

required by these sciences lead to 'the most radical reduction possible: the qualitative differences of concrete reality are reduced to precisely measurable quantities'.[145] The sciences insist on laws of 'general validity' in order to transcend 'the mere classification of appearances';[146] the laws must be '*strict* and mathematically self-evident'.[147] This leads to a complete stripping away of all perceptual content. 'In the last analysis, the product of these sciences is a set of absolutely non-qualitative – and therefore absolutely imaginary – conceptual entities which undergo changes that can only be described quantitatively.'[148]

The historical sciences by contrast do not aim for formulae of universal validity but rather for a causal understanding of what is unique and peculiar in the individual event. Their explanatory concepts will be qualitative and close to the 'inexhaustible manifold' that is empirically given. They will thus be neither precise nor unambiguous. 'All qualities as such . . . necessarily have the character of relative "vagueness"', says Weber.[149] In his discussion of how we understand the behaviour of others in the essay on Knies, he stresses this ambiguity: '[T]he interpretation of mental processes has the same status as any science which does not completely abstract from qualities: it employs concepts which are in principle not absolutely unambiguously definable.'[150] We see here the same contrast that Neurath drew. The abstract concepts of exact science are both precise and unambiguous; the concrete concepts of the historical sciences, close to everyday empirical reality, have the character of *Ballungen*. But before exploring further the connections of Weber's and Neurath's ideas, we should conclude our discussion of Weber and the *Methodenstreit*.

We have said that Weber ascribed different aims to the nomological and the historical sciences and that this provides them with different methodologies. At the most general level however even the aims are the same. Both kinds of science look for causal knowledge. Historical sciences want causal knowledge of the characteristics of an event that make it unique; nomological sciences, of characteristics that are universal. This places an unexpected restriction on the domain of the natural sciences: the nomological method is appropriate just for 'problems in which the *essential features* of phenomena – the properties of phenomena which are worth knowing – are identical with their *generic features*'.[151] In describing the historical sciences Weber talked repeatedly of causal

---

[145] Weber 1903–6 [1975], p. 56. [146] Ibid., p. 56. [147] Ibid. [148] Ibid. [149] Ibid., p. 178.
[150] Ibid., pp. 178–9. [151] Ibid., p. 57.

knowledge. In his discussion of Knies he claims that history has a different 'theoretical purpose' from nomological science. It aims not for the '"discovery of laws"... but rather the causal explanation of cultural-historical facts'.[152] He put it somewhat differently in the essay on Roscher, but still with the emphasis on causality: the historical sciences attempt 'to order phenomena into a universal system of concrete "causes" and "effects" which are immediately and intuitively understandable'.[153] Both kinds of science are in equal pursuit of causal knowledge then. But we may fail to see this, Weber warned, if we mistakenly identify causality with nomological regularity.[154]

How though is a genuine causal explanation to be recognised? Hume taught that causes must be regularly associated with their effects. This demand is echoed for modern readers in C.G. Hempel's deductive-nomological model of explanation.[155] According to Hempel causes and effects have a law-like relation that allows the effect to be deduced from the cause. Weber altogether rejected deducibility on account of his 'anti-realist' view of the concepts of historical explanation. Hegel is a contrast. Hegel took general concepts describing perceptual reality to be '"metaphysical realities" that comprehend and imply individual things and events as their instances of *realisation*'. This, Weber argued, supplied Hegel with two different conclusions at once: 'On the one hand, reality can be deduced from the general concepts. On the other hand reality is comprehended in a thoroughly perceptual fashion: with the *ascent* to concepts, reality loses none of its perceptual content.'[156]

A number of historians of the time, Roscher among them, rejected Hegel's emanationism but nevertheless hoped to retain deducibility. For Weber this is impossible. He accuses Roscher of the mistake of thinking that 'because the general concepts are formed by abstracting from reality, so... it must in turn be possible to deduce reality from these concepts'. Here is where Weber's anti-realism about concepts enters. The general concepts we use to describe the dense manifold of perceptual reality do not designate distinct properties that make up that reality. They are artificial constructs made up by us to serve certain purposes for us. As we have seen, concepts of this kind can never comprehend reality completely. '*For this reason*', Weber argues, 'total reality can never *actually* be deduced from concepts of this sort. The historical event, as an object of knowlege, lacks *necessity*.'[157]

---

[152] Ibid., p. 142.   [153] Ibid., p. 57.   [154] Ibid., p. 58.   [155] Hempel 1965.
[156] Weber 1903–6 [1975], p. 70.   [157] Weber 1904 [1949], p. 106.

This leaves unanswered the question of what justifies the choice of one feature among others as the cause of a given effect. That is the question Weber needed to answer in order to defend historical explanation as scientific. Weber's answer had two parts. One attempted to validate the choice of concepts to be used, and the other to validate the specific explanations offered. Carl Menger had made clear that in the deductive sciences the two solutions support one another.[158] The appropriate explanatory concepts are those that represent the essences of phenomena and one knows one has described the essences when one has something from which the effects can be deduced. Both Weber and Menger agreed that it will never be possible to deduce the actual effects that occur in the world around us. Menger's reasons seem to be much like those of John Stuart Mill: the causes are usually too numerous and too short lived for us to identify.[159] On any particular occasion we are very likely to have omitted a number of significant factors from our inventory, hence the effects we deduce may be radically mistaken. This argument relies on what we think is best labelled the *complexity* of reality: there are too many relevant features for us to deal with adequately. Sometimes Weber too worries about complexity. General laws treat separate facts whereas what history needs to explain is why a particular bundle of factors appears together at a particular place and time. All laws can do in this regard is to trace a given configuation back to an earlier one and that to an earlier still, without ever coming to grips with the question of what caused the configuration. Weber's central concern, however, is not with complexity but with density. Perceptual reality is not composed of distinct recurring properties that can serve in lists of causes; it is instead a 'homogenous' and 'fluid' manifold. Since reality is not predivided into factors to begin with, what justifies the choice of one set of concepts over another in studying it? In natural science the answer is much like Menger's. We choose concepts that can figure in exact laws. But in social science?

Weber's answer is well known. The choice of the concepts to treat in social inquiry depends on a determination of what aspects of the episode in question are culturally significant. For example the cultural significance of exchange in a money economy 'can be the fact that it exists on a mass scale as a fundamental component of modern culture'.[160] Strict laws like those aimed for in the natural sciences will be

[158] Menger 1883.  [159] Mill 1836, see also Cartwright 1994.
[160] Weber 1904 [1949], p. 77.

practically useless in explaining facts like these since, as we have seen, the cost of universal validity is lack of content. Weber summarises: 'The conclusion which follows from the above is that an "objective" analysis of cultural events which proceeds according to the thesis that the ideal of science is the reduction of empirical reality to "laws," is meaningless.'[161] Nevertheless the explanations that are offered must be objectively valid; they must tell a correct causal story. '[C]ultural science in our sense involves "subjective" presuppositions insofar as it concerns itself only with those components of reality which have some relationship, however indirect, to events to which we attach cultural *significance*. Nonetheless, it is entirely *causal* knowledge exactly in the same sense as the knowledge of significant concrete (*individueller*) natural events which have qualitative character.'[162] We still need an account of what the validity of this kind of causal knowledge consists in.

Regularities play a role in Weber's answer: '[A] *valid* imputation of any individual effect [to particular concrete causes] without the application of *"nomological" knowledge* – i.e., the knowledge of recurrent causal sequences – would in general be impossible.'[163] So regularities matter. But they need not be strict nor need they literally 'cover' the case at hand in the sense required by the hypothetico-deductive model. To evaluate a given hypothesis about what caused a given effect, it helps to know what we can '*generally* expect'[164] from that cause even though the circumstances may seldom be the same from case to case. Weber tells us that to defend specific causal imputations we must use 'general knowledge', causal relationships expressed in rules (presumably what we now mean by 'rules of thumb') and the 'application of the category of "objective possibility"'.[165] These suggestions, though, sound more methodological than metaphysical. They still do not tell us what makes a causal imputation objectively valid. But we have said already as much as is necessary about Weber to make clear the theses we want to develop about Neurath. So we shall not pursue further the thorny issue of whether Weber can answer the metaphyscial question – or indeed whether he should try to. For Neurath of course the question does not make sense. What is required for science is not a theory of truth – an account of what in the world makes our scientific claims true, but rather an account of acceptability – what are the best methods of rational control that maximise the chance of getting scientific conclusions that will do what we want.

---

[161] Ibid., p. 80.  [162] Ibid., p. 82, original italics.  [163] Ibid., p. 79, original italics.
[164] Ibid., original italics.  [165] Ibid., p. 80.

Before returning to Neurath we should mention one more important figure. Weber published his criticisms of the historical school in 1903, 1905 and 1906. In 1906 he also published a criticism of the methodology of Eduard Meyer, the eminent and influential historian of classical antiquity. Meyer's importance has been described in this way: 'It is difficult to grasp how [Meyer's] more open and comprehensive standpoint influenced historical thinking in Germany and how its concerns opened up new directions and problems. In any case, parts of the late works of Max Weber read as though they could have been written on the canvas of universal history that Eduard Meyer had prepared.'[166]

In 1907 Meyer published the second edition of his monumental history of antiquity[167] with a gigantically expanded introduction which included a lengthy defense of his own views against Weber. Two kinds of factors govern historical events: general laws that are true for all events and specific causes that give the event its individuality. Sciences like anthropology, psychology or the natural sciences try to grasp the general form of the world and subsume its particular events under concepts that have their own inner law. The specific causes of a singular event that are not governed by general laws are always infinite and the choice among them is relative and subjective. But it is not just the number and complexity of causes that introduce subjectivity. 'It is only through historical consideration that the individual occurrence becomes an historical event.'[168] A different essay on theory and method of history written about the same time is explicit in its praise of Rickert and its criticism of Lamprecht for believing that historical science can proceed in exactly the same manner as natural science. The essay opens on the first page with a description of the 'infinite diversity of particulars' that we sum up as a fact. The essay also takes up the issue of free will and materialism that sparked off the debate in *Der Kampf* that would be so important to Neurath in 1931. There is one other similarity with concerns of Neurath that is worth noting. In the discussion of chance and necessity in this essay Meyer argues that conceiving of all the causal chains that lead to a single event as belonging to one system is a mere construction, 'just an idea'.[169] This claim is surprisingly close to Neurath's own view that the Laplacean mind that knows all the initial conditions is pure metaphysics.[170]

And what about Otto Neurath? We should remember that in 1905

---

[166] Tenbruck 1987, pp. 224–5.  [167] Meyer 1907.  [168] Meyer 1907, p. 199.
[169] Meyer 1910, p. 19.  [170] Neurath 1931c [1973], p. 404.

Neurath was studying in Berlin where Schmoller and Meyer taught and in 1906 these two men approved his dissertation on antique economies. Neurath was active in Schmoller's Verein für Sozialpolitik and he participated in, and indeed spoke at, the famous Vienna meeting of 1909 in which both Max and Alfred Weber spoke vehemently about introducing value judgement into research conclusions, or more accurately, about claiming to derive value judgements from them. The young Neurath – ultimately more a sociologist than anything else if we try to clarify his discipline – was an associate of Weber and the other members of the 'new generation' of the *Verein*. He was an insider in this discussion.

One of the locations for the debate over values in the Verein in 1909 was a session on productivity containing among others a paper by Karl Ballod. Neurath spoke at the meeting against the notion of productivity as part of his campaign against the usefulness of the concept of money in economics. About the concept of income that speakers had been using, Neurath argues: 'It is questionable now whether the concept of income in the narrow definition, or even in the wider conception it has today, is sufficient to handle successfully the complex of questions it is cut out to solve.'[171] Neurath complained that, for example, focus on income makes it sound as if this is what career choice depends on, leaving out issues of artistic or religious freedom or the honour accompanying various careers. Imagine the choice between jobsite $A$ and $B$. In $A$ the quantity of housing and food is better, in $B$ the prestige is higher. Can we make a calculus and compare? 'Out of the question',[172] says Neurath, arguing a view familiar to us from his attacks on false quantification. He also reveals his clear presumption that the process cannot reasonably be studied by conceptually dismantling it into separate constitutents: '*We can only consider such a complex as a whole.*'[173] As with works of art, we consider each as a totality.

Neurath acknowledged the roots of the view: 'This is a chief point in the programme of the historical school.' He himself endorsed it in clear terms: 'Above all it is questionable whether it is possible to fix the value of a single object in social life without simultaneously considering the entire social life.'[174] Nevertheless we know that Neurath, like Weber, believed in the usefulness, the importance, the ineliminability of the abstract concepts of exact science, concepts that pick out at best aspects of events and not in a way that truly pictures them. Neurath expressed that view in

---

[171] Neurath 1910b, p. 599.  [172] Ibid., p. 600.  [173] Ibid., p. 600, original italics.
[174] Ibid., p. 602.

a mild way at the Verein discussion. 'It is not sufficient to consider a complex piece by piece, but on that account one should not dismiss the undertaking of considering the causal connections which stand between the parts. It is not a matter here of either-or.'[175]

These early remarks of Neurath point to the conclusion of our next section: Neurath shared with Weber the ontology of congestion. He did not of course use this ontology as Weber did to drive a wedge between the methodology of the natural and the social sciences. According to Weber the natural sciences produce 'strict and mathematically self-evident' laws of 'general validity'; political economy does not do so because it studies the individual event and the properties that make it unique. For Neurath each of the features Weber uses to map out two distinct kinds of scientific endeavour are shared by the disciplines on both sides of the contested divide. Both must develop strict laws and both must be turned to in order to understand and predict the individual event.

Neurath it may seem was more optimistic than Weber. For Weber natural science could work as it did because the features 'worth knowing' in its domains are the 'generic' features – the ones that figure in strict laws. For Neurath that will be true in social inquiry as well. Of course that does not commit him to the view that strict laws can be found for all the features we would like to study. The fact that we select an aspect of a phenomenon as an object of study and interest does not mean that it can be made an object of scientific law. But this does not rule out the hunt for laws about social and political characteristics. Neurath complained about just this assumption in Weber and Rickert: 'A great part of the trains of thought which are linked to Rickert and related thinkers', Neurath complains in 1931, 'produce no scientific laws even where they can be physically interpreted.'[176]

But this attribution of greater optimism is somewhat misleading, for the gain that is bought at the level of theory is paid for at the level of application. For Neurath every application requires judgement, local knowledge and free decision. Nothing follows from the body of sciences simply as a matter of course, but every concrete prediction is a matter of construction. That is indeed what Popper could not tolerate about Neurath. There was no laid-down methodology with fixed rules: Neurath 'paves the way for any arbitrary system to set itself up as "empirical science".'[177]

This fact does not, however, tell against social science. Sociology and

---

[175] Popper 1935 [1959], p. 97.   [176] Neurath 1932a [1983], p. 77.   [177] Popper 1935 [1959], p. 97.

political economy are not unscientific because predictions of individual events cannot be made by their laws alone. As Neurath argues: 'The course of a leaf in the wind cannot be predicted either, though kinematics, climatology, meteorology are quite well developed sciences. It is not an intrinsic property of a developed science that it should be able to predict any individual event.'[178]

Yet Neurath was more optimistic – vastly more optimistic. For he believed we can construct exact concepts and strict laws throughout the sciences and that these laws can be equally controllable by empirical data whether they are in political economy or in hydrodynamics. The successes of application in technology set a model. Deductivity fails: the equations of the physicist must be combined with the lessons of material engineering; decisions must be made as to whether to use the wave theory of light or the incompatible particle theory; auxiliaries are up for grabs; and once *Ballungen* are taken fully into account there are not even any proper logical connections between the concepts. Nevertheless precise and accurate predictions are achieved. Good (or not) as they may be at providing understanding of the historical particular, Weber's ideal types can not offer us anything like this promise for social inquiry. But Neurath believed that standard scientific method, statistics and physicalism 'embraced with vigour across all domains of scientific inquiry' can.

How in this optimistic account did Neurath hope to solve the problem of legitimising the concepts of social science, the problem for which Weber invoked values and interests? The defence of social science quoted above follows the description of a case even more difficult to predict, it seems, than the course of the forest fire that figured so centrally in our discussions in section 3.2:

> For the prediction of behaviour of a group in some respect it is often necessary to know the whole life of the group. The individual ways of behaviour that can be lifted out of the totality of events, the construction of the machine, the building of temples, the rules of marriage are, in their changes not autonomously computable; they have to be regarded as parts of the complex that is being investigated at the time. In order to know how the building of temples will change, we have to be acquainted with the manner of production, the social order, the kinds of religious behaviour at the initial moment in time, we have to know to which modification *all this together* will be subjected.[179]

This passage reminds us of the answers we already know. Scientific concepts owe the legitimacy they have to their utility in the tool box of

---

[178] Neurath 1932a [1983], p. 77.   [179] Ibid., pp. 75–6.

unified science: their usefulness with the concepts and laws of other sciences at the point of action. Of course sometimes we are lucky. Not all kinds of events are 'equally difficult to approach' for prediction. From some conditions we can often 'roughly deduce' what will happen, for instance, 'From the modes of production of one age to the modes of production and social order of the next, from which we can try to get more predictions about religious behaviour and the like.' (On the other hand the opposite – from religious behaviour to modes of production – 'doesn't succeed, as experience shows'.)[180]

Here we see the common thread that we promised between our two stories, the one of unified science and the other of the attack on method. Neurath abandoned hope for a single theoretical account – a system of science from whose strict laws predictions about individual events can be deduced – in part for the same reason that he renounced the possibility of any scientific method that could unambiguously legitimate the choice among the sets of scientific concepts and laws. No system of scientific laws can predict the events we study since, it seems we have learned, the processes that bring them about are not in fact composed of distinct causes whose behaviour can be recorded in laws. For the same reason the history of such events cannot vote in an unambiguous and determinate way 'yes' or 'no', for or against a set of laws and concepts. The renunciation of the unified picture and the rejection of method are, it turns out, obverse consequences of the same phenomenon – the congestion of events.

### 3.4.3 The density of concepts

We have made a long detour through the work of Weber and even Rickert, whose neo-Kantian point of view was much at odds with Neurath's own. We did so because we see the doctrine of the congestion of empirical events that Weber struggled with reflected in the shift from Neurath's first Boat of 1913 to his second of 1921, emerging at last in the third Boat of 1932 as the motive force for Neurath's attack on method. The detour has been long because the terms of the debate that Weber was engaged in, like the formal sociology of Simmel and the semiotics of Tönnies, are not the stock-in-trade of philosophy of science nowadays nor part of the standard education about the Vienna Circle. So there is a danger they may not seem real to the contemporary reader. But they

[180] Ibid., p. 76.

were trenchant for the political economists and sociologists whom Neurath lived among.

The second Boat, we should recall, occurs in *Anti-Spengler*, which Neurath wrote when again in contact with Weber, who submitted a testimonial on his behalf for the trial proceedings at the time; and he was again thinking about Weberian themes. Rickert is referred to at various places in the attack on Spengler and despite his differences with Rickert it is clear that here Neurath shares with him and Weber a belief in the density of events and an attitude towards concepts that we have labelled, by contrast with Mill, 'anti-realist'. Indeed it is this stress on density that brings to the second Boat a focus on a kind of holism not associated with the first. The first Boat, we have seen, was primarily opposed to presuppositionless foundations: '[I]t is of no use to . . . renounce knowledge already gained'; '[T]he current state of knowledge has been presupposed'; '[O]ur thinking is of necessity full of tradition'; and finally 'We are like sailors who are forced to reconstruct . . . with beams they carry along.'

It is in the second Boat that the idea of empirical reality as 'an indefinite confusion that is tangled up in the most varied ways' becomes significant. In the passage leading up to the second Boat recall the miner's attempts to grasp the 'plenitude' and the god who is trying to cope with the same 'plenitude' (see section 2.3.3.) The miner fashions plans and sketches to do this; we as scientists use 'conceptually shaped results.' It is clear that these concepts do not pick out factors that literally make up the manifold, as Mill supposed. Our concepts are rather constructs that (somehow) contribute to our endeavour 'to gain some yield'. The second Boat lays out the same problematical relation between concepts and the manifold they describe that exercised Weber and Rickert, even if their solutions are at odds. Weber legitimated the choice of concepts by reference to values whereas Neurath refused any such comforting legitimation and insisted we recognise the role of voluntary decision. The solutions differ, but the ontology that creates the need for solution is the same. The view that the events of empirical reality are dense and cannot be reduced to simpler causally significant elements was not new to Neurath after the war. As we noted in the Introduction, 'The phenomena that we encounter are so much interconnected that they cannot be described by a one-dimensional chain of statements.' That is from 1913. But the kinds of problems this can create about the relation between concepts and reality were not actually disturbing to Neurath at that time. Before the war Neurath could talk about a system 'that coincides with reality in certain points' without remark even while at the same time noting both that 'an infinite number of systems can be indicated' and also that 'a complete mastery of the whole multiplicity seems an impossibility to us.'[181]

We know that within the Vienna Circle Neurath was deeply opposed to metaphysics. By all accounts he was the member of the Circle most sensitive on this matter. Recall the anecdote retold in our Introduction of Neurath's constant humming of 'M-m-m-m'. And we know that this was not a new concern but one that worried him even before the war. So it is important to notice that Neurath held his views on density not as a metaphysical doctrine but rather as a scientific one. We can look at Rickert to see the contrast. Rickert worked in the tradition of Kant and Hegel and he was prepared to make general statements about the fundamental nature of empirical reality as such. These kinds of claims were the basis for his views about the boundaries of scientific law. His arguments rest on the assumption that scientific concepts provide only 'the kind of representation of reality in which reality as unique, preceptual and individual, and thus *as reality itself*, disappears'.[182] This of course was in no way peculiar to Rickert. The *Methodenstreit* was after all triggered by Carl Menger's claims that 'full empirical reality' does not have the kind of character that allows it to be described by 'exact concepts' nor brought under 'strict laws'.[183]

The kinds of observations we find in Neurath's work that lead us to claim that he too found the events of the historical processes he studied dense and irreducible are different. They are specific claims with specific content. 'The indefinite tangle' that leads up to the second Boat does not, we should recall, appear in a text on philosophy but rather in an attack on a theory about human history, an attack that rests for the most part on the grounds that Spengler had gotten his facts wrong. For instance Neurath, as we have seen, argued against Spengler's relativism by laying claim to 'the existence of some common features' shared among all people.[184] Whether true or false, plausible or implausible, this is an empirical claim about a particular domain of study. In this case it is all human expressions that are 'woven into structures' not all of empirical reality and the weaving is a matter of fact not a result of the very nature of empirical reality. In Spengler's vision humanity is carved up into cultures and all cultures follow the same trajectory, eventually

---

[181] Neurath 1916 [1983], pp. 25, 24.   [182] Rickert 1896–1902 [1927], p. 41, our italics.
[183] Menger 1883.   [184] Neurath 1921a [1973], p. 199.

ending in destruction. Neurath argued over and over against both Spengler's general claim and against his specific examples. The complex details of the real causal stories do not bear out Spengler's histories. That is in large part because of 'the cover of the earth'. Cultures are not isolated entities, each propelled from within through the prescribed stages of development. Spengler's template fits so badly, Neurath argued, because all human activity is interlocked.

As a last example consider one further discussion of Neurath's, written at about the same time on a very different topic:

> The true market economy was already born under this co-operation of farsighted thinkers... It would be a mistake to say that the Manchester teachings about competition, teachings of *laissez faire*, were the cause of the change, but equally so it would be a mistake to say the change, which in any case was going on in all areas, was the cause of these teachings. It would be far better to designate these teachings along with other conditions as causes of the whole run of events, since in general the intuition should prevail that the fullness of the occurrences at one time point is the cause of the occurrences at the next. We must avoid viewing a part of these occurrences as cause of another part, the production processes as causes of the religious, moral, or political.[185]

Once more we see Neurath urging the density of events, but again not as a doctrine of metaphysics but rather as a concrete empirical claim about a particular historical episode.

From 1931 onward Neurath broadened his views. He took the congestion that he had repeatedly found characteristic of the historical processes he studied and made with it the famous linguistic turn of the Vienna Circle. The view that the historical event is dense and defies segmentation and that the concepts of science can represent but not picture the event is an ontological doctrine. That is metaphysics and Neurath would have no truck with it. Weber claimed to transform it into an epistemological doctrine. But that will not help. Epistemology is no better than metaphysics. Neither constitute positive knowledge, hence neither have any claim to knowledge at all. Neurath's view by contrast is neither about the world nor about the relations between us and the world; it is rather about our descriptions of the world, the nature of the concepts we use. When Neurath broadens his doctrines about the density of events beyond particular historical episodes they remain empirical claims – but now claims to be defended by evidence about our language and its use.

[185] Neurath 1920c.

Think again about the contrast between Mill and Rickert. For Mill the complex singular event is really made up of a number – albeit a very large number – of separable factors that our science aims to picture. For Rickert it seems that this is not the case. There is first one single globular event that gives rise in time to another and then another; we devise scientific concepts to represent these events with some kind of values in view. But what is this talk of events being composed or not of separable factors that are pictured by our scientific concepts? How can such talk be verified? Neurath did not indulge in it. He tells us about our claims and our concepts: 'Complex messy statements of little cleanliness – *Ballungen* – are the basic materials of the sciences.'[186] And they are not composed of more elementary precise concepts – concepts that are not dense – 'In this form they are even a protest against elementary statements.' For Neurath there are no elementary concepts in our starting base. If we want to introduce such concepts they will inevitably have an entirely artificial connection with our everyday empirical ones. This is easier to see if one thinks about how Carnap's *Aufbau* works. If you started with *Ballungen*-style concepts you could not construct from them the concepts of physics or geometry as Carnap tried to do.

Do not be misled by our tracing the roots of *Ballungen* to ontological doctrines. The linguistic turn of the Vienna Circle must be taken seriously, as must Neurath's agreement with Tönnies that we always start within a natural language. Neurath was not trying to mirror the world surreptitiously while talking only about relations between concepts and sentences. He was talking about all there is to talk about and the claims he made are true, if they are, because our concepts are as he said they are, not because events in the world are composed in one way or another. The method of verification and argumentation must follow in train. Neurath's claims about the nature of our empirical concepts and their relations must be taken to be empirical claims subject to empirical confirmation. And that is how he intended it. There is only positive knowledge; Neurath genuinely did not want to do philosophy.

A distinction made by Carnap can provide us with a good device for summarising the central point of this section – the distinction between the formal and the material mode. Section 2.5.2 described Carnap's thesis about the two: claims commonly made in the material mode that seem to lack genuine empirical content – claims about causality or

---

[186] Neurath 1935b [1983], p. 128.

determinism, for example – are often unperspicuous ways of formulating claims that make very good sense when rendered in the formal mode – as, for instance, claims about deductive relations among statements in our theories. From the First World War onwards Neurath argued in favour of the density of events in the historical and sociological matters he studied. But to take on this view as a general thesis about the nature of empirical reality, as did Weber and the earlier protagonists of the *Methodenstreit*, would be to fall into metaphysics, which Neurath would not be tempted to do. In 1931, however, he hit upon the concept of *Ballungen* and a number of important consequences about the relations of theory to data followed – consequences similar to views about the relation of science to reality that underlay the *Methodenstreit*, only now by Neurath cast acceptably in the formal mode. In 1932 Carnap articulated the formal/material mode distinction; already one year before Neurath had exemplified it. We do not wish to suggest though that Neurath turned back to the old issues of the *Methodstreit* in formulating his views of 1931. Rather he developed them in a context far more sympathetic from his Marxist and physicalist perspective – he developed his ideas about *Ballungen* in tandem with his efforts to understand how to think about history from a Marxist and materialist – for Neurath, physicalist – point of view. The section after next will tell the story of how this happened.

Before turning to this story we should like to make two remarks. The first is about the terms we have been using. Neurath often talked about the complexity of events but this can have a misleading connotation. We often think of a complex as made up of distinct pre-existing parts each playing its own distinguishable role. But as we have seen this does not seem to be what Neurath had in mind when he talked of 'the infinite confusion that is tangled up in the most varied ways'. We prefer to talk of 'the density of concepts' and the 'congestion of events' – but we will nevertheless persist with 'complexity' when Neurath himself uses it. Our second remark concerns an apparent counter-example to Neurath's repeated insistence on the interconnectedness of all human activity. We shall take it up in the next section.

### 3.4.4 *The separability of planning and politics*

There seems to be a startling lacuna in Neurath's belief in the tangle of events and the intertwining of causes and effects, a lacuna which is especially troubling because it threatened his entire programme for social change. Throughout the course of the Bavarian revolution and for a long time afterwards Neurath clung to the claim that the economic arrangements for full socialisation could be insulated from the volatile forces of politics and power. As we saw (section 1.3.3), at his trial he defended his service across all three Socialist governments with the claim that he saw himself as an economic technician without commitment to any particular political form. With the second Soviet Republic he worked 'essentially only organizationally';[187] he did not consider the political question of the legitimacy of either the Hoffmann or the Soviet governments. He explained: 'I felt neutral about both governments and saw my duties as consisting only in pursuing the interests of the state and proleteriat in the area of the original socialisation plans.'[188] It seems he believed that some kind of wall could be constructed between the economic and political forces at work in society.

The belief in the separability of the economic plans for socialisation and their implementation from the political process is surprising given Neurath's tendency to picture all aspects of social life as bearing on one another; and it is a belief for which he has often been criticised. Otto Bauer made much of this in his statement at Neurath's trial proceedings. He repeatedly urged that Neurath was concerned not with 'the constellations of political power that the realisation of socialism requires but with the technical and organisational means for its realisation'.[189] For instance: 'It is characteristic for Neurath that he never understood his problem politically but always – according to his own favourite expression – "*gesellschaftstechnisch*" (i.e., as a social engineer)';[190] 'Neurath recommends a planned order and rearrangement of economic life from above through an administration standing over the society, and it is not interesting to him whether that is an Imperial-and-Royal supreme command, a democratic parliamentary government or a soviet dictator';[191] he served 'in succession an Imperial-and-Royal war ministry, a bourgeois-socialist coalition government and a Soviet republic. For he imagined that each of these regimes could allow for his "socio-technical" plans to be realised as well as any other'.[192] Finally Bauer closed with the criticism, noted in section 1.3.3: 'A Marxist can accuse him of not understanding that all social reforms are secured through the political power relations under which they are carried

---

[187] Neurath's defence statement at the *Standgericht*, State Archives Munich. See also 1920f.
[188] Neurath's defence statement at the *Standgericht*, State Archives Munich.
[189] Otto Bauer's statement for Neurath's *Standgericht*, State Archives Munich, p. 051790.
[190] Ibid., p. 051788.    [191] Ibid., p. 051791.    [192] Ibid., p. 051793.

out.'¹⁹³ Serious as the mistake is, Bauer concluded, it is nevertheless not legally punishable.

Weber too found Neurath sincere about the separation, but thought him foolish in this regard (see section 1.3.3 again). More recently an East German author put the case thus:

> Neurath, who never wanted to hurt anyone, overlooked the fact that the political and economic conditions of the socialist plan could only be created once the working class had seized political power, once the capital ownership of the means of production had been made secure.[194]

The same critic poked fun at what he considered other aspects of Neurath's political naivety as well: Neurath's Central Planning Agency was so busy socialising the universities and the press that it never occurred to anyone to disarm the bourgeois until late in the first Soviet republic.

One place where the separation between politics and economic planning played a clear role in Neurath's thought and in his own efforts was in the issue of the timing of the full socialisation programme. Neurath had predicted that socialism could be achieved in Bavaria within five to ten years and he did not abandon this prediction even after the Bavarian revolution was crushed. In the 1920 article in which he explained his views on the relationship between the political and economic structures, he maintained, 'Today socialism stands before the door.'[195] This was a repeat of the claim he had made from prison just after Leviné's trial and before his own and Toller's.[196] Then he explained that his plans for full socialisation had failed for lack of time. What happened was primarily a result of the bourgeois war, strikes, wild wages and the encroachment of the individual industrial councils. Socialism could still be achieved in a few years – if only the political conditions were right.

But they weren't. Erich Mühsam like many others, both critics and supporters of the revolution, agreed with Neurath that the revolution had failed because of the shortness of time:

> The bourgeois world always had a big advantage on its side: time. It escapes, agitates and waits. Conflicts within the revolution, the awakening fears, dissatisfaction of parts of the population (which is never totally involved and committed), the 'state of the world', politics, and so forth allow the future to grow so that time 'will heal the wounds'. So the theme frequently appears among the Munich Revolutionaries: if we had time . . . a couple of months . . . if only we

---

[193] Ibid., p. 051793.  [194] H.Beyer 1957, p. 63; see also his 1982, p. 61.
[195] Neurath 1920e, p. 224.  [196] Neurath 1919e.

were left undisturbed in our work . . . During the first workday as a People's deputy Landauer wrote to his friend, 'If I am only left some week's time, I hope to achieve something; but it is easily possible that it is only a couple of days, and then it was only a dream.'[197]

Mühsam, unlike most others, believed that the time was ripe politically. He thought that the actions of Munich would open up the revolution for all of Germany.[198] As we saw, the Communists did not consider the situation at all opportune and resisted the attempt to establish the first Soviet Republic. Leviné's wife, Rosa Leviné-Meyer, relates an amusing story about their efforts:

All of the available speakers were mobilised and went speeding from factory to factory, from meeting to meeting, to elucidate the views of the party. My hard-earned savings in the Ukraine stood the party in good stead: they were spent, almost to the penny, in taxi fares.[199]

It might have been the case that Mühsam was right about the political possibilities but history seems to have borne out Leviné and the Communists. Neurath's confidence in the possibilities for implementing full socialisation would not have been so surprising had his arguments been in line with Mühsam's rather than Leviné's. Even the reverse would not have been so puzzling. But Neurath's estimate that socialisation could be achieved within five to ten years was based entirely on his views about the state of scientific and technological knowledge at the time. He deliberately restricted himself to predictions made just on these grounds, as if they could operate in isolation; correlatively he seemed to have had no view about the power structure and its effects.

The same surprising gap in Neurath's thought, a kind of optimistic neglect of the kinds of interconnections that he usually stressed in other work, comes out in another concrete example. Kurt Eisner had first asked the well-known liberal economist and politician Lujo Brentano to become Minister of Trade and Industry in his government. But Brentano refused. Failure of socialisation was inevitable, he believed, given the run-down conditions that existed in Bavaria at the close of the First World War.[200] According to Ernst Toller, Brentano appeared before the first meeting of Eisner's socialisation council on 22 January 1919 and 'made a pronouncement which caused the industrial magnates to sit up; "Industry can only be socialised", he said, "when it is already in existence, and in Germany today it is non-existent."'[201]

[197] In Viesel 1976, p. 8.   [198] Mühsam 1929.   [199] Leviné-Meyer 1973, p. 91.
[200] Brentano 1931.   [201] Toller 1934, pp. 142–3.

### Where 'Ballungen' come from

Neurath and Brentano disagreed over a large number of issues. We have seen that Neurath argued for a moneyless economy of 'payment in kind'. Brentano made fun of him. Neurath was 'an enraptured Utopian' with an economic plan 'like the kind that might have existed in ancient Egypt, where everyone lives directly from the King'.[202] They differed on more concrete questions, too. Brentano was a longstanding advocate of free trade – why make shirts in Germany at three times the price they cost to make in America? Neurath answered with his familiar objection that the profit motive was not sensitive to people's needs: 'So long as there are hungry people and a need for clothes we should mandate the production of food or clothes for internal use.'[203]

It would not be surprising given this context that Neurath would have objected to Brentano's appointment. But that was not his chief reaction. Rather, he was puzzled: 'How can it be explained that a man like Eisner, who wanted socialisation with every fibre of his heart, put this important position in the hands of a liberal who probably did not want socialisation?'[204] But Eisner's motive was clear. He tried to appoint Brentano for the same kind of reasons that moved Neurath to keep the current entrepreneurs at the heads of industries and banks to implement his plans for socialisation. The organisation of production, transport, credit and the like required experts; and the Bavarian economy could not be rebuilt without them. Brentano had established contact and respect with Bavaria's capitalist powers. These were by Neurath's very own arguments essential if socialisation were to have any hope of succeeding.

Last of all let us look at one more area in which Neurath has been accused of inappropriately segregating intertwined elements, and the defence he gave for doing so. One of the significant non-Marxist strands of Neurath's thought that is sometimes pointed to is his focus on the arrangements for consumption rather than production.[205] This is apparent in his views about the nationalisation of industry. We know that he did not agree with Otto Bauer that production was a central issue. In his 1919 lecture and pamphlet 'The Character and Course of Socialisation' Neurath remarked that there had already been a great deal of Marxist propaganda concerning the socialisation of industry. What was lacking was attention to a different, but equally Marxist concern, that of the need for central planning. According to the newspaper accounts this was one of the topics he stressed during the two and a half hour performance

---

[202] Brentano 1931, p. 364.   [203] Neurath 1919e, p. 227.
[204] Neurath 1919h [1973], p. 18, translation altered.   [205] H. Beyer 1957.

before the Council of Ministers that resulted in his appointment as head of the Central Economic Administration: nationalisation could only be of help if planned utilisation of resources to serve people's needs were already in place.[206] In 1920 he argued the same point with telling examples:

> Nationalising basic production . . . and the ripe industries does not help the starving. It is all the same to the workers whether they burn state or private coal in their stoves; what matters to them is looking after the proper distribution so that the coal is not used for luxury purposes.[207]

Nationalisation was just one issue. Neurath's insulation of the production and distribution plan from questions of ownership and political organisation was thoroughgoing. We must never, he urged, confuse *Sozialisierung* – remodelling of the order of life – with *Vergesellschaftung* – changing the power relationships.[208]

Just after the destruction of the Bavarian Socialist governments Neurath did try to explain his belief that the economic and political orders could be kept separate. Political forms are not irrelevant, he admitted, and economic structures do depend on them – but not at every step. It is essential for socialism to enforce this independence since a socialisation plan would take years to carry out and it must be kept stable across shifts of political power that are very likely to occur during the period of transition. The economic plan must be safeguarded and considered to be a 'closed whole'. In 1930, it seems, Neurath still held similar views. In a discussion in May he presented a systematic overview of the development of Marxism. During the review he praised the Russians for their commitment to total planning. They treated themselves as a closed whole and embarked for ten years on an autonomous economic course.[209] Although things may be more expensive in the short run in the totally closed-in economy in Russia, after ten years' life will be far better as a result of the systematic planning. Already England had had three crises while the Russians had had none. In 1932, in a description to Carnap of his work in Moscow, Neurath again painfully noted the unhappy consequences for social planning when it cannot be insulated from politics: in Moscow it is especially important that he work entirely as a technocrat. Any ideology leads to disagreement – especially when

---

[206] *Münchner Neueste Nachrichten*, 25 March 1919.
[207] Neurath 1919g, pp. 269–70. This example also appears in his 1920f.   [208] Neurath 1920c.
[209] 'Diskussion: Boltzmanngasse 8 May 1930', RC 029-18-01 ASP.

the official line changes.[210] Whether Neurath was right or not about the possibility of the economic 'closed whole', these beliefs show at least that he did not hold his claims about the interconnection of events as a necessary feature of empirical reality, that is, as a piece of a priori metaphysics.

### 3.4.5 How Marxists think of history

We have already pointed out that Neurath's first public use of the concept of *Ballungen* seems to have been on 4 March 1931. The idea was developed as one of many connected themes in two papers that we know from Neurath himself were written in 1931: 'Physicalism' and 'Sociology in the Framework of Physicalism'.[211] The first mention of *Ballungen* in the published literature occurs in the latter. Throughout that year Neurath was deeply engaged with the problems of protocol sentences. But he was equally concerned with the issue of sociology as an empirical science and with Marxism. Not only did he write the book *Empirical Sociology*; during this period he was also following a debate in *Der Kampf* to which he contributed at the end of the year. The debate involved another left-wing member of the Vienna Circle, Edgar Zilsel. The starting point was a review by Zilsel of Max Adler's 'Textbook' on the materialist conception of history.[212] It is significant for our story about the genesis of *Ballungen* that the idea first appeared in print in 'Sociology in the Framework of Physicalism', where Neurath brought together his main concerns from the protocol sentence debate and from his own physicalist version of Marxist materialism.

Max Adler was the intellectual leader of the social democracy and one of the visible neo-Kantian heads of the Austro-Marxist movement. He was also a co-founder of *Der Kampf* along with Neurath's friend Otto Bauer. In his book Adler had pictured the classical philosophical mind–body problem as parallel to the Marxist distinction between ideology – or superstructure – and the economic basis – or substructure – and used it to argue against a materialistic philosophy that reduces spirit to matter. His view on the mind–body problem provided Adler with grounds for defending the compatibility of Marxism and religion. Zilsel on his part argued that Marxism is both scientific and materialist and that the mind–body problem is at any rate only a fuzzy philosophical

---

[210] Letter to Carnap, 1 October 1932, RC 029-12-08 ASP.   [211] Neurath 1931b, 1932a.
[212] Zilsel 1931a, M. Adler 1930–2.

problem that has nothing to do with real questions of how social and economic structures come about.

The debate between Adler and Zilsel continued in a second article by Zilsel[213] with an intervention in defence of the compatibility of Marxism with religion by Wilhelm Frank[214] and a final article by Neurath '*Weltanschauung und Marxismus*' ('Worldview and Marxism').[215] Neurath's ideas in this article were the same as ones he presented in *Empirical Sociology*, as well as those formulated in the two articles where *Ballungen* first appear. Neurath's principal point was that Marxism is no *Weltanschauung*, it is – as he had long held – a science. Nor is Marxism compatible, or for that matter incompatible, with any *Weltanschauung*, for there is no such thing as a *Weltanschauung*. There is only science, a unified science in which empirical sociology is contained; all the rest is empty talk – and likely to be politically insidious. As Neurath argued in 'Physicalism': 'The work on unified science replaces all former philosophy. At this point, science without a world-view confronts "world-views" of all kinds, philosophies of all kinds.'[216] (It is interesting to note that here it is not just metaphysics, but all of philosophy that physicalism is to combat.) Physicalism is the key to Neurath's own views on historical materialism, which are not only his central topic in *Der Kampf* but are also discussed in *Empirical Sociology* and in 'Sociology in the Framework of Physicalism'. Briefly what is crucial for Neurath is that the interaction between the economic basis on the one hand and ideological factors on the other must be cast into the framework of strict physicalism. But 'physicalism is perfectly monistic'[217] and this applies to Marxist sociology as well.

How does this connect with Neurath's simultaneously developing thoughts about *Ballungen*? To trace this influence we have first to consider the question of historical materialism in the context of a much broader and earlier discussion. If the little debate in *Der Kampf* of 1931 was the locus for Neurath's thinking about *Ballungen*, the big debate in Marxism from the 1890s was the source. Its main protagonists were Friedrich Engels, the German revisionist Eduard Bernstein, the Russian orthodox G. Vladimir Plekhanov, and the Italian heterodox Antonio Labriola. Max Adler serves as starting reference and intellectual link between the two debates. In his book of 1930[218] Adler argued from the standpoint of Kantian idealism against the materialist philosophy commonly attrib-

---

[213] Zilsel 1931b.   [214] W. Frank 1931.   [215] Neurath 1931d.
[216] Neurath 1931b [1983], p. 56.   [217] Ibid.   [218] M. Adler 1930–2.

uted to Marx. It is impossible, he claimed, to derive consciousness from physical motion; nor is the causal explanation of social phenomena in conflict with the existence of a human will. He explicitly called into question the traditional distinction between 'material' and 'spiritual' (ideological) factors in the historical process.

Adler referred to Engels' doctrine of the 'mediate causes' to give textual support to his interpretation. Specifically, he quoted from two famous letters by Engels, one to Joseph Block, dated 21–22 September 1890, the other to Heinz Starkenburg, dated 25 January 1894. In the extract from the first one, Engels remarks that

> there are innumerable intersecting forces, an infinite group of parallelograms of forces which give rise to one resultant – the historical event. This may again be viewed as the product of a force which, taken as a whole, works *unconsciously* and without volition. For what each individual wills is obstructed by everyone else, and what emerges is something that no one willed.[219]

In the second letter Engels points out that 'political, juridicial, philosophical, religious, literary, artistic, etc., development is based on economic development. But all these react upon one another and also on the economic base.'[220] These letters were the two triggers of the ensuing discussions and various interpretations given within the Marxist community of the thesis of historical materialism.

Adler also referred to a pamphlet by Bernstein of March 1889 and warned against Bernstein's interpretation of Engels' letters as grounds for dualism. In an intermediate position between Bernstein and Adler himself, Adler incidentally mentions Plekhanov, who, precisely because he is 'not really a materialist, has always determined to work out the human nature of economic relationships'.[221] The immediate question is, what did Plekhanov have to say about Engels' letters?

In 1908 Plekhanov published his book *Fundamental Problems of Marxism*.[222] Adler reviewed the German translation in *Der Kampf* in 1910;[223] later that year he wrote again about Plekhanov and his views of the dialectic interaction between thought and object. In between is a brief letter of Karl Kautsky's discussing Plekhanov and answering a question about one of Neurath's favourite philosophers, Ernst Mach: are Mach's teachings compatible with Marxism? Kautsky's answer is exactly the thesis of Neurath's 1931 paper. Of Marxism, Kautsky

---

[219] Marx and Engels 1934, p. 476.   [220] Ibid., p. 517.
[221] Adler 1930–2 [1964], footnote, p. 103.   [222] Plekhanov 1908.   [223] M. Adler 1910b.

affirms, 'I understand it not as a philosophy, but as an experiential science, a particular conception of history. The conception is certainly incompatible with idealist philosophy, but not incompatible with Machian epistemology.'[224] His later remark about Marx in the same letter is also close to Neurath's own thought: 'Marx is no philosopher; rather he has announced the end of all philosophy.'[225]

Neurath did not contribute to this series of articles in 1910–11. But he certainly read them and had them to turn back to in 1931 along with Bauer's 'World-View of Capitalism' which we described in section 2.5.1. Looking in detail at both its content and its language, Neurath's contribution to *Der Kampf* in 1931 would be far more appropriately placed in the 1910 series than in that of 1931. Let us just point to the most striking feature. Neurath's 1931 piece, we have seen, is about Marxism as a *Weltanschauung*. But this concept played little part in the 1931 exchanges between Adler and Zilsel. It was central in 1910. Adler's discussion then, entitled 'Marxism and Materialism',[226] focused on Plekhanov's opening statement – 'Marxism is an entire *Weltanschauung*. It is, briefly expressed, the modern materialism, which represents the most highly developed *Weltanschauung* to date.'[227] The term *Weltanschauung* does play a role in Wilhelm Frank's 1931 contribution, unlike Adler's and Zilsel's. But his very title, 'Is the Marxist Conception of History Materialistic', like Zilsel's, directs the attention of the 1931 reader back to the topic of the big debate of the teens.

Plekhanov's own work on that subject in 1897 was 'The Materialist Conception of History', a review of a newly published essay under a cognate title by Antonio Labriola, whom he viewed as the first true Italian Marxist.[228] In his essay Labriola attacked the 'theory of factors'. His view was largely a critique of what he considered vulgar interpretations of Marxism as the theory that the 'economic factor' dominates in history. Like Engels (as he interpreted him), Labriola rejected philosophical materialism. He believed instead in the interrelations of human activities. The historical process, he argued, develops organically and pluralistically as a 'complex'; multiple factors intervene that cannot be disregarded nor considered autonomously. In an earlier essay of 1887 Labriola had already complained that Hegel's monistic philosophy of history had become a Procrustean bed for the historical sciences, which needed to concern themselves with highly differentiated forms of social

---

[224] Kautsky 1909, p. 452.  [225] Ibid.  [226] M. Adler 1910b.  [227] Quoted ibid., p. 564.
[228] Plekhanov 1897, Labriola 1896.

life such as law, language, and art: 'The original centres of civilisation are many in number and cannot be reduced by any sleight of hand ... the consideration of so many separate series of events, so many factors that resist simplification, so many unintended coincidences, ... makes it seem highly improbable ... to suppose that there is at the root of everything a real unity.'[229]

In 'Historical Materialism' Labriola warned that these factors are not to be taken as real. Like Weber and unlike Mill, for Labriola and Plekhanov the factors studied in science are not real elements separately operating in concrete events. Laws about them do not truly govern what happens, for 'these concurrent factors, which abstract thought conceives and then isolates, have never been seen acting each for itself'.[230] It is in this sense that factors represent 'much less than the truth, but much more than a simple error'.[231] In Labriola's account the study of history begins from the *complexus* as the primitive datum:

The narrator finds himself, in a word, confronted with a complexus of accomplished facts and of facts on the point of being produced, which in their *totality* present a certain aspect ... Yet into this complexus he must introduce a certain degree of analysis, resolving it into groups and into aspects of facts, or into concurrent elements [factors], which afterwards appear at a certain moment as independent categories.[232]

In his review Plekhanov stressed this aspect of the 'doctrine of factors':

A historical factor is an *abstraction*, and the idea of it originates as a result of a process of abstraction. Thanks to the process of abstraction, various *sides* of the social *complex* assume the form of separate *categories*, and the various manifestations and expressions of the activity of social man – morals, law, economic forms, etc. – are converted in our minds into separate forces which appear to give rise to and determine this activity and to be its ultimate causes ... a multiform reflection ... of the single indivisible history.[233]

The density aspect of *Ballungen* missing from Duhem's views on scientific abstraction is the exact analogue at the linguistic level of this complexus that Labriola introduced at the conceptual level. As we have seen Neurath's early views on the make-up of empirical reality were already in line with those of Labriola and Plekhanov, especially with regard to his belief in the interconnectedness of basic empirical facts and his talk

---
[229] Labriola, *Problems of the Philosophy of History*, quoted in Kolakowski 1978, p. 287.
[230] Labriola 1896 [1904], p. 147.  [231] Ibid., p. 151.  [232] Ibid., pp. 141–2, our italics.
[233] Plekhanov 1897 [1969], p. 108, our italics.

240                    *Unity on the earthly plane*

of their complexity or congestion. This is especially clear in a remark of 1911:

Scientific progress in economics obtains once empirical *complexes* prompt *abstractions* which lead to combinations whose reality or realisability one can investigate.[234]

In 1931, at just the time that *Ballungen* began to play a crucial role in Neurath's attack on pseudorationalism, these views re-emerged, and in much the same terms as those used by Labriola in his criticism of Hegel. In *Empirical Sociology* Neurath writes:

The materialistic conception of history begins with the total process of life. If one continues to use the traditional delimited terms such as 'religion', 'art', 'science', 'law', and so on, these formations appear as interwoven into the total social process. This interweaving might perhaps be described in this way, one might of course predict the course of modes of production and social changes, but it would be hopeless to write an autonomous 'history' of religion, art, mathematics and so on, such histories could be written only within the framework of an historical account of the total process.[235]

A similar defence of the *Ballungen* aspect of events – their 'smunched together' character – appeared in 'Sociology in the Framework of Physicalism', where Neurath argues:

The individual ways of behaviour that can be lifted out of the totality of events, the construction of machines, the building of temples, the rules of marriage are, in their changes, not 'autonomously' computable; they have to be regarded as parts of the complex that is being investigated at the time.[236]

It was a familiar German tradition to treat nations or peoples as organisms and to devise theories of the stages of development that a people pass through. This is the tradition that Spengler was tacking onto in his story of the decline of the west. Neurath disapproved. We have already looked at his criticisms of Spengler in 1921. He returned to his attack on the underlying assumptions in 1931. To the separation of the world into peoples Neurath opposed 'the cover of the earth'. The relevant empirical object of study is 'a "human and cultural cover" which is stretched over the whole earth'.[237] Scientists 'delimit parts of the human cover as against other parts';[238] yet 'each "state" is part of the human cover, a texture of citizens, judges, soldiers, policemen, civil servants, prisons, schools, houses, streets, etc. are joined together'.[239] On this view,

---
[234] Neurath 1911, p. 82, our italics.   [235] Neurath 1931c [1973], p. 352.
[236] Neurath 1932a [1983], pp. 75–6.   [237] Neurath 1931c [1973], p. 332.   [238] Ibid., p. 368.
[239] Ibid.

'the notion of a single people as a continuous entity will simply be invalid'.[240] The task is 'just to compare partial changes in the great object "humanity"'.[241]

The precision and cleanliness of the scientific concepts we construct contrasts radically with the complex, rough character of the descriptions we must use when dealing with real history: 'If we wish to predict what peoples, states or organisations will do . . ., [w]e must be satisfied to consider certain rough facts of a complex character.'[242] The density of events surfaces in every empirical fact the sociologist and the historian pay heed to. In 'Sociology in the Framework of Physicalism' Neurath puts this thought in the much stronger form of a definition: 'historical period = non-analysed complex of conditions'.[243]

The similarities between the ideas of Neurath in 1931 and those of Labriola and Plekhanov run deep. All three reject the metaphysical speculations of Hegelian idealism. More generally, all, as Labriola urges, 'look for *gradual* progress in the *gradual* abandonment of every sort of metaphysical hypothesis in the works of human thought'.[244] It is precisely on these grounds that they all criticised the distinction between the natural and the social sciences. Related disciplines like sociology and history must be pursued scientifically or rejected altogether.[245] The materialist conception of history itself was offered by Labriola as an example of this kind of unification. According to Labriola, 'The materialist conception marks the culminating tendency in the investigation of the historic-social laws.'[246] Neurath wrote in similar terms that 'the example of Marxism shows us how sociological correlations can be investigated and how laws of relations can be established'.[247] As we have seen, the members of the historical school in Germany denied the possibility of producing general and abstract social and historical laws. Plekhanov attacked them for this.[248] Neurath too was explicit in his rebukes, even of his contemporaries and advocates: '[Weber] tries, like Rickert, to carry out a division in the pursuit of science which makes impossible any universal connection of all law-like features that are required by the empirical sciences in their forecast.'[249]

Did Neurath read Plekhanov's essay on Labriola? If not, it is quite surprising that both began their presentation of the materialist conception

---

[240] Ibid., p. 332.  [241] Ibid.  [242] Ibid., p. 404.  [243] Neurath 1932a [1983], p. 85.
[244] See Labriola 1896 [1904], p. 156.
[245] See ibid., p. 225, Neurath 1932a [1983] p. 82 and section 5.
[246] Labriola 1896 [1904], p. 156.  [247] Neurath 1932a [1983], p. 82.
[248] See Plekhanov 1908, p. 96 and the editor's note on p. 62.  [249] Neurath 1931c [1973], p. 357.

of history with exactly the same remark, almost verbatim: the label 'historical materialism' must be preferred to the much used 'economic materialism' on the grounds that the former emphasises the contrast with unwelcome idealism.[250] In any event there can be no doubt that Neurath was familiar with the works of Labriola and Plekhanov. Plekhanov, who had been described by Lenin as the most educated intelligence of socialism, was a personal friend of Viktor Adler, the first leader of the Austrian Socialists and father of Friedrich Adler, translator of Duhem. He was moreover, as we have seen, involved in debates with Max Adler in *Der Kampf*.

Labriola too was well known in the Austrian socialist group of which Neurath was a part. Wilhelm Ellenbogen was one tie. Ellenbogen was an economic expert and a friend of Viktor Adler who worked unsuccessfully with Otto Bauer to try to push through a number of socialisation projects. Like Neurath, Ellenbogen wrote regularly for *Der Kampf* throughout the twenties. Although he and Neurath would stand on the same side of the Social Democratic Party discussions in favour of making coalitions with the bourgeoisie to stop Fascism, they were opposed on much else, such as the need for religion, the importance of materialism, and the possibilities for achieving full socialisation in the near future, a topic in which Neurath attacked Ellenbogen in *Der Kampf* in 1922.[251] By the 1890s Ellenbogen and Labriola were close friends and corresponded regularly. Indeed Ellenbogen's intimate knowledge of contemporary Italian thought was a great asset in his early recognition and analysis of the dangers of Fascism.[252] Labriola's doctrines were also debated in works by both well-known socialists and sociologists. Ernst Untermann's book *Dialektisches*, published in 1907, contained a comparison between Labriola's conception of historical materialism and dialectical monism,[253] and Emile Durkheim, whose work Neurath as a sociologist would have known, published a review in 1897 of Labriola's essay on the materialist conception of history.[254]

When the Labriola–Plekhanov idea of the density and complexity of the empirical objects of study reappeared in Neurath's work in 1931, the Vienna Circle was full sail in the middle of its 'linguistic turn'. Neurath applied the doctrine at one and the same time to the basic concepts and sentences of the empirical sciences and to the basic facts of sociology

---

[250] See Plekhanov 1908, p. 105 and Neurath 1931c [1973], p. 349.   [251] Neurath 1922a.
[252] See Ellenbogen 1981.
[253] Our thanks to Conrad Wiedemann for finding the Untermann book.
[254] Durkheim 1897 [1982].

and history. Recall Neurath's summary: '"historical period" = non-analysed complex of conditions'.[255] As we have seen, the exact correlate of this view on the linguistic level is there in the same article. The basic statements of the empirical sciences are complex. That is on account of *Ballungen* – those 'imprecise, *unanalysed* terms'[256] that enter protocol sentences from the language of every day life. The complexity of the basic observation-statements of the sciences continued as a main issue for Neurath, especially in later criticism of Schlick[257] and Popper.[258] One notable remark that can serve to summarise Neurath's deep and continuing objections to false ideals and false precisions appears in a 1934 article against Schlick, where he made explicit reference to his earlier denunciations of the Laplacean myths: 'Though we can co-ordinate precise mathematical formulations with our imprecise observation statements, the assumption that one would have to arrive at precise elementary statements if one only had sufficient intelligence at one's disposal, leads to a fiction that resembles that of Laplace's spirit – a perfectly metaphysical notion.'[259]

We see then a striking resemblance between Neurath's doctrines on *Ballungen* and certain Marxist doctrines about how to study history – which he was writing about simultaneously, and that at the same time as his drive to defend the public nature of science and language in the protocol sentence debate and to insist on physicalism as the right criterion for marking out the domain of what is knowable. One might ask, did Neurath's philosophical ideas spring from his Marxism; alternatively was the articulation of this set of Marxist ideas a consequence of the philosophy he was defending in the Thursday evening meetings of the Vienna Circle? Not surprisingly, we should like to use Neurath's own doctrines about the tangle of events to reject the question. Neurath came to the Vienna Circle with a number of views about politics, economics and history and about how to study them. He was responsive to certain problems that set the background to the *Methodenstreit* and to its treatment at the hands of Weber, though not to Weber's solution. He was a Marxist and a physicalist, and we do not know in what order. Like his science, his metascience was practice oriented. It could no more deny the apparent necessity for decision and negotiation at each stage of choice than it could accept that science might be a personal construction or that it could start with data whose admissibility did not depend on

---

[255] Neurath 1932a [1983], p. 85.   [256] Neurath 1932d [1983], p. 91, our italics.
[257] Neurath 1934.   [258] Neurath 1935b.
[259] Neurath 1934 [1983], p. 107, cf. 1931c [1973], p. 404.

assessments about the reliability of the individuals and methods used to collect it. He argued about protocols, he defended empirical sociology, he sat in on discussions of imprecision and he read *Der Kampf*. Out of this tangle he forged the idea of *Ballungen* and developed the attack on method.

### 3.5 NEGOTIATION, NOT REGULATION

Neurath may not have cared about truth but he did care about effectiveness. That generates a serious problem. How are we to make choices that are likely to work? Here we must come to terms with a general frustration. As we have stressed throughout, Neurath insisted that there is no universal method to tell us how to choose between competing claims. We are expected to act, but there is no unified theory from which to predict the consequences of our actions. We are supposed to unify the sciences at the point of action. But how are we supposed to do that? In Neurath's view theories can be evaluated and actions can be chosen, but in different ways for different times and different contexts. And he will not tell you how. How could he? It takes detailed local knowledge to make a good decision. That is the task of the expert.[260]

Neurath's philosophical strategy was identical to his strategy for central planning. We have already described his plan for an Experts Section to provide specialist advice to the Controlling Council (section 1.2.4). He did not expect to solve problems himself but rather to create structures in which others with special knowledge and special skills could solve them. What after all was the Kranold–Neurath–Schumann plan for socialisation? Let us look again at figure 1.1. Its central feature is that it is no plan at all; it is a metaplan. It is a blue print for a centralised structure of bureaus, commissions and committees; *their* job is to think problems through and propose changes. The lack of detail in his schemes is equally apparent in his suggestions about the kinds of laws that are to be worked out by the Central Planning Agency. Neurath laid out the overall aims:

1. Production and use are to be centrally planned so as to prevent the recurrent crises, the mass unemployment and the depressions characteristic of the free market economy.

---

[260] Should the philosopher provide rules for the expert? No, only another expert can do that. We may enquire, 'How do we know that a particular proposal will work?' The answer must be a scientific one and hence an uncertain one. We may want more but we can not have it. We must be careful not to expect Neurath to enter into the Kantian project of assuring us that knowledge is possible. That, for him, is abhorrent metaphysics.

2. The income of society is to be divided in such a way that personal advantages ('food, clothing, housing, pleasure possibilities, and education') and personal property will be distributed according to individual achievements and characteristics (such as age, sex and state of health), without any group privileges ( such as privileges of birth, station or rights of inheritance).[261]

His advice about how to achieve the aim was, as always, to set up a committee of experts to figure it out.

This abstract advice was easy to deride. Here, for example, is a description from an on-the-spot observer:

> Under the flood of new socialisation plans, Munich acquired at that time a new Central Planning Agency under the direction of a demagogue called from Austria, Dr Neurath. The Central Planning Agency had dozens of control centres, nationalisation and organisation centres and special councils without end: centres on the Russian model for socialisation of the press, of housing, of food, of clothes, of education for pleasures. But in the city, theft, robbery, and mugging increased alarmingly.[262]

Neurath would reply that one cannot socialise without such a structure. It is necessary if society is to accomplish the dual tasks of creating detailed workable plans while ensuring overall co-ordination. Two issues need to be kept separate: the question of whether Bavaria should have been attempting to socialise at this time, versus the question of the level of detail and the actual structure of whatever socialisation plan was to be undertaken.

Despite Neurath's commitment to the creation of large-scale, long-term structures, when he was convinced that he had to act immediately it seems he was prepared to do so, and often he was successful in what he did, as in his wartime work as a manager of provisions and rations and his period as a district administrator for Radzwillow. Joseph T. Simon (not the Simon of the Bavarian socialist government), whose father was a friend of Neurath's at the *Neue Handelsakademie* in Vienna, wrote with admiration of Neurath's efforts during the war. In Simon's view, Otto Neurath 'was the brain behind the care of the troops on the Italian front – his unequalled intelligence and powers of dedication allowed him to set up a regulated supply of food and munitions for the fighting troops from resources both scarce and difficult to obtain'.[263] This was of course in time of emergency. Socialisation was for the long run. That takes

---

[261] Neurath 1919d and 1919c.
[262] von Müller 1954, p. 320. Hillmayr 1974 also associates plundering during the food searches and while opening the safes with Neurath's naivety. [263] Simon 1979.

serious planning and planning in Neurath's view is always best carried out by those with detailed knowledge and actual experience.

This is true not only in the high level of abstraction of his general scheme for socialising but also in his more day-to-day decision-making. Let us consider, as one example, the heated question of the socialisation of the press discussed in section 1.3.2. Socialisation of the press came to a head under the Hoffman regime. Three days after Neurath was appointed, the Central Planning Agency called a meeting in the Ministry of Commerce, Trade and Industry of representatives of all levels and areas of press workers – writers, editors, printers, setters and so forth. Josef Simon was away and Neurath took charge. As we have seen, he stressed both then and at other times that there was no intention to limit freedom of expression in any way, but rather the converse.[264] The upshot of the meeting was a proposal for a law mandating a 'space-exchange' in which newspapers must give space to each party. At the time Neurath announced that he had no already-set plan. In the spirit of joint effort that characterised Neurath's career, he wanted the press to make its own arrangements – only if that failed would the government take a hand.[265] The result was typical of Neurath's efforts: not a plan for socialisation of the press, but rather, a committee to form a plan.

One of Neurath's first projects when he took over the Central Planning Agency, and probably his favourite, was the socialisation and reform of the polytechnics. This may seem a crazy place to start at a time when credit, food, supplies, raw materials, transportation and markets were all lacking. But his behaviour was not so fatuous, given his overall view. The construction of the new social order required that young technicians learn new subjects, subjects that would be of help in social planning. Neurath had in mind very practical subjects like workers' psychology or geography for the study of traffic patterns.[266]

The introduction of more experts with a better mix of practical and theoretical training for the kinds of concrete needs the sciences are supposed to serve will of course not remove the worries Popper had. The experts will still come to problems with different background assumptions and they will confront data that points in different directions. How are they to settle on a single programme? Popper wants a rule; Neurath wants more information, more discussion, more science;[267] but ulti-

---

[264] Of course not everyone believed in this intention. Cf. Stadtrat Max Gerstl, 1919.
[265] See *München-Augsburger Abendzeitung*, 147, 31 March 1919.
[266] *München-Augsburger Abendzeitung* 159, 7 April 1919; see also Dr Lindner's statement of 13 June in Neurath's trial proceedings.   [267] Cf. Neurath 1935b [1983], p. 123.

## Negotiation, not regulation 247

mately, a decision. You may feel you need a rule to bring yourself to a decision. Neurath proposed that you flip a coin. As we saw in section 2.4.2, this was Neurath's advice in his first published attacks on certainty: 'Today it is already of actual importance for the wise man who is conscious of the incompleteness of his insight, who refuses superstition, and who nevertheless wants to act decisively. Only the auxiliary motive can strengthen his will without demanding the sacrifice of his honesty.'[268] The auxiliary motive, explains Neurath,

> appears in its purest form as a drawing of lots. If a man is no longer able to decide on the basis of insight which of several actions to prefer, he can draw lots, or, equally well, declare vaguely that he will just do 'something or other', or that he will wait and see which resolution, after some hesitation will come out on top, as if leaving the decision to exhaustion, or at any rate to an agent quite outside the motive in question, that belongs to the category of the parrot who draws the 'planet'.[269]

It was regularly his advice about choosing between different social orders. Recall 'In general it is not possible to create an order of life that takes equal account of different views . . . Perhaps struggle will decide . . . perhaps . . . the choice will be made with the help of an inadequate metaphysical theory or in some other way. Tossing a coin would be much more honest.' Nothing will fix what we should do. There are no ultimate rules. There are only hard decisions and hard scientific tasks, not arbitrary, but reasoned. Yet the reasons themselves are supported only by more hard decisions and hard scientific work.

How then are we to resolve conflicts about what hypotheses to accept or what programmes to adopt? Many look for some Archimedean point outside the terms of the debate from which they can assure themselves that they are right and their opponents wrong.[270] They look for the concept of moral good or objective truth; God; a categorical imperative; or perhaps the more welfare-oriented utilitarian rule: 'Act so as to promote the greatest good for the greatest number.' Neurath made fun of these rules. Where is the voice of God? In his own writings he never urged anyone to be a Marxist or to be concerned with human welfare because it is right or just to do so. He framed his recommendations differently: '*One who loves Socialism* can see it realised only through the victory of the working class; *one who loves the working class* can catch sight of its victory only through the realisation of socialism.'[271] Philosophic views found the same expression. 'To one

---

[268] Neurath 1913b [1983], p. 10. [269] Ibid., pp. 4–5.
[270] Thanks to Harry Collins for help with this point. [271] Neurath 1922b, p. 84 our italics.

who holds the scientific attitude, statements are only means to predictions.'²⁷² Neurath praised Marx and Engels for a similar attitude:

> [W]hen Marx and Engels try to prove the assertion that the present order must perish and give way to one less full of suffering, they never argue with references to the 'injustice' of our order, or to the claim that the 'just cause' or 'truth' must win. They show strictly empirically how under certain conditions existing men behave or will behave. They know no ideal order, by which certain sufferings are judged to be disturbances.²⁷³

For Neurath when there is conflict, you persuade, you educate, you negotiate. This is no start for a more detailed or sophisticated philosophic position, but rather a programme for life; indeed a programme for Neurath's own life. Let us consider persuasion. Neurath intended to persuade and he often succeeded. Of all those tried for their work in the Bavarian revolution, Neurath was the only non-military defendant who was not a minister; he was technically a civil servant. Part of the explanation of why Neurath was pursued and tried lies in his strength as a public speaker. He was often accused of being a demagogue. J.M. Segitz, the Minister of the Interior under Hoffmann, is supposed to have said that he had never experienced anything like Neurath in forty years of political practice.²⁷⁴ It has also been claimed that Neurath won his case for setting up the Central Planning Office by threatening to appeal to the masses saying that within three days he would have 200,000 workers marching in the streets behind him.²⁷⁵ Schumann described Neurath's power in talking to workers' gatherings more favourably, but made the same point. At the time that the Kranold–Neurath–Schumann programme was formulated, 'Neurath began to lecture at mass meetings. He was extremely fascinating and made an enormous impression, especially in the mining areas of Southern Saxony where his lecturing tours were like triumphal processions.'²⁷⁶

---

²⁷² Neurath 1931c [1973], p. 326.
²⁷³ Neurath 1931c [1973], p. 346. This we think is the appropriate reply to Hegselmann's (1979) worries that Neurath's positivism is inconsistent with his drive to improve society. Neurath expects people to care about human misery, and he hopes to educate them to do so where they do not. In the end he thinks that without the blinders of metaphysics, enough people will do so that it will be possible to set up the new socialist economy. But he never makes the claim that one should do so because it is *right* or *good*. The fully planned economy will eliminate crises, and that is what we aim to do. It is not necessary to tack on the idle addendum 'and crises are wrong'. Neurath's attitude here reflects his repudiation of the entire Kantian framework, in which he resembled Reichenbach. For both Neurath and Reichenbach the idea was repellent that we should prefer people who try to redistribute coal because it is their duty rather than because doing so alleviates human misery. (For Reichenbach's views, cf. his 1951.)
²⁷⁴ See Peter Kritzer 1969.   ²⁷⁵ Cf. Kritzer, ibid.   ²⁷⁶ Schumann 1973, p. 17.

Neurath similarly carried his views about education into his own life. He believed that a scientific education for the proletariat was crucial and he did his part in providing it. We have seen that his engagement in adult education for workers was central to his active political life in Red Vienna. Neurath taught at municipal adult-education institutes as well as at the party education institute of the Social Democratic Workers Party.[277] Even more well known was his work from 1925 to 1934 for the Social and Economic Museum in Vienna, where the Vienna method of picture statistics was invented and widely applied. As we have seen, the goal of the Vienna Method was to facilitate visually knowledge of socio-economic statistical facts; that is, to enable workers to comprehend the 'statistical age'. Neurath argued vigorously for the importance of workers' access to statistics. An early discussion is in his review of Wladimir Wohtinsky.[278] In this review Neurath praised Wohtinsky's attempt to make statistics more readable. It had faults, Neurath admitted, but was a milestone in the development of statistics for the proletariat. Neurath stressed that the proletariat needs to learn statistics, both to mount their own political and economic measures and to criticise the capitalist economic order. From 1925 onwards Neurath wrote often on this issue.[279]

Persuasion and education of course will not always produce agreement. What do we do then? There is a well-known answer, which Imre Lakatos often cited. One of these occasions was at a history and philosophy of science seminar in Cambridge: 'That is the point at which we get out our guns.' Richard Braithwaite was in the audience. Braithwaite sat up with a start and exclaimed, 'No. No. No. That is the point at which we retire to our gardens.'[280] Lakatos' solution would have been abhorrent to Neurath; in 1942 he wrote to Carnap that 'The most terrible thing would be if people got power to bully other people.'[281] But Braithwaite's solution will not do either. Social reform calls, after all, for common action. The option that remains is co-operation and compromise. Neurath knew that even before the Bavarian revolution:

> The application of the auxiliary motive [to choose one plan from among many] needs a prior high degree of organisation; only if the procedure is more or less common to all will the collapse of human society be prevented. The traditional uniformity of behaviour has to be replaced by conscious co-operation.[282]

---

[277] See Stadler 1989; for a discussion of Neurath's pedagogical activities in the context of the socialist politics in Red Vienna, see J. Cat, N. Cartwright and H. Chang, forthcoming.
[278] Neurath 1926.   [279] Neurath 1933e [1991].
[280] N.C. observed this incident, probably in 1971.
[281] Neurath to Carnap, 17 July 1942 RC 102-57-08 ASP.   [282] Neurath 1913b [1983], p. 10.

Nor had Neurath the illusion that the need for co-operation and compromise would come to an end. Neurath's friend when he moved to Oxford, the English socialist G.D.H. Cole, had claimed that a good system of production demanded good people. Neurath objects: 'We must fashion the new order in its requirements so that good and bad men can be mixed with each other in roughly unchanged measure.'[283]

The recognition of the need to work co-operatively with those with whom you disagree and the willingness to curb your own opinions in the service of the joint effort was a part of Neurath's character. One often-remarked sign of this is that Neurath wrote, sometimes by himself and sometimes with others, a number of the 'Vienna Circle' and 'Unity of Science' manifestos. What he wrote expressed a kind of compromise position often far from his own. Nevertheless the work went under his name.

In 1935 and 1936 Neurath urged this view on Popper as well. Popper quarrelled with the Vienna Circle when in Neurath's opinion they should have been joining together to fight larger, more dangerous enemies. Neurath urged: 'This overstressing of the littlest difference is certainly not good for co-operation.'[284] Popper replied that he had an aversion to all 'narrow scientific fraternities: scientific friends can and should objectively dispute'.[285] Neurath thought differently:

To the Fraternity, in my opinion, belongs the NET DISCUSSION. One fights over what one cannot previously unite on. But one should not carry on the NET DISCUSSION in public. Really. That is why many of us place our papers with each other before publishing them. That helps eliminate many superficial differences. That is community in a good sense – it seems to me.[286]

The point mattered deeply to Neurath and six years later it generated a quarrel that almost fractured his friendship with Carnap. Carnap removed himself from editorial responsibility for Neurath's contribution to the 'Foundations of the Unity of Science' series, which they were jointly editing with Morris. Neurath was deeply hurt. He could not understand why. 'Withdrawing a name is so serious an action, that I assume that you can tell me what detrimental effects for our movement, readers, mankind as a whole you expected from your name being on that paper.'[287]

[283] Neurath 1922c, p. 337.
[284] Popper-Neurath correspondence, 24 August 1935, Vienna Circle Archives.
[285] Popper-Neurath correspondence, 7 October 1935, Vienna Circle Archives.
[286] Popper-Neurath correspondence, 4 February 1936, Vienna Circle Archives.
[287] Neurath to Carnap, 24 September 1945, RC 102-55-14 ASP.

Neurath's commitment to compromise was clear throughout his active political life. During the period of the Bavarian revolution he carried on with his socialisation programme under three different regimes and in no case was he in agreement with the majority of any. And outside his official position Neurath made personal efforts to lessen antagonism, to co-operate, to negotiate. He knew that one could not socialise industrial Saxony without an agrarian base, or primarily agrarian Bavaria by itself, and he and Schumann went back and forth between Munich and Dresden trying to get some joint plan accepted.[288]

Neurath knew too that Munich could not carry on if it were isolated from the countryside. But the farmers tended to be staunchly Catholic and conservative, opposed to socialist ideas. Neurath's work with Sebastian Schlittenbauer, the head of the conservative Bavarian People's Party, which was described in section 1.3, was undertaken entirely on his own without instructions from any party. Together they devised a non-coercive plan for uniting the farmers. On 24 March Neurath spoke at a meeting of the party. Schlittenbauer introduced him laughingly; how surprised his colleagues must be to find themselves addressed by a Communist.[289] The evening edition of the *München-Augsburger Abendzeitung* on 26 March records a report from Dr Schlittenbauer at a meeting called by Simon that the Bavarian People's Party was at that time in agreement with Neurath. But the coalition was shaky and with damaging consequences, especially for the first Soviet Republic. According to Mühsam, 'The meetings of the Central Council were mostly taken up with useless discussion and much time was lost over the resistance of the farmers to Neurath's designs for socialisation.'[290] Their earlier agreement to go along with Neurath depended on very restrictive conditions (for instance that only very large farms would be socialised). On Neurath's own account Schlittenbauer made a statement much later endorsing an economic dictatorship and a planned economy.[291] But by then the socialist government was long dead.

The list of Neurath's efforts for co-operation goes on. Christians should not be excluded from socialist concerns;[292] the unions must stop fighting among themselves;[293] it is crucial to work quickly in the new republic to organise the councils so that they do not compete among themselves and destroy the possibility of full socialisation. And so forth. We need only recall Neurath's remark from section 2.5.1 about avoiding

[288] Neurath 1920b.   [289] Ibid.   [290] Mühsam 1929, p. 60.   [291] Neurath 1920b.
[292] Ibid., and Neurath 1931d.   [293] Neurath 1920f.

offending the churches to summarise: 'I spend my life on the most different levels of neutrality at the same time.'

The general topics of this section are the familiars of philosophic life, though philosophers are apt to put the problems more abstractly. How can we reason about how to reason? How can we rationally choose the terms and rules of the discussion? How can we remove the tormenting suspicion that the task cannot be achieved and our entire structure of claims and defences rests not on reason but on arbitrary choice? These are the traditional questions of epistemology. Classical philosophy offered Neurath transcendental arguments in answer. Neurath would have no truck with transcendental argument, nor with epistemology. But he did not dismiss the questions, for they are questions that matter for our common lives together. How could one leave a matter of such importance to pure reason and abstract argument? We begin with a philosophical problem: how are choices to be made when claims compete? We are given a practical answer: persuade, educate, negotiate; in the end, decide and act. The point is, it is an answer. Neurath worked throughout his life to show that it would work; indeed must work, for it is all we have. We must think about our problems, negotiate with our opponents, make our choices and see to it that they are carried through.

# Conclusion

'There is no scientific method. There are only scientific methods. And each of these is fragile; replaceable, indeed destined for replacement; contested from decade to decade, from discipline to discipline, even from lab to lab.' We are apt to take these as special insights of our own postmodern period, the final rejection of the foundationalist myth of a clean, clear and certain scientific edifice which was spawned by the Positivists and nurtured by the Vienna Circle. In this book we have shown how Otto Neurath, working from the heart of Logical Positivism in the first and second Vienna Circles, launched an all-out attack on foundationalism and the myth of the mechanical method. There are, he declared, no unrevisable, incorrigible or undeniable givens. Neither are there guaranteed means of epistemic ascent, nor methods that can get us from what we already accept to new truths or to new falsifications without the need for on-the-spot judgements. Neurath made a radical break with what since has been called the 'spectator view of knowledge', the view that knowledge is the reflection of independent reality and that truth consists in some correspondence relation between signifier and the object signified.

To understand Neurath's distinctive alternative it is important to note again the peculiar nature of his self-confessed 'scientism'.[1] It excelled in the scientific *attitude*. The scientific attitude stands to scientific method as the scientific world-conception stands to philosophical world-views: it

> always starts from the individual which it joins with what is similar into greater, and clearly surveyable complexes. It recognises no 'world' as a whole, it does not aim at comprehending a mighty world-picture in its totality, at a world-view. If one speaks of a scientific world-'conception' in contradistinction to a philosophical world-'view', 'world' is not to indicate a definitive whole, but the daily growing sphere of science. This conception is derived from individual research work, which one wants to incorporate into a unified science.[2]

[1] Neurath 1935a [1983], p. 115.   [2] Neurath 1930a [1983], pp. 32–3.

Deeply aware of the need for abstract concepts and logical generalisations in scientific theorising, the scientific world conception nevertheless embraces the concrete – it 'absorbs everything that can be experienced'[3] – and abjures a priori starting points. From the point of view of the scientific attitude, the scientific world conception is something we construct.

Neurath's conception of science and its metatheory places us squarely, if oddly, in the field of 'post-modern' debate. Yet in one respect Neurath remained resolutely modernist. Is the convergence of his reasoning with ideas forwarded by contemporary theorists of science undercut by his Enlightenment orientation? Did he not 'buy into' the metanarrative of emancipation over the course of history? That depends on how one thinks of continuing the 'unfinished project' of Enlightenment. As Habermas describes it:

> The project of modernity formulated in the 18th century by the philosophers of the Enlightenment consisted in their efforts to develop objective science, universal morality and law, and autonomous art according to their inner logic. At the same time, this project intended to release the cognitive potentials of each of these domains from their esoteric forms. The Enlightenment philosophers wanted to utilise this accumulation of specialised culture for the enrichment of everyday life – that is to say, for the rational organisation of everyday social life.[4]

That was then. What about Enlightenment now? Having lately discovered the need for 'post-metaphysical' thinking, critical theorists like Habermas deem themselves to have advanced beyond anything the Vienna Circle could conceive of.[5] But recall Neurath's long-held view: for reason to fulfil its Enlightenment promise it has to be *reconceptualised*.[6] Moreover, if 'post-modernism' represents not a blanket denouncement of rationality but rather a programme to seek a new and better understanding of what reason is and can do, then Neurath's drive for rationality without foundations is not at odds with it either. Neurath's progressivism did not recapitulate in thought the order nature planned. Nor did his engagement for emancipation derive from the objective telos of history. It derived from decision and from action, consciously planned.

Still, Neurath's belief in proletarian solidarity, internationalist socialism and total socialisation is sustained by one residual universalism that

---

[3] Ibid., p. 32.   [4] Habermas 1981 [1983], p. 9.   [5] Habermas 1988, pp. 14, 35.
[6] Encouragingly, this longstanding conviction of Neurath converges with the insight Wellmer 1985 and Schnädelbach 1979 [1987] retrieve from Horkheimer and Adorno's later *Dialectic of Enlightenment* (1947), written during the Second World War.

## Conclusion

was never given up throughout the variations in his image of knowledge. It is always 'one' boat in which we all sit. Perhaps we should follow Hilary Putnam's suggestion to change the metaphor to a *fleet* of boats?

The people in each boat are trying to reconstruct their own boat without modifying it so much at any one time that the boat sinks, as in Neurath's image. In addition, people are passing supplies and tools from one boat to another and shouting advice and encouragement (or discouragement) to each other. Finally, people sometimes decide they don't like the boat they're in and move to a different boat altogether. (And sometimes a boat sinks or is abandoned.) It's all a bit chaotic; but since it is a fleet, no one is ever totally out of signalling distance from all the other boats. There is, in short, both collectivity and individual responsibility.[7]

Note that Putnam's suggestion that 'the whole of culture' be 'put . . . in the boat'[8] was anticipated by Neurath himself, as the context of his last Boat makes clear. Nor, given his own views about the social nature of language, can Putnam claim these individual boats to be self-sufficient. So his pluralism cannot exceed Neurath's, unless he were to speak of different fleets – which raises the very spectre of relativism that he so vigorously denounced elsewhere at the time. Nevertheless two things distinctive to Neurath are lost in Putnam's version of the Boat. Not only does Putnam substitute the new pathos of individualism for the old idea of collectivism. But also his fleet of little boats is more reminiscent of an afternoon on the Charles River than (to use Carnap's allusion to Neurath's metaphor) 'the boundless ocean of unlimited possibilities',[9] to say nothing of the turbulent seas across which Neurath in 1919 tried to 'steer full sail'.[10]

That it is indeed the *same* boat in which we *all* sit is important for Neurath's 'anti-philosophy'. Neurath insisted it is *one* not due to unacknowledged unversalist designs nor to the demands of conversational convenience:

The unity we have before us, as a goal for the encyclopedism of logical empiricism, is based on the actual store of expressions which people have in common all over the world. Its evolution would be based on conventions which could never be definite or authoritative as far as the aspirations of conscientious logical empiricists are concerned. Pluralism is the aura of this scientific world community of the common man. The encyclopedism of logical empiricism . . . competes with no philosophy, and is anti-totalitarian through and through.[11]

---

[7] Putnam 1981 [1983], p. 204.  [8] Ibid.  [9] Carnap 1934 [1937], p. xv.
[10] Neurath 1919b [1973], p. 155.  [11] Neurath 1946 [1983], p. 242.

For Neurath we are all in the same boat because it is one world in which we act. As Marx taught, we must participate in renewal. Conscious efforts to transform society, Neurath insisted, are not contradicted by the insight that 'both the willed and the willing are part of the development':[12]

> Germany is ravaged by civil war. Hunger, epidemics, and murder are at work, the Horsemen of the Apocalypse. Who can resist them? Will and cognition.[13]

But not the will and cognition of anyone alone; we must will and act together:

> Through the November revolution the urban proletariat and its friends found themselves with much power. They lacked, however, a clear picture of the future which could have guided the will.[14]

The Horsemen of the Apocalypse ravaged not only the Central Europe of the 1920s, 1930s and 1940s. Everywhere will and cognition are required so that the world will 'stand before us as a whole again'. But as Neurath learnt – and as we can learn from him – that world is unified only in action.

---

[12] Neurath 1920a, pp. 44–5.    [13] Ibid, p. 44.    [14] Ibid, p. 45.

# References

## UNPUBLISHED SOURCES

Staatsarchiv München, Akte no. I 2139 [State Archives Munich], concerning the trial against Otto Neurath at the Standgericht München, 1919
Letter, Otto Neurath to Josef Frank, The Hague, 8 July 1939 (in possession of Marie Neurath)
Letter, Paul Neurath to Lola Fleck, New York, 9 April 1978
Letter, Österreichisches Staatsarchiv, Abteilung Kriegsarchiv [War Archive] to Lola Fleck, Vienna, 28 November 1978
Letter, Universitätsarchiv Heidelberg [University Archive] to Lola Fleck, 1 June 1977
Conversation, Lola Fleck with Heinrich Neider, Vienna, December 1975
Conversations, Lola Fleck with Marie Neurath, Rechberg near Graz, June and September 1977
Conversation, Lola Fleck with Franz Rauscher, Vienna, December 1977
Conversation, Lola Fleck with Grete Schütte-Lihotzky, Vienna, February 1977
Correspondence, Rudolf Carnap, Archives of Scientific Philosophy (ASP), Dept of Special Collections, Hillman Library, University of Pittsburgh
Correspondence, Otto Neurath, Vienna Circle Archives, Rijksarchief Noord-Holland, Haarlem
Correspondence, Hans Reichenbach, Archives of Scientific Philosophy (ASP), Dept. of Special Collections, Hillman Library, University of Pittsburgh
Correspondence, Ferdinand Tönnies, Tönnies Nachlaß, Schleswig-Holsteinische Landesbibliothek, Kiel

## PUBLISHED MATERIAL

Adler, Friedrich, 1918, *Ernst Machs Überwindung des mechanischen Materialismus*, Vienna: Wiener Volksbuchhandlung
Adler, Max, 1904, *Kausalität und Teleologie im Streite um die Wissenschaft*, Marx-Studien 1, pp. 195–433
   1910a, *Der Sozialismus und die Intellektuellen*, Vienna: Wiener Volksbuchhandlung
   1910b, 'Marxismus und Materialismus', *Der Kampf*, 3
   1913, *Marxische Probleme: Beiträge zur Theorie der materialistischen Geschichtsauffassung und Dialektik*, Stuttgart: Dietz 1913

1930-2, *Lehrbuch der materialistischen Geschichtsauffassung*, 2 vols., Berlin: Laub; rev. edn *Soziologie des Marxismus*, 3 vols., Vienna: Europa Verlag, 1964

Ariew, Roger, 1984, 'The Duhem Thesis', *The British Journal for the Philosophy of Science* 35, pp. 313-25

Atlanticus (pseud.), 1898, *Produktion und Konsum im Sozialstaat*, Stuttgart: Dietz; 2nd rev. edn K. Ballod, *Der Zukunftsstaat – Produktion und Konsum im Sozialstaat*, 1919; 3rd rev. edn 1920

Avenarius, Richard, 1892, *Der menschliche Weltbegriff*, Leipzig: Reisland, 4th enlarged edn. 1927

Ay, E., 1969, 'Zur Rolle Otto Neuraths in der Novemberrevolution', *Wissenschaftliche Zeitschrift der Universität Greifswald, Gesellschafts- und sprachwissenschaftliche Reihe* 18, pp. 259-64

Ayer, Alfred J., ed., 1959, *Logical Positivism*, New York: Free Press

Ballod, Karl (*see* Atlanticus)

Barrett, R.B. and R.F. Gibson, eds., 1990, *Perspectives on Quine*, Oxford: Blackwell

Bauer, Helene, 1923, 'Geld, Sozialismus und Otto Neurath', *Der Kampf* 16, pp. 195-202

Bauer, Otto, 1907a, *Die Nationalitätenfrage und die Sozialdemokratie*, Marxstudien 2, Vienna: Wiener Volksbuchhandlung

1907b, 'Bücherschau', *Der Kampf* 1 (March issue)

1907c, 'Die Arbeiterbibliothek', *Der Kampf* 1 (October issue)

1910, 'Eine Parteischule für Deutschösterreich', *Der Kampf* 4 (January issue)

1919, *Der Weg zum Sozialismus*, Vienna: Wiener Volksbuchhandlung, repr. in Bauer, *Werkausgabe*, vol. II, Vienna, 1976

1920, 'Die Sozialisierungsaktion im ersten Jahre der Republik', Vienna: Wiener Volksbuchhandlung, repr. in Bauer, *Werkausgabe*, vol. II, Vienna, 1976

1923, *Die österreichische Revolution*, Vienna: Wiener Volksbuchhandlung; abridged trans. *The Austrian Revolution*, New York: Burt Franklin, 1925

1924, 'Das Weltbild des Kapitalismus', in O. Jenssen, ed., *Der lebendige Marxismus. Festgabe zum 70. Geburtstag von Karl Kautsky*, Jena: Thüringer Verlagsanstalt, pp. 407-64, excerpts trans. by P. Goode as 'The World View of Organized Capitalism' in Bottomore and Goode 1978, pp. 208-17

Baumgarten, Eduard, ed., 1964., *Max Weber. Werk und Person*, Tübingen: Mohr

Beck, Hermann, 1919, *Sozialisierung als organisatorische Aufgabe*, repr. from *Wege und Ziele der Sozialisierung. Protokoll der 1. sozialistischen Wirtschaftskonferenz des Bundes Neues Vaterland 27. 12. 1918-2.1.1919*, Berlin

Belke, Ingrid, 1978, *Die sozialreformerischen Ideen von Josef Popper-Lynkeus (1838-1921) im Zusammenhang mit allgemeinen Reformbestrebungen des Wiener Bürgertums um die Jahrhundertwende*, Tübingen: Mohr

Berghel, H., A. Hübner, and E. Köhler, eds., 1979, *Wittgenstein, der Wiener Kreis under der kritische Rationalismus*, Vienna: Hölder-Pichler-Tempsky

Bergmann, Hugo, 1971, 'Itelson, Gregorius', in *Encyclopedia Judaica*, vol. IX, Jerusalem: Keter Publishing, p. 1147

Bergström, Lars, 1982, 'Interpersonal Utility Comparisons', in Haller 1982b, pp. 282-312

Bernstein, Eduard, 1897, *Die Voraussetzungen des Sozialismus und die Aufgaben der Sozialdemokratie*, Stuttgart: Dietz, excerpts trans. as *Evolutionary Socialism*, repr. New York: Schocken, 1961
Beyer, Hans, 1957, *Von der Novemberrevolution zur Räterepublik in München*, Berlin: Rütten und Loening
  1982, *Die Revolution in Bayern 1918/1919*, Berlin: Deutscher Verlag der Wissenschaften
Beyer, Wilhelm Raymund, 1983, 'Otto Neuraths allgemeiner Arbeitgeber', in P. Damerow, P. Furth and W. Lefèbre, eds., *Arbeit und Philosophie*, Bochum: Germinal, pp. 231–7
Bickel, Cornelius, 1991, *Ferdinand Tönnies. Soziologie als skeptische Aufklärung zwischen Historismus und Rationalismus*, Opladen: Westdeutscher Verlag
Blackmore, John, ed., 1992, *Ernst Mach – A Deeper Look*, Dordrecht: Kluwer
Blum, M., 1985, *The Austro-Marxists (1890–1918): A Psychological Study*, Lexington, KY: University of Kentucky Press
Blumenberg, Hans, 1979, *Schiffbruch mit Zuschauer. Paradigma einer Daseinsmetapher*, Frankfurt: Suhrkamp
  1987, *Die Sorge geht über den Fluß*, Frankfurt: Suhrkamp
Boltzmann, Ludwig, 1905, 'Über eine These Schopenhauers', in *Populäre Schriften*, Leipzig: Barth, pp. 385–402, trans. by P. Foulkes as 'On a Thesis of Schopenhauer's', in *Theoretical Physics and Philosophical Problems*, ed. by B. McGuinness, Dordrecht: Reidel, 1974, pp. 185–98
Bool, Flip, 1982, 'Figurativer Konstruktivismus und Kritische Grafik von 1924 bis 1971', in Stadler 1982a, pp. 219–26
Born, Max and Albert Einstein, 1971, *The Born–Einstein Letters*, ed. and trans. by Irene Born, London: Macmillan 1971
Bottomore, Tom, 1978, 'Introduction', in Bottomore and Goode 1978, pp. 1–44
Bottomore, Tom and P. Goode, eds., 1978, *Austro-Marxism*, Oxford: Clarendon Press
Botz, G., H. Hautmann, H. Konrad, and J. Weidenholzer, eds. 1978, *Bewegung und Klasse. Studien zur österreichischen Arbeitergeschichte*, Vienna
Brentano, Lujo, 1931, *Mein Leben*, Jena: Eugen Diederichs
Broos, Kees, 1982, 'Bildstatistik Wien-Moskau-Den Haag von 1928 bis 1965', in Stadler 1982a, pp. 214–18
Buek, Otto, 1926, 'Gregorius Itelson †', *Kant Studien* 31, pp. 428–30
Cahnmann, Werner J., ed., 1973, *Ferdinand Tönnies. A New Evaluation*, Leiden: Brill
Capek, Milič, 1968, 'Ernst Mach's Biological Theory of Knowledge', *Synthese* 29, 171–91, repr. in R.S. Cohen and R.J. Seeger, eds., 1970, *Ernst Mach: Physicist and Philosopher*, Dordrecht: Reidel, 1970
Carnap, Rudolf, 1928, *Der logische Aufbau der Welt*, Berlin: Bernary, trans. by R.A. George as *The Logical Structure of the World*, Berkeley: University of California Press, 1967
  1930a, 'Die alte und die neue Logik', *Erkenntnis* 1, pp. 12–26, trans. by I. Levi, 'The Old and the New Logic', in Ayer 1959, pp. 133–46

1930b, 'Einheitswissenschaft auf physischer Basis', *Erkenntnis* 1, p. 77

1932a, 'Die physikalische Sprache als Universalsprache der Wissenschaft', *Erkenntnis* 2, pp. 432–65, rev. edn. trans. by M. Black as *The Unity of Science*, London: Kegan, Paul, Trench Teubner and Co., 1934

1932b, 'Psychologie in physikalischer Sprache', *Erkenntnis* 3, pp. 107–42, trans. by G. Schick as 'Psychology in Physicalist Language', in Ayer 1959, pp. 165–98

1932c, 'Über Protokollsätze', *Erkenntnis* 3, pp. 215–28, trans. by R. Creath and R. Nollan as 'On Protocol Sentences' in *Nous* 21 (1987), pp. 457–70

1934a, *Logische Syntax der Sprache*, Vienna: Springer, rev. edn. trans. by A. Smeaton, as *The Logical Syntax of Language*, London: Kegan, Paul, Trench Teubner and Co, 1937, repr. Paterson, NJ: Littlefield, Adams and Co., 1959

1934b, 'Theoretische Fragen und praktische Entscheidungen', *Natur und Geist* 2, pp. 257–60, repr. in H. Schleichert, ed., *Logischer Empirismus – Der Wiener Kreis*, Munich: Fink, 1975, pp. 173–6

1936/7, 'Testability and Meaning', *Philosophy of Science* 3, pp. 419–71, 4, pp. 1–40, rev. edn. repr. New Haven: Yale Graduate Philosophy Club, 1954

1961, 'Vorwort zur zweiten Auflage', *Der logische Aufbau der Welt*, Hamburg: Meiner, trans. by R.A. George as 'Preface to 2nd edition' in trans. of Carnap 1928, pp. v–xi

1963, 'Intellectual Autobiography' and 'Comments and Replies', in Schilpp 1963, pp. 3–84, 859–1016

Carnap, Rudolf, Hans Hahn and Otto Neurath, 1929, *Wissenschaftliche Weltauffassung – Der Wiener Kreis*, Vienna: Wolf, repr. in Neurath 1981, pp. 299–336, trans. in Neurath 1973, pp. 299–318

Cartwright, Nancy, 1983, *How the Laws of Physics Lie*, Oxford: Oxford University Press

1994, 'Mill and Menger: Ideal Elements and Stable Tendencies,' *Poznan Studies in the Philosophy of Science and the Humanities*, 38, pp. 171–88

Cat, Jordi, forthcoming, 'The Popper-Neurath Debate and Neurath's Attack on Scientific Method', in *Studies in the History and Philosophy of Science*

Cat, Jordi, Nancy Cartwright and Hasok Chang, 1991, 'Otto Neurath: Unification as the Way to Socialism', in J. Mittelstrass, ed., *Einheit der Wissenschaften*, Berlin: de Gruyter

forthcoming, 'Otto Neurath: Politics and Unity of Science', in P. Galison and D. Stump, eds., *The Disunity of Science*, Stanford: Stanford University Press

Chaloupek, Günther K., 1990, 'The Austrian Debate on Economic Calculation in a Socialist Economy', *History of Political Economy* 22, pp. 659–75

Cherniak, Christopher, 1986, *Minimal Rationality*, Cambridge, MA.: MIT Press

Coffa, Alberto, 1985, 'Idealism and the Aufbau', in N. Rescher, ed., *The Heritage of Logical Positivism*, New York: University Press of America, pp. 133–56

1991, *The Semantic Tradition from Kant to Carnap. To the Vienna Station*, ed. by L. Wessels, Cambridge: Cambridge University Press

Cohen, Robert S., 1963, 'Dialectical Materialism and Carnap's Logical Empiricism', in Schilpp 1963, pp. 99–158

1967, 'Neurath, Otto', in P. Edwards ed., *Encyclopedia of Philosophy*, New York: Macmillan and Free Press, vol. VI, pp. 477–9
1968, 'Ernst Mach: Physics, Perception and the Philosophy of Science', *Synthese* 18, pp. 132–79, repr. in Cohen and Seeger 1970
Cohen, Robert S. and R.J. Seeger, 1970, eds., *Ernst Mach: Physicist and Philosopher*, Dordrecht: Reidel
Comte, Auguste, 1864 *Cours de Philosophie Positive*, Paris: J.B. Baillier et Fils, repr. 1853 as *The Positive Philosophy of Auguste Comte*, trans. by Harriet Martineau, London: John Chapman
Creath, Richard, 1990, 'Carnap, Quine and the Rejection of Intuition', in Barrett and Gibson 1990, pp. 55–66
D'Acconti, Alessandra, 1986, 'Note Sull'Olismo Neurathiano', *Annali della Facoltá di Lettere e Filosofia, Universitá di Bari*, 29, pp. 409–41
Dahme, Heinz-Jürgen and Otthein Rammstedt, eds., 1984, *Georg Simmel und die Moderne. Neue Interpretationen und Materialien*, Frankfurt: Suhrkamp
Dahms, Hans-Joachim, 1985, ed., *Philosophie, Wissenschaft, Aufklärung*, Berlin: de Gruyter
  1990, 'Die Vorgeschichte des Positivismusstreits: von der Kooperation zur Konfrontation. Die Beziehungen zwischen Frankfurter Schule und Wiener Kreis 1936–1942', in *Jahrbuch für Soziologiegeschichte 1990*, ed. by H.-J. Dahme, C. Klingemann, M. Neumann et al., Opladen: Leske and Budrich, pp. 9–79
  1994, *Positivismusstreit*, Frankfurt: Suhrkamp
Dahms, Hans-Joachim and Michael Neumann, 1994, 'Sozialwissenschaftler und Philosophen in der Münchner Räterepublik', in *Jahrbuch für Soziologiegeschichte 1992*, ed. by C. Klingemann, M. Neumann, K.-S. Rehberg et al., Opladen: Leske und Budrich, pp. 115–46
D'Alembert, J., 1929, *Discours préliminaire de l'Encyclopédie*, Paris: Colin
Dewey, John, 1938, 'Unity of Science as Social Problem', in Neurath, Bohr, Dewey *et al.* 1938, pp. 29–38
  1939, *Theory of Valuation* (International Encyclopedia of Unified Science, vol. II, no. 4), Chicago: University of Chicago Press
Dorst, Tankred, ed., 1969, *Die Münchner Räterepublik. Zeugnisse und Kommentar*, Frankfurt: Suhrkamp
Du Bois-Reymond, Emil, 1872, *Die Grenzen des Naturerkennens*, 6th edn, Leipzig: von Veit and Co., 1884
Duhem, Pierre, 1906, *La Théorie Physique: Son Objet Sa Structure*, 2nd edn 1914, trans. by P. Wiener as *The Aim and Structure of Physical Theory*, 1954, Princeton: Princeton University Press, 2nd edn. 1962, New York: Atheneum
Durkheim, Emile, 1897, Review of French trans. of Labriola 1897, *Revue Philosophique* 44, pp. 645–51, repr. and trans. by W.D. Halls as 'Marxism and Sociology: The Materialist Conception of History', *Durkheim: The Rules of Sociological Method*, 1982, ed. Steven Lukes, London: Macmillan
Dvorak, Johann, 1982, 'Otto Neurath und die Volksbildung – Einheit der

Wissenschaft, Materialismus und umfassende Aufklärung', in Stadler 1982a, pp. 149–56, trans. by T.E. Uebel as 'Otto Neurath and Adult Education – Unity of Science, Materialism, and Comprehensive Enlightenment', in Uebel 1991a, pp. 265–74

Eisner, Kurt, 1919, 'Das Regierungsprogramm', in K. Eisner, *Die Neue Zeit* 1st series, Munich: Georg Müller, pp. 20–9

Elkana, Yehuda, 1981, 'A Programmatic Attempt at an Anthropology of Knowledge', in E. Mendelsohn and Y. Elkana, eds., *Sciences and Cultures. Sociology of Science Yearbook* 5, Dordrecht: Reidel, pp. 1–76

Ellenbogen, Wilhelm, 1981, *Menschen und Prinzipien. Erinnerungen Urteile und Reflexionen eines Kritischen Sozialdemokraten*, ed. by F. Weissenstein, Vienna: Böhlau, 1981

Elsenhans, Th., ed., 1909, *Bericht über den 3. Internationalen Kongress für Philosophie zu Heidelberg, 1–5 September 1908*, Heidelberg: Carl Winters Universitätsbuchhandlung

Engels, Friedrich, 1886, *Ludwig Feuerbach under der Ausgang der klassischen deutschen Philosophie*, trans. as *Ludwig Feuerbach and the Outcome of Classical German Philosophy*, New York: International Publ., 1935

Eschbach, Achim, 1988, 'Karl Bühler und Ludwig Wittgenstein', in A. Eschbach, ed., *Karl Bühler's Theory of Language*, Amsterdam: John Benjamins, pp. 385–407

Farrell, James T., 1959, 'Contribution', in Lamont 1959, p. 13

Feldbauer, Peter and Wolfgang Hösl, 1978, 'Die Wohnverhältnisse der Wiener Unterschichten und die Anfänge des genossenschaftlichen Wohn- und Siedlungswesen', in Botz, Hautsmann, Konrad and Weidenholzer, 1978

Fine, Arthur, 1993, 'Fictionalism', *Midwest Studies in Philosophy* 18, Notre Dame, IN: University of Notre Dame Press

Fischer, Heinz, 1968, ed., *Zu Wort gemeldet: Otto Bauer*, Vienna

Fleck, Christian, 1990, *Rund um Marienthal. Von den Anfängen der Soziologie in Österreich bis zu ihrer Vertreibung*, Vienna: Verlag für Gesellschaftskritik

Fleck, Karola (= Lola), 1979, 'Otto Neurath. Eine biographisch-systematische Untersuchung', Doctoral dissertation, Karl-Franzens-Universität Graz, rev. edn trans. in this volume

1982a, 'Otto Neurath's Beitrag zur Theorie der Sozialwissenschaften', in Stadler 1982a, pp. 100–3, trans. by T.E. Uebel as 'Otto Neurath's Contribution to the Theory of the Social Sciences', in Uebel 1991a, pp. 203–8

1982b, 'Anna Schapire', in Stadler 1982a, pp. 229–30

Frank, Philipp, 1917, 'Die Bedeutung der physikalischen Erkenntnistheorie Ernst Machs für das Geisteslebens unserer Zeit', *Die Naturwissenschaften* 5, pp. 65ff., trans. as 'The Importance for Our Times of Ernst Mach's Philosophy of Science', in Frank 1941b and Frank 1949b, pp. 61–79

1938, 'Ernst Mach – The Centenary of his Birth', *Erkenntnis* 7, pp. 247ff., trans. as 'Ernst Mach and the Unity of Science', in Frank 1941b and 1949b, pp. 79–89

1941a, 'Introduction: Historical Background', in Frank 1941b, pp. 3–16
1941b, *Between Physics and Philosophy*, Cambridge MA: Harvard University Press
1949a, 'Historical Introduction', in Frank 1949b, pp. 1–51
1949b, *Modern Science and its Philosophy*, Cambridge, MA: Harvard University Press
Frank, Wilhelm, 1931, 'Ist die marxistische Geschichtsauffassung materialistisch?', *Der Kampf* 24, pp. 163–6
Freud, Sigmund, 1900, *Die Traumdeutung*, Vienna, trans.: *The Interpretation of Dreams*, in *Standard Edition* 4–5, London: The Hogarth Press
Freudenthal, Gideon, 1989, 'Otto Neurath: From Authoritarian Liberalism to Empiricism', in M. Drascal and O. Gruengard, eds., *Knowledge and Politics*, Boulder: Westview, pp. 207–40
Freund, Julian, 1978, 'German Sociology in the Time of Max Weber', in *A History of Sociological Analysis*, ed. by T. Bottomore and R.A. Nisbet, New York: Basic Books, pp. 149–86
Friedman, Michael, 1987, 'Carnap's *Aufbau* Reconsidered', *Nous* 21, pp. 521–45
1992, 'Epistemology in the *Aufbau*', *Synthese* 93, pp. 15–58
Frisby, David, 1986, *Fragments of Modernity: Theories of Modernity in the Work of Simmel, Kracauer and Benjamin*, Cambridge, MA: MIT Press
Galison, Peter, 1990, 'Aufbau/Bauhaus: Logical Positivism and Architectural Modernism', *Critical Inquiry* 16, pp. 709–52
1993, 'The Cultural Meaning of *Aufbau*,' in Stadler 1993, pp. 75–94
Gerstl, Max, 1919, *Die Münchner Räterepublik*, Munich: Verlag politische Zeitfragen
Gesellschafts- und Wirtschaftsmuseum Wien, 1930, *Gesellschaft und Wirtschaft*, Leipzig: Bibliographisches Institut
Giedymin, Jerzy S., 1976, 'Duhem's Instrumentalism and its Critique: A Reappraisal', in *Essays in Memory of Imre Lakatos*, ed. by R.S. Cohen, P.K. Feyerabend, M.W. Wartofsky, repr. in J.S. Giedymin, *Science and Convention*, Oxford: Pergamon, 1982, pp. 90–108
Gillies, Donald, 1993, *Philosophy of Science in the Twentieth Century*, Blackwell: London
Glaser, Ernst, 1976, 'Otto Neurath und der Austromarxismus', *Die Zukunft* (April), pp. 16–22
Goodman, Nelson, 1951, *The Structure of Experience*, Cambridge, MA: Harvard University Press; 3rd edn. Dordrecht: Reidel, 1978
Gorges, Imela, 1980, *Sozialforschung in Deutschland 1872–1914: gesellschaftliche Einflüsse auf Themen- und Methodenwahl des Vereins für Sozialpolitik*, Berlin: Schriften des Wissenschaftszentrums; repr. Frankfurt: Hain, 1986
Greenaway, Frank, forthcoming, *Science Universal*
Grunberger, Richard, 1973, *Red Rising in Bavaria*, London: A. Barker
Gulick, Charles, 1948, *Austria from Habsburg to Hitler*, Berkeley and Los Angeles: University of California Press
Habermas, Jürgen, 1981, *Die Moderne – ein unvollendetes Projekt*, in *Kleine politische*

*Schriften IV*, Frankfurt: Suhrkamp, 1981, trans. by S. Ben-Habib: 'Modernity vs. Postmodernity', *New German Critique* 22 (1981), repr. in H. Foster, ed., *The Anti-Aesthetic*, Seattle: Bay Press, 1983

1988, *Nachmetaphysisches Denken*, Frankfurt: Suhrkamp, trans. by W.H. Hohengarten as *Postmetaphysical Thinking*, Cambridge, MA: MIT Press, 1992

Haller, Rudolf, 1966, 'Der Wiener Kreis und die analytische Philosophie', in F. Sauer, ed., *Forschung und Fortschritt*, Graz, pp. 33–46, repr. in Haller 1979b, pp. 79–98

1977, 'Österreichische Philosophie', *Conceptus* 28–30, pp. 56–66, repr. in Haller 1979b, pp. 5–22

1979a, 'Über Otto Neurath', in Haller 1979b, pp. 99–106, trans. by T.E. Uebel as 'On Otto Neurath', in Uebel 1991a, pp. 25–32

1979b, *Studien zur Österreichischen Philosophie*, Amsterdam and Atlanta, GA: Rodopi

1979c, 'Geschichte und wissenschaftliches System bei Otto Neurath', in Berghel, Hübner and Köhler, 1979, pp. 302–7, trans. by T.E. Uebel as 'History and the System of Science in Otto Neurath' in Uebel 1991a, pp. 33–40

1982a, 'Das Neurath-Prinzip', in Stadler, ed., 1982a, pp. 79–87, in Dahms, ed., 1985, pp. 205–20, repr. in Haller 1986b, pp. 108–24, trans. by T.E. Uebel as 'The Neurath Principle: Its Grounds and Consequences', in Uebel 1991a, pp. 117–30

1982b, ed., *Schlick und Neurath – ein Symposion, Grazer Philosophische Studien* 16/17

1985, 'Der erste Wiener Kreis', *Erkenntnis* 22, pp. 341–58, trans. by T.E. Uebel as 'The First Vienna Circle', in Uebel 1991a, pp. 95–108

1993, *Neopositivismus: Eine historische Einführung in die Philosophie des Wiener Kreises*, Darmstadt: Wissenschaftliche Buchgesellschaft

1995, 'Dirt and Crystal: Neurath on the Language of Science', in K. Gavroglu, J. Stachel and M. Wartofsky, eds., *Physics, Philosophy and the Scientific Community*, Dordrecht: Kluwer, pp. 287–300

Haller, Rudolf and Heiner Rutte, 1977, 'Gespräch mit Heinrich Neider, Wien: Persönliche Erinnerungen an den Wiener Kreis', *Conceptus* 11, nos. 28–30, pp. 21–40

Haller, Rudolf and Friedrich Stadler, 1988, eds., *Ernst Mach: Werk und Wirkung*, Vienna: Hölder-Pichler-Tempsky

Hansen, Reginald, 1968, 'Der Methodenstreit in den Sozialwissenschaften zwischen Gustav Schmoller und Karl Menger: Seine wissenschaftshistorische und wissenschaftstheoretische Bedeutung', in A. Diemer, ed., *Beiträge zur Entwicklung der Wissenschaftstheorie im 19. Jahrhundert*, Meisenheim a. Glan: Hain, pp. 137–73

Harding, Sarah G., ed., 1976, *Can Theories Be Refuted? Essays on the Duhem-Quine Thesis*, Dordrecht: Reidel

Hegselmann, Rainer, 1979, 'Otto Neurath – Empiristischer Aufklärer und Sozialreformer', in Neurath 1979, pp. 7–78

Heidelberger, Michael, 1985, 'Zerspaltung und Einheit: vom logischen Aufbau der Welt zum Physikalismus', in Dahms 1985a, pp. 144–89
Hempel, Carl Gustav, 1965, *Aspects of Scientific Explanation and Other Essays* New York: Free Press
   1982, 'Schlick und Neurath: Fundierung vs. Kohärenz in der Wissenschaftlichen Erkenntnis', in Haller 1982b, pp. 1–18
Hennis, Wilhelm, 1987, 'A Science of Man: Max Weber and the Political Economy of the German Historical School', in Mommsen and Osterhammel 1987, pp. 25–58
Herz, Rudolf and Dirk Halfbrodt, 1988, *Revolution und Fotographie*, Berlin: Nishen
Hillmayr, Heinrich, 1974, *Roter und weisser Terror in Bayern nach 1918*, Munich: Nusser
Hoffmann, Robert, 1978, 'Entproletarisierung durch Siedlung? Die Siedlerbewegung in Österreich 1918–1938', in Botz, Hautmann, Konrad and Weidenholzer et al. 1978, pp. 713–42
Hofmann-Grüneberg, Frank, 1988, *Radikal-empiristische Wahrheitstheorie. Eine Studie über Otto Neurath, den Wiener Kreis und das Wahrheitsproblem*, Vienna: Hölder-Pichler-Tempsky
Holton, Gerald, 1992, 'Ernst Mach and the Fortunes of Positivism in America', *Isis* 83, pp. 27–69, repr. in Holton, *Science and Anti-Science*, Cambridge, MA: Harvard University Press, 1994
Hookway, Christopher, 1988, *Quine*, Palo Alto: Stanford University Press
   1990, *Scepticism*, London: Routledge
Horkheimer, Max and Adorno, Theodor W., 1947, *Dialektik der Aufklärung*, Amsterdam, 7th edn, Frankfurt: Fischer, 1980, trans. by J. Cumming as *Dialectic of Enlightenment*, New York: Herder & Herder, 1972
Iggers, Georg G., 1969, *The German Conception of History*, Middletown, CT: Wesleyan University Press
Jacoby, E.G, 1971, *Die moderne Gesellschaft im sozialwissenschaftlichen Denken von Ferdinand Tönnies*, Stuttgart: Enke
Jay, Martin, 1973, *The Dialectical Imagination. A History of the Frankfurt School and the Institute of Social Research*, Boston: Little, Brown and Co.
Jerusalem, Wilhelm, 1909, 'Soziologie des Erkennens', *Die Zukunft* 15 May, repr. in W. Jerusalem, *Gedanken und Denker. Gesammelte Aufsätze, Neue Folge*, Vienna: Braunmüller, 1925, pp. 140–53
   1924, 'Die soziologische Bedingtheit des Denkens und der Denkformen', in M. Scheler, ed., *Versuche zu einer Soziologie des Wissens*, Leipzig: Duncker and Humblot, pp. 182–227, repr. in Meja and Stehr 1982, pp. 27–56
Johnston, William M., 1972, *The Austrian Mind. An Intellectual and Social History 1848–1938*, Berkeley: University of California Press
Käsler, Dirk, 1983, 'In Search of Respectability: The Controversy over the Destination of Sociology during the Conventions of the German Sociological Society', in *Knowledge and Society: Studies in the Sociology of Culture Past and Present*, vol. IV, New York: JAI Press, pp. 227–72

1984, *Die frühe deutsche Soziologie 1909 bis 1934 und ihre Entstehungsmilieus*, Opladen: Westdeutscher Verlag
Kaufmann, Felix, 1936, *Methodenlehre der Sozialwissenschaften*, Vienna: Springer
Kautsky, Karl, 1909, 'Ein Brief über Marx und Mach', *Der Kampf* 2, pp. 451–2
Keuzenkamp, Hugo A., 1994 'Probability, Econometrics and Truth, a Treatise on the Foundations of Econometric Inference', Dissertation, Center and Department of Econometrics, Tilburg University
Keynes, J.M., 1921, *A Treatise on Probability*, London: Macmillan
Kinross, Robin, 1979, 'Otto Neurath's Contribution to Visual Communication, 1925–1945', Dissertation, University of Reading
Kitching, Gavin, 1988, *Karl Marx and the Philosophy of Praxis*, London: Routledge
Kleene, Stephen C., 1939, 'Review: Carnap, *The Logical Syntax of Language*', *Journal of Symbolic Logic* 4, pp. 82–7
Klose, Olaf, Eduard Georg Jacoby and Irma Fischer, eds., 1961, *Ferdinand Tönnies, Friedrich Paulsen: Briefwechsel 1876–1908*, Kiel: Hirt
Köhler, Eckehart, 1982, 'Zu einigen Schriften von Otto Neurath über Logik, Ethik und Physik', in Stadler 1982a, pp. 107–11, trans. by T.E. Uebel 'On Neurath's Writings on Logic, Ethics and Physics' in Uebel 1991a, pp. 109–16
Kolakowski, L., 1978, *Main Currents of Marxism*, vol. II, Oxford: Oxford University Press
Koppelberg, Dirk, 1987, *Die Aufhebung der Analytischen Philosophie*, Frankfurt: Suhrkamp
1990, 'How and Why to Naturalize Epistemology', in Barrett and Gibson 1990, pp. 200–11
Korsch, Karl, 1919, *Was ist Sozialisierung?*, Hanover: Verlag Freies Deutschland, repr. in K. Korsch, *Schriften zur Sozialisierung*, ed. by E. Gerlach, Frankfurt: Europäische Verlagsanstalt, 1969, pp. 15–69
Kritzer, Peter, 1969, *Die bayerische Sozialdemokratie und die bayerische Politik in den Jahren 1918 bis 1923*, Munich: Neue Schriftenreihe des Stadtarchivs
Krüger, Dieter, 1987, 'Max Weber and the "Younger" Generation in the Verein für Sozialpolitik', in Mommsen and Osterhammel 1987, pp. 71–86
Kruntorad, Paul, ed., 1991, *Jour Fixe der Vernunft. Der Wiener Kreis und die Folgen*, Vienna: Hölder-Pichler-Tempsky
Labriola, Antonio, 1896, *Del materialismo storico. Dilucidazione preliminare*, Rome: Loescher, repr. in Labriola, *La concezione materialistica della storia*, Rome: Loescher, 1902, trans. by C.H. Kerr as 'Historical Materialism' in Labriola, *Essays on the Materialistic Conception of History*, Chicago: C.H. Kerr & Co, 1904
Lamont, Corliss, ed., 1959, *Dialogue on John Dewey*, New York: Horizon
Laplanche, J. and J-B. Pontalis, 1967, *Vocabulaire de la Psychanalyse*, Paris: Presses Universitaires de France, trans. D. Nicholson-Smith as *The Language of Psycho-Analysis*, London: Hogarth Press, 1973
Leichter, Otto, 1923, *Die Wirtschaftsordnung in der sozialistischen Gesellschaft*, Marx Studien 5.1, Vienna: Wiener Volksbuchhandlung
Lenin, Vladimir I., 1908, *Materialism and Empirio-Criticism*, trans. 1927, New York: International Publishers, repr. 1972

Lepenies, Wolf, 1985, *Die drei Kulturen*, Munich: Hanser, trans. by R.J. Hollingdale as *Between Literature and Science: The Rise of Sociology*, Cambridge: Cambridge University Press 1988

Lepsius, Rainer M., 1977, 'Max Weber in München', *Zeitschrift für Soziologie* 6, pp. 103–18

Leviné-Meyer, Rosa, 1973, *Leviné: The Life of a Revolutionary* Farnborough: Saxon House

Liebersohn, Harry, 1988, *Fate and Utopia in German Sociology 1870–1932*, Cambridge, MA.: MIT Press

Lindenlaub, Dieter, 1967, *Richtungskämpfe im Verein für Sozialpolitik*, *Vierteljahresschrift f. Sozial- u. Wirtschaftsgeschichte* Beiheft (monograph series) 52, Wiesbaden: Steiner

Luxemburg, Rosa, 1898, 'Sozialreform oder Revolution', repr. in R. Luxemburg, *Schriften zur Theorie der Spontaneität*, Reinbeck bei Hamburg, 1970

McGuinness, Brian, 1987, ed., *Unified Science*, Dordrecht: Reidel

Mach, Ernst, 1883, *Die Mechanik in ihrer Entwicklung*, Leipzig: Brockhaus, 4th edn. trans. by T.J. McCormick as *The Science of Mechanics*, 2nd edn. 1902, Chicago: Open Court

1886, *Beiträge zur Analyse der Empfindungen*, Jena: Fischer, trans. by C.M. Williams, rev. by S. Waterlow, as *The Analysis of Sensations*, New York: Dover, 1959

1905, *Erkenntnis und Irrtum*, Leipzig: Barth; 5th edn. 1926 trans. by T.J. McCormack and P. Foulkes: *Knowledge and Error*, Dordrecht: Reidel, 1976

Mandel, Ernest, 1969, *Marxist Economic Theory*, New York: Monthly Review Press

Martinez-Alier, Juan (with Klaus Schlüpmann) 1987, *Ecological Economics. Energy, Environment and Society*, Oxford: Basil Blackwell

Marx, Karl, 1845, 'Thesen zu Feuerbach', first publ. in F. Engels, 1886, *Ludwig Feuerbach und das Ende der klassischen deutschen Philosophie*, trans.: 'Theses on Feuerbach', in Marx and Engels 1978, pp. 143–5

Marx, Karl and Friedrich Engels, 1846, 'Die Deutsche Ideologie. Erster Teil: Feuerbach. Gegensatz von materialistischer und idealistischer Anschauung', ed. by D. Riazanov, *Marx-Engels Archiv* I (1926), pp. 233–307; rev. edn. in Marx and Engels, *Werke*, vol. III, trans. 'The German Ideology (Part I)', in Marx and Engels 1978, pp. 146–99

1848, *Manifest der Kommunistischen Partei*, repr. in Marx and Engels, *Werke*, vol. IV, trans. as 'Manifesto of the Communist Party', in Marx and Engels 1978, pp. 469–500

1934, *Correspondence, 1846–1895* ed. and trans. by Dona Torr, London: Martin Lawrence

1978, *The Marx–Engels Reader*, ed. by R.C. Tucker, 2nd edn., New York: Norton

Mauthner, Fritz, 1901, *Beiträge zu einer Kritik der Sprache*, vol. 1, Berlin, 2nd edn 1906, repr. Frankfurt: Ullstein 1982.

Meja, Volker and Nico Stehr, eds., 1982, *Der Streit um die Wissenssoziologie*, 2 vols., Frankfurt: Suhrkamp

Menger, Carl, 1883, *Untersuchung über die Methode der Socialwissenschaften und der Politischen Oekonomie insbesondere*, Vienna, trans. by F.J. Nock: *Problems of Economics and Sociology*, Urbana, IL.: University of Illinois Press, 1963

Mergner, Gottfried, ed., 1971, *Gruppe Internationaler Kommunisten Hollands*, Reinbeck bei Hamburg

Meyer, Eduard, 1907, *Geschichte des Altertums*, 2nd edn, Stuttgart and Berlin: J.G. Cotta'sche Buchhandlung Nachfolger

1910, *Kleine Schriften zur Geschichtstheorie und wirtschaftlichen und politischen Geschichte des Altertums*, Halle

Mill, John S., 1836, 'On the Definition of Political Economy,' in J.M.Robson, ed., *The Collected Works of John Stuart Mill*, Toronto: University of Toronto Press, vol. IV, 1967

Mitchell, Allan, 1965, *Revolution in Bavaria 1918–1919. The Eisner Regime and the Soviet Republic*, Princeton: Princeton University Press

Mohn, Erich, 1978, *Der Logische Positivismus. Theorien und politische Praxis seiner Vertreter*, Frankfurt: Campus

1985, 'Die politische Praxis Otto Neurath's während der Räterepublik in Bayern', in Dahms 1985, pp. 30–9

Mommsen, Wolfgang J. and Jürgen Osterhammel, 1987, eds., *Max Weber and his Contemporaries*, London: Unwin Hyman

Morentz, Ludwig and Erwin Münz, 1968, *Revolution und Räteherrschaft in München. Aus der Statdchronik 1918/19*, Munich

Morgan, Mary, 1990, *The History of Econometric Ideas*, Cambridge: Cambridge University Press

Mormann, Thomas, 1991, 'Neuraths Enzyklopädismus: Entwurf eines radikalen Empirizismus', *Journal for General Philosophy of Science* 22, pp. 73–100

Mühsam, Erich, 1929, *Von Eisner bis Leviné*, Berlin-Britz: Fanal

Müller, Karl H., 1982, 'Planwelten. Die ökonomischen Beiträge von Otto Neurath', in Stadler 1982a, pp. 104–6

1991a, *Symbole-Statistik-Computer-Design. Otto Neurath's Bildpädagogik im Computerzeitalter*, Vienna: Hölder-Pichler-Tempsky

1991b, 'Neurath's Theory of Pictorial-Statistical Representation', in Uebel 1991a, pp. 223–43

Müller, Richard, 1925, *Vom Kaiserreich zur Republik. Ein Beitrag zur Geschichte der revolutionären Arbeiterbewegung während des Weltkrieges*, Vienna, repr. Berlin, 1974

von Müller, Karl Alexander, 1954, *Mars und Venus. Erinnerungen 1914–1919*, Stuttgart: Gustav Klipper

Müller-Meiningen, Ernst, 1923, *Aus Bayerns schwersten Tagen*, Berlin/Leipzig: Vereinigung wissenschaftlicher Verleger

Musil, Robert, 1908, *Beitrag zur Beurteilung der Lehren Machs*, Berlin, repr. Reinbeck bei Hamburg: Rowohlt, 1980, trans. by K. Mulligan as *On Mach's Theories*, Munich: Philosophia, 1982

Nagel, Ernest, 1959, 'Contribution' in Lamont 1959, pp. 10–13
  1961, *The Structure of Science*, New York: Harcourt Brace and World
Neider, Heinrich, 1977, 'Gespräch mit Heinrich Neider', ed. by R. Haller and R. Rutte, *Conceptus* 28–30, pp. 21–42
Nemeth, Elisabeth, 1981, *Otto Neurath und der Wiener Kreis: Wissenschaftlichkeit als revolutionärer politischer Anspruch*, Frankfurt: Campus
  1982a, 'Die Einheit der Planwirtschaft und die Einheit der Wissenschaft', in Haller 1982b, pp. 437–49, trans. by T.E. Uebel as 'The Unity of Planned Economy and the Unity of Science', in Uebel 1991a, pp. 275–84
  1982b, 'Otto Neurath's Utopien – Der Wille zur Hoffnung', in Stadler, ed., 1982a, pp. 94–9, trans. by T.E. Uebel as 'Otto Neurath's Utopias – The Will to Hope', in Uebel 1991a, pp. 285–94
  1990, 'Wissenschaftliche Weltanschauung zwischen Sozialrevolution und Sozialreform', *Conceptus* 24 no. 61, pp. 73–90
  1991, 'Zur emanzipatorischen Perspektive in Otto Neurath's Begriff wissenschaftlichen Wissens', *Mesotes* 1, pp. 56–65
  1992, 'Nietzsche, Neurath und die Reflexivität der Moderne', in P. Muhr, P. Feyerabend, and C. Wegeler, eds., *Philosophie, Psychoanalyse, Emigration. Festschrift für Kurt Rudolf Fischer*, Vienna: Universitätsverlag, pp. 251–66
  1994, 'Empiricism and the Norms of Scientific Knowledge: Some Reflections on Otto Neurath and Pierre Bourdieu', in H. Pauer-Studer, ed., *Norms, Values and Society*, Vienna Circle Institute Yearbook 2, Dordrecht: Kluwer, pp. 23–32
Neumann, G., 1973, 'Military Life', in Neurath 1973, pp. 7–11
Neurath, Marie, 1973a, 'Preface', in Neurath 1973, p. xiii
  1973b, 'University Days', in Neurath 1973, p. 7
  1973c, 'Otto's Last Day', in Neurath 1973, pp. 78–80
Neurath, Otto, 1903a, 'Eine soziale Niederlassung (Settlement) in Wien', *Der Arbeiterfreund* 41, pp. 238–42
  1903b, 'Sozialwissenschaftliches von den Ferial-Hochschulkursen in Salzburg', *Der Arbeiterfreund* 41, pp. 271–9
  1903c, 'Wolframs Faust', *Goethe-Jahrbuch* 24, pp. 233–5
  1904a, 'Geldzins im Altertum', *Plutus* 1, pp. 569–73
  1904b, 'Die Arbeiterfrage im Lichte der Kulturgeschichte', *Der Arbeiterfreund* 42, pp. 161–6
  1906a, *Zur Anschauung der Antike über Handel, Gewerbe und Landwirtschaft*, Jena: Gustav Fischer (see also Neurath 1906–7)
  1906b, 'Ludwig Hermann Wolframs Leben, als Einleitung zu seinem Faust', in Neurath 1906c, pp. E1–518
  1906c, ed., F. Marlow (L.H. Wolfram), *Faust. Ein dramatisches Gedicht in drei Abschnitten*, Berlin: Frensdorf
  1906/1907, 'Zur Anschauung der Antike über Handel, Gewerbe und Landwirtschaft', *Jahrbücher für Nationalökonomie und Statistik*, 3rd. series, 32, pp. 577–606, 34, pp. 145–205

1909, *Antike Wirtschaftsgeschichte*, Leipzig: Teubner, 2nd rev. edn. 1918; 3rd edn. 1926

1910a, 'Zur Theorie der Sozialwissenschaften', *Jahrbuch für Gesetzgebung, Verwaltung und Volkswirtschaft im Deutschen Reich* 34, pp. 37–67, repr. in Neurath 1981, pp. 23–46

1910b, 'Diskussionsbeitrag über die Produktivität der Volkswirtschaft', Verhandlungen des Vereins für Sozialpolitik, Vienna, 27–29 September 1909, *Schriften des Vereins für Sozialpolitik* 132, pp. 599–602

1910c, 'Die Kriegswirtschaft', *Jahresbericht der Neuen Wiener Handelsakademie 1909*, Vienna, pp. 5–54, repr. in Neurath 1919a, pp. 6–42

1910d, *Lehrbuch der Volkswirtschaftslehre*, Vienna: Hölder

1911, 'Nationalökonomie und Wertlehre', *Zeitschrift für Volkswirtschaft, Sozialpolitik und Verwaltung* 20, pp. 52–114

1912, 'Das Problem des Lustmaximums', *Jahrbuch der Philosophischen Gesellschaft an der Universität zu Wien*, Vienna, pp. 45–59, trans. as 'The Problem of the Pleasure Maximum' in Neurath 1973, pp. 113–22

1913a, 'Probleme der Kriegswirtschaftslehre', *Zeitschrift für die gesamte Staatswissenschaft* 69, pp. 438–501

1913b, 'Die Verirrten des Cartesius und das Auxiliarmotiv (Zur Psychologie des Entschlusses)', *Jahrbuch der Philosophischen Gesellschaft an der Universität zu Wien 1913*, pp. 45–59, trans. as 'The Lost Wanderers and the Auxiliary Motive (On the Psychology of Decision)', in Neurath 1983, pp. 1–12

1913c, 'Über die Stellung des sittlichen Werturteils in der wissenschaftlichen Nationalökonomie', Interne Diskussionspapiere des Vereins für Sozialpolitik zur Werturteilsdiskussion 1913, repr. in Neurath 1981, pp. 69–70

1913d, 'Das neue Statut der österreichisch-ungarischen Bank und die Theorie der Zahlung', *Zeitschrift für die gesamte Staatswissenschaft* 69, pp. 51–84

1913e, 'Die Kriegswirtschaftslehre als Sonderdisziplin', *Weltwirtschaftliches Archiv* 1, pp. 342–8, repr. in Neurath 1919a, pp. 1–5, trans. as 'The Theory of War Economy as a Separate Discipline', in Neurath 1973, pp. 125–30

1914, 'Einführung in die Kriegswirtschaftslehre', *Mitteilungen aus dem Intendanzwesen*, Vienna, repr. in Neurath 1919a, pp. 42–133

1915, 'Prinzipielles zur Geschichte der Optik', *Archiv für die Geschichte der Naturwissenschaften und Technik* 5, pp. 371–89, trans. as 'On the Foundations of the History of Optics', in Neurath 1973, pp. 101–12

1916, 'Zur Klassifikation von Hypothesensystemen', *Jahrbuch der Philosophischen Gesellschaft an der Universität Wien 1914 und 1915*, pp. 39–63, trans. as 'On the Classification of Systems of Hypotheses', in Neurath 1983, pp. 13–31

1917a, 'Das Begriffsgebäude der Wirtschaftslehre und seine Grundlagen', *Zeitschrift für die gesamte Staatswissenschaft* 73, pp. 484–520, repr. in Neurath 1981, pp. 103–29

1917b, 'Kriegswirtschaft, Verwaltungswirtschaft, Naturalwirtschaft', *Europäische Staats- und Wirtschaftszeitung* 2, pp. 966–9, repr. in Neurath 1919a, pp. 147–51

1918a, 'Josef Popper-Lynkeus, seine Bedeutung als Zeitgenosse', *Neues Frauenleben* 20, no. 3, pp. 33-8, repr. in Neurath 1981, pp. 131-6

1918b, *Die Kriegswirtschaftslehre und ihre Bedeutung für die Zukunft*, Veröffentlichungen des Deutschen Kriegswirtschaftsmuseum in Leipzig 4, Leipzig: Brandstetter

1919a, *Durch die Kriegswirtschaft zur Naturalwirtschaft*, Munich: Callwey

1919b, 'Die Utopie als gesellschaftstechnische Konstruktion', in Neurath 1919a, pp. 228-31, trans. as 'Utopia as a Social Engineer's Construction', in Neurath 1973, pp. 150-5

1919c, *Die Sozialisierung Sachsens*, Chemnitz: Verlag des Arbeiter- und Soldatenrats im Industriebezirk Chemnitz

1919d, *Wesen und Weg der Sozialisierung, Gesellschaftliches Gutachten vor dem Münchner Arbeiterrat 25 Januar 1919*, Munich: Callwey, repr. in Neurath 1919a, pp. 209-20, trans. as 'Character and Course of Socialisation', in Neurath 1973, pp. 135-50

1919e, 'Sozialisierung und Räterepublik', *Neue Erde* 1, no. 14 (June), pp. 226-9

1919g, 'Die Sozialisierung und die wirschaftlichen Räte', *Neue Erde* 1, no. 17 (July), pp. 269-73

1919h, 'Bayerische Sozialisierungserfahrungen', *Neue Erde* 1, no. 4 (March), pp. 41-4, nos. 5-6 (April), pp. 57-60, 2, nos. 8-9 (June), pp. 117-20, no. 10 (July), pp. 131-4, rev. and repr. in Neurath 1920b

1920a, 'Ein System der Sozialisierung', *Archiv für Sozialwissenschaft und Sozialpolitik* 48, pp. 44-73

1920b, *Bayerische Sozialisierungserfahrungen*, Aus der sozialistischen Praxis 4, Vienna: Neue Erde Verlag, excerpts trans. in Neurath 1973, pp. 18-28

1920c, *Vollsozialisierung*, Deutsche Gemeinwirtschaft 15, Jena: Diederichs

1920d, *Betriebsräte-Lehrerschule*, Reichenberg: Runge and Co

1920e, 'Wirtschaftsplan, Planwirtschaft, Landesverfassung und Völkerordnung', *Der Kampf* 13, pp. 224-7

1920f, 'Die wirtschaftlichen Räte im Programm der bayrischen Vollsozialisierung', *Der Kampf* 13, pp. 136-41

1921a, *Anti-Spengler*, Munich: Callway, repr. in Neurath 1981, pp. 139-96, excerpts trans. as 'Anti-Spengler', in Neurath 1973, pp. 158-213

1921b, 'Das Generallohnsystem', *Der Kampf* 14, pp. 35-6

1921c, *Betriebsräte, Fachräte, Kontrollrat und die Vorbereitung der Vollsozialisierung*, Berlin: Hoffmann

1921d, 'Das Forschungsinstitut für Gemeinwirtschaft', *Die Waage* 25 June, Vienna

1922a, 'Vollsozialisierung und gemeinwirtschaftliche Anstalten', *Der Kampf* 15, pp. 54-60

1922b, 'Österreichs Baugilde und ihre Entstehung', *Der Kampf* 15, pp. 84-9

1922c, 'Reichsgilden', *Der Kampf* 15, pp. 330-7

1923a, 'Geld und Sozialismus', *Der Kampf* 16, pp. 145-7

1923b, 'Geld, Sozialismus, Marxismus', *Der Kampf* 16, pp. 283-8

1924, 'Kirche und Proletariat', *Die Sozialistische Erziehung* 3, no. 8, pp. 177–84
1925a, *Wirtschaftsplan und Naturalrechnung*, Berlin: Laub
1925b, 'Sozialistische Nützlichkeitsrechnung und kapitalistische Reingewinnrechnung', *Der Kampf* 18, pp. 391–5
1925c, 'Gesellschafts- und Wirtschaftsmuseum in Wien', *Österreichische Gemeindezeitung* 2, no. 16, pp. 1–15, repr. in Neurath 1991, pp. 1–17
1926, 'Rezension', *Der Kampf* 19, pp. 91–3
1927, 'Bildliche Darstellung gesellschaftlicher Tatbestände', *Die Quelle* 77, pp. 130–6, repr. in Neurath 1991, pp. 118–25
1928a, *Lebensgestaltung und Klassenkampf*, Berlin: Laub, repr. in Neurath 1981, pp. 227–93, excerpts trans. as 'Personal Life and Class Struggle' in Neurath 1973, pp. 249–98
1928b, 'Rezension: R. Carnap, *Der Logische Aufbau der Welt und Scheinprobleme der Philosophie*', *Der Kampf* 21, pp. 624–6, repr. in Neurath 1981, pp. 295–7
1929a, 'Wissenschaftliche Weltauffassung', *Arbeiterzeitung*, Vienna, 13. October, p. 17, repr. in Neurath 1981, pp. 345–7
1929b, 'Bertrand Russell, der Sozialist', *Der Kampf* 23, pp. 234–8, repr. in Neurath 1981, pp. 337–44
1930a, 'Wege der wissenschaftlichen Weltauffassung', *Erkenntnis* 1, pp. 106–25, trans.: 'Ways of the Scientific World Conception', in Neurath 1983, pp. 32–47
1930b, 'Bürgerlicher Marxismus', *Der Kampf* 23, pp. 227–32, repr. in Neurath 1981, pp. 349–56
1931a, 'Physicalism: The Philosophy of the Vienna Circle', *The Monist* 41, pp. 618–23, repr. in Neurath 1983, pp. 48–51
1931b, 'Physikalismus', *Scientia* 50, pp. 297–303, trans.: 'Physicalism' in Neurath 1983, pp. 52–7
1931c, *Empirische Soziologie. Der wissenschaftliche Gehalt der Geschichte und Nationalökonomie*, repr. in Neurath 1981, pp. 423–527, partly trans. as 'Empirical Sociology', in Neurath 1973, pp. 319–421
1931d, 'Weltanschauung und Marxismus', *Der Kampf* 24, pp. 447–51, repr. in Neurath 1981, pp. 407–12
1931e, 'Kommunaler Wohnungsbau in Wien', *Die Form* 6, no. 3
1931f, 'Das Gesellschafts- und Wirtschaftsmuseum in Wien', *Minerva Zeitschrift* 7, no. 9/10, pp. 153–6, repr. in Neurath 1991, pp. 192–6
1931g, 'Bildhafte Pädagogik im Gesellschafts- und Wirtschaftsmuseum in Wien', *Museumskunde*, new series 3, no. 3, pp. 125–9, repr. in Neurath 1991, pp. 197–206
1932a, 'Soziologie im Physikalismus', *Erkenntnis* 2, pp. 393–431, trans. as 'Sociology in the Framework of Physicalism' in Neurath 1983, pp. 58–90
1932b, 'Die "Philosopie" im Kampf gegen den Fortschritt der Wissenschaft', *Der Kampf* 25, pp. 385–9, repr. in Neurath 1981, pp. 571–6
1932c, 'Sozialbehaviourismus', *Sociologicus* 8, pp. 281–8, repr. in Neurath 1981, pp. 563–70
1932d, 'Protokollsätze', *Erkenntnis* 3, pp. 204–14, trans. as 'Protocol Statements', in Neurath 1983, pp. 91–9

1932e, 'Einheitswissenschaft und Empirismus der Gegenwart', *Erkenntnis* 2, pp. 310–11

1932f, 'Das Fremdpsychische in der Soziologie', *Erkenntnis* 3, pp. 105–6

1933a, *Einheitswissenschaft und Psychologie*, Vienna: Gerold, trans. by H. Kraal as 'Unified Science and Psychology', in B. McGuinness 1987, pp. 1–23

1933b, 'Soziale Aufklärung nach Wiener Methode', *Mitteilungen der Gemeinde Wien* 1933 no. 100, pp. 25–33, repr. in Neurath 1991, pp. 231–9

1933c, 'Die pädagogische Weltbedeutung der Bildstatistik nach Wiener Methode', *Die Quelle* 1933 no. 3, pp. 209–12, repr. in Neurath 1991, pp. 240–3

1933d, 'Museums of the Future', *Survey Graphic* 22, 458–63, repr. in Neurath 1973, pp. 218–23

1933e, *Bildstatistik nach Wiener Methode in der Schule*, Vienna and Leipzig: Deutscher Verlag für Jugend und Volk, repr. in Neurath 1991, pp. 265–336

1934, 'Radikaler Physikalismus und "wirkliche Welt"', *Erkenntnis* 4, pp. 346–62, trans. as 'Radical Physicalism and "the Real World"' in Neurath 1983, pp. 100–14

1935a, 'Einheit der Wissenschaft als Aufgabe', *Erkenntnis* 5, pp. 16–22, trans. as 'The Unity of Science as a Task', in Neurath 1983, pp. 115–20

1935b, 'Pseudorationalismus der Falsifikation', *Erkenntnis* 5, pp. 353–65, trans. as 'Pseudorationalism of Falsification', in Neurath 1983, pp. 121–31

1935c, '1. Internationaler Kongress für Einheit der Wissenschaft in Paris 1935', *Erkenntnis* 5, pp. 377–406, repr. in Neurath 1981, pp. 649–72

1935d, *Was bedeutet rationale Wirtschaftsbetrachtung?*, Vienna: Gerold, trans. by H.Kraal as 'What Is Meant by a Rational Economic Theory?', in B. McGuinness 1987, pp. 67–109

1936a, *Le développement du Cercle de Vienne et l'avenir de l'Empiricisme logique*, Paris: Hermann and Cie, trans. by B. Treschmitzer and H.G. Zilian as 'Die Entwicklung des Wiener Kreises und die Zukunft des Logischen Empirismus', in Neurath 1981, pp. 673–703

1936b, *International Picture Language*, London: Kegan Paul

1936c, 'Physikalismus und Erkenntnisforschung', *Theoria* 2, pp. 97–105, 234–7, trans. as 'Physicalism and the Investigation of Knowledge', in Neurath 1983, pp. 159–67

1936d, 'Une encyclopédie internationale de la science unitaire', *Actes du Congrès Internationale de Philosophie Scientifique, Sorbonne, Paris 1935*, Facs. II 'Unité de la Science', Paris: Herman and Cie, pp. 54–9, trans. as 'An International Encyclopedia of Unified Science' in Neurath 1983, pp. 139–44

1936e, 'L'encyclopédie comme "modèle"', *Revue de Synthèse* 12, pp. 187–201, trans. as 'Encyclopedia as Model', in Neurath 1983, pp. 145–58

1936f, 'Einzelwissenschaften, Einheitswissenschaft, Pseudorationalismus', *Actes du Congrès Internationale de Philosophie Scientifique, Sorbonne, Paris 1935*, Facs. I 'Philosophie Scientifique et Empirisme Logique', Paris: Herman &

Cie, pp. 57–64, trans. as 'Individual Sciences, Unified Science, Pseudorationalism', in Neurath 1983, pp. 132–8

1937a, 'Unified Science and its Encyclopedia', *Philosophy of Science* 4, pp. 265–77, repr. in Neurath 1983, pp. 172–82

1937b, 'Inventory of the Standard of Living', *Zeitschrift für Sozialforschung* 6, pp. 140–51

1937c, 'Die neue Enzyklopädie des wissenschaftlichen Empirismus', *Scientia* 62, pp. 309–20, trans. as 'The New Encyclopedia of Scientific Empiricism', in Neurath 1983, pp. 189–99

1937d, 'The Departmentalization of Unified Science', *Erkenntnis* 7, pp. 240–6, repr. in Neurath 1983, pp. 200–205

1937e, 'Prognosen und Terminologie in Physik, Biologie und Soziologie', *Travaux du IXe Congré International de Philosophie, IV. Unité de la Science*, Paris: Hermann, pp. 77–85, repr. in Neurath 1981, pp. 787–94

1938a, 'Unified Science as Encyclopedic Integration', in Neurath, Bohr, Dewey, et al, 1938, pp. 1–27

1938b, 'The Departmentalization of Unified Science', *Erkenntnis (Journal of Unified Science)* 7, pp. 240–6, repr. in Neurath 1983, pp. 200–205

1939, *Modern Man in the Making*, New York: Knopf

1941a, 'Universal Jargon and Terminology', *Proceedings of the Aristotelian Society* new series 41, pp. 127–48, repr. in Neurath 1983, pp. 213–29

1941b, 'The Danger of Careless Terminology', *New Era* 22, pp. 145–50

1942, 'International Planning for Freedom', *New Commonwealth Quarterly* (April), pp. 281–92, (July), pp. 23–8, repr. in Neurath 1973, pp. 442–70.

1944, *Foundations of the Social Sciences*, International Encyclopedia of Unified Science, vol. II no. 1, Chicago: University of Chicago Press

1946a, 'The Orchestration of the Sciences by the Encyclopedism of Logical Empiricism', *Philosophy and Phenomenological Research* 6, pp. 496–508, repr. in Neurath 1983, pp. 230–42

1946b, 'Visual Education: Humanisation versus Popularisation', in Neurath 1973, pp. 227–48

1973, *Empiricism and Sociology*, ed. by Marie Neurath and Robert S. Cohen, trans. by P. Foulkes and M. Neurath, Dordrecht: Reidel

1979, *Wissenschaftliche Weltauffassung, Sozialismus und Logischer Empirismus*, ed. by R. Hegselmann, Frankfurt: Suhrkamp

1981, *Gesammelte philosophische und methodologische Schriften*, ed. by R. Haller and H. Rutte, Vienna: Hölder-Pichler-Tempsky

1983, *Philosophical Papers 1913–1946*, ed. and trans. by Robert S. Cohen and Marie Neurath, Dordrecht: Reidel

1991, *Gesammelte bildpädagogische Schriften*, ed. by R. Haller and R. Kinross, Vienna: Hölder-Pichler-Tempsky

1994, *Die Einheit von Wissenschaft und Gesellschaft*, ed. by P. Neurath and E. Nemeth, Vienna: Bölau

Neurath, Otto and Olga Hahn, 1909a, 'Zum Dualismus in der Logik', *Archiv für*

*Philosophie, 2. Abteilung: Archiv für systematische Philosophie*, new series 15, pp. 149–62, repr. in Neurath 1981, pp. 5–16

1909b, 'Zur Axiomatik des logischen Gebietskalküls', *Archiv für Philosophie, 2. Abteilung: Archiv für systematische Philosophie*, new series 15, pp. 345–7

1910, 'Über die Koeffizienten einer logischen Gleichung und ihre Beziehung zur Lehre von den Schlüssen', *Archiv für Philosophie, 2. Abteilung: Archiv für systematische Philosophie*, new series 16, pp. 149–76

Neurath, Otto and A. Schapire-Neurath, 1910a, *Lesebuch der Volkswirtschaftslehre*, Leipzig: Klinkhart, 2nd edn Leipzig: Glöckner, 1913

1910b, 'Vorwort', in Francis Galton, *Genie und Vererbung*, trans. by O. Neurath and A. Schapire-Neurath, Leipzig: Klinkhardt, pp. i–vii

Neurath, Otto and Wolfgang Schumann, 1919, *Können wir heute sozialisieren?, Eine Darstellung der sozialistischen Lebensordnung und ihres Werdens*, Leipzig: Klinkhardt

Neurath, Otto, Niels Bohr, John Dewey, Bertrand Russell, Rudolf Carnap and Charles W. Morris, 1938, *Encyclopedia and Unified Science*, International Encyclopedia of Unified Science, vol. 1 no. 1, Chicago: University of Chicago Press

Neurath, Paul, 1973, 'Memories of my Father', in Otto Neurath 1973, pp. 29–41

1982, 'Otto Neurath und die Soziologie', in Haller 1982b, pp. 223–40

1994, 'Otto Neurath (1882–1945): Leben und Werk', in Otto Neurath 1994, pp. 12–95

Neurath, Wilhelm, 1878, 'Darwinismus und Social-Oekonomie', in Wilhelm Neurath 1880b, pp. 165–238

1880a, 'Die Funktion des Geldes', in Wilhelm Neurath 1880b, pp. 313–521

1880b, *Volkswirtschaftliche und socialphilosophische Essays*, Vienna: Faesy and Frick

1880c, 'Abriß einer Selbstbiographie', Universitätsarchiv Tübingen, excerpts trans. as 'Autobiographical Sketch', in Otto Neurath 1973, pp. 2–4

1891, *Moral und Politik*, Vienna, repr. in Wilhelm Neurath 1902, pp. 183–201

1892, *Die wahren Ursachen der Überproduktionskrisen sowie der Erwerbs- und Arbeitslosigkeit*, Vienna, repr. in Wilhelm Neurath 1902, pp. 202–27

1902, *Gemeinverständliche nationalökonomische Vorträge*, ed. by E.O. v. Lippmann, Braunschweig: Vieweg

Niekisch, Ernst, 1958, *Gewagtes Leben. Erinnerungen eines Revolutionärs*, Cologne: Keipenhauer & Witsch, repr. 1974

Nyiri, J.C., 1989, 'Collective Reason: Roots of a Sociological Theory of Knowledge', in W. Gombocz, H. Rutte and W. Sauer, eds., *Traditionen und Perspektiven in der analytischen Philosophie. Festschrift für Rudolf Haller*, Vienna: Hölder-Pichler-Tempsky, pp. 600–18, repr. in J.C. Nyiri, *Tradition and Individuality*, Dordrecht: Kluwer, 1992, pp. 25–38

Oaks, Guy, 1988, *Weber and Rickert. Concept Formation in the Cultural Sciences*, Cambridge, MA: MIT Press

Oberdan, Thomas, 1990, 'Positivism and the Theory of Observation', in *PSA 1990*, vol. 1, ed. by A. Fine, M. Forbes, and L. Wessel, East Lansing: Philosophy of Science Association, pp. 25–37

Österreichische Akademie der Wissenschaften, 1976ff., *Österreichisches biographisches Lexikon 1815–1950*, Vienna: Verlag der österreichischen Akademie der Wissenschaften

Peirce, Charles Sanders, 1887, 'The Fixation of Belief', *Popular Science Monthly* 12, pp. 1–15, repr. in H.S. Thayer, ed., *Pragmatism. The Classic Writings*, New York: New American Library, 1970, pp. 61–78

Pick, Käthe, 1920, 'Wissenschaftlicher Sozialismus und Utopistik', *Der Kampf* 13, pp. 193–7

Plekhanov, G.V., 1897, 'Review of Labriola, Essais sur la conception materialiste de l'histoire, Paris 1897', *Novoye Slovo* (in Russian), trans. in German edn of Plekhanov 1908 (1910) and as 'The Materialist Conception of History' in repr. of Plekhanov 1908 [1969], pp. 103–38

   1908, *Fundamental Problems of Marxism* (in Russian), var. edns, trans. by J. Katzer, London: Lawrence & Wishart, 1969

Poincaré, Henri, 1902, *Science et Hypothèse*, Paris, trans. by W.J. Greenstreet as *Science and Hypothesis*, London: Scott, 1905, repr. New York: Dover, 1952, also trans. by G.B. Halsted in Poincaré, *The Foundations of Science*, Lancaster, PA: The Science Press, 1946

Popper, Karl, 1935, *Die Logik der Forschung*, Vienna: Springer, trans. with addenda by author as *The Logic of Scientific Discovery*, London: Hutchinson, 1959, repr. New York: Harper Torchbooks

Popper-Lynkeus, Josef, 1878, *Das Recht zu leben und die Pflicht zu sterben*, repr. New York: Johnson Reprint Corp., 1972

   1912, *Die allgemeine Nährpflicht als Lösung der sozialen Frage*, Dresden: Reisner

Proust, Joelle, 1986, *Questions de Forme*, Paris: Gallimard, trans. by A.A. Brenner: *Questions of Form. Logic and the Analytic Proposition from Kant to Carnap*, Minneapolis: University of Minnesota Press, 1989

Putnam, Hilary, 1981, 'Philosophers and Human Understanding', in A.F. Heath, ed., *Scientific Explanation*, Oxford: Clarendon Press, repr. in H. Putnam, *Realism and Reason, Philosophical Papers*, vol. III, Cambridge: Cambridge University Press, 1983, pp. 184–204

Quine, W.V.O., 1951, 'Two Dogmas of Empiricism', *Philosophical Review* 60, pp. 20–43, repr. in Quine, *From a Logical Point of View*, 1953, rev. edn. 1980, Cambridge, MA.: Harvard University Press, pp. 20–46

   1969, 'Epistemology Naturalized', in W.V.O. Quine, *Ontological Relativity and Other Essays*, New York: Columbia University Press, pp. 69–90

Rabehl, Bernd, 1978, 'Über Gewalt und Terrorismus. Kritik des falschen Avantgardismus', *Kritik. Zeitschrift für sozialistische Diskussion* no. 15/16

Rabinbach, A., 1983, *The Crisis of Austrian Socialism: From Red Vienna to the Civil War (1927–1934)*, Chicago: University of Chicago Press

Rammstedt, Otthein, 1988, ed., *Simmel und die frühen Soziologen. Nähe und Distanz zu Durkheim, Tönnies und Weber*, Frankfurt: Suhrkamp

Reichenbach, Hans, 1951, *The Rise of Scientific Philosophy*, Berkeley: University of California Press

Reisch, G., 1994, 'Planning Science: Otto Neurath and the International

Encyclopedia of Unified Science', *British Journal for the History of Science* 27, pp. 153–75

Rey, Abel, 1906, *La théorie de physique chez les physiciens contemporains*, Paris, trans. by R. Eissler: *Die Theorie der Physik bei den modernen Physikern*, Leipzig: Klinkhardt, 1908

Richardson, Alan, 1990, 'How Not to Russell Carnap's *Aufbau*', in *PSA 1990*, vol. 1, ed. by A. Fine, M. Forbes, and L. Wessel, East Lansing: Philosophy of Science Association, pp. 3–14

Rickert, Heinrich, 1896–1902, *Die Grenzen der naturwissenschaftlichen Begriffsbildung: eine logische Einleitung in die historischen Wissenschaften*, 2 vols. 4th edn., 1927, Tübingen: Mohr; 5th edn, trans., abridged and ed. by G. Oakes as *The Limits of Concept Formation in Natural Science: A Logical Introduction to the Historical Sciences*, Cambridge: Cambridge University Press, 1986

Ringer, Fritz K., 1968, *The Decline of the German Mandarins. The German Academic Community 1890–1933*, Cambridge, MA.: Harvard University Press

Rossi-Landi, Ferrucio, 1968, *Il linguaggio come lavoro e come mercato*, Milan, trans. M. Adams et al others *Language as Work and Trade. A Semiotic Homology for Linguistics and Economics*, South Headley, MA.: Bergin and Garvey, 1983

Rothschild, Kurt W., 1961, 'Wurzeln und Triebkräfte der Entwicklung der österreichischen Wirtschaftsstruktur', in *Österreichs Wirtschaftsstruktur*, ed. by Wilhelm Weber, Berlin

Rühle, Otto, 1971, *Baupläne für eine neue Gesellschaft*, Reinbeck bei Hamburg

Rutte, Heiner, 1977, 'Positivistische Philosophie in Österreich', *Conceptus* 28–30, pp. 43–56

1982a, 'Der Philosoph Otto Neurath', in Stadler 1982a, pp. 70–8, trans. as 'The Philosopher Otto Neurath', in Uebel 1991a, pp. 81–94

1982b, 'Über Neurath's Empirismus und seine Kritik am Empirismus', in Haller 1982e, pp. 366–84, trans. by T.E. Uebel as 'On Neurath's Empiricism and his Critique of Empiricism', in Uebel 1991a, pp. 175–90

Ryckman, Thomas A., 1991, 'Designation and Convention. A Chapter of Early Logical Empiricism', in *PSA 1990*, ed. A. Fine, M. Forbes and L. Wessel, East Lansing: Philosophy of Science Association, vol. II, pp. 149–58

Schäfer, Lothar, 1974, *Erfahrung und Konvention*, Stuttgart: Fromm-Holzbog

Schilpp, Paul Arthur, 1963, ed., *The Philosophy of Rudolf Carnap*, LaSalle: Open Court

Schleichert, Herbert, ed., 1975, *Logischer Empirismus – der Wiener Kreis*, München: Fink

Schleier, Hans, 1988, 'Der Kulturhistoriker Karl Lamprecht, der "Methodenstreit" und die Folgen', in Karl Lamprecht, *Alternative zu Ranke*, ed. by H. Schleier, Leipzig: Reclam jnr., pp. 7–45

Schlick, Moritz, 1917, 'Raum und Zeit in der gegenwärtigen Physik', *Die Naturwissenschaften* 5, 3rd enlarged edn, Berlin: Springer, 1920, trans. by H.L. Brose as *Space and Time in Contemporary Physics*, Oxford: Oxford

University Press, repr. (with trans. of rev. in German 4th edn.) in Schlick, *Philosophical Papers, Vol.* I *(1909–1922)* ed. by H.L. Mulder and B. van de Velde-Schlick, Dordrecht: Reidel, 1979, pp. 207–69

1918/25, *Allgemeine Erkenntnislehre*, Berlin: Springer, 1918, 2nd rev. edn. 1925, trans. by H. Feigl and A. Blumberg as *General Theory of Knowledge*, Lasalle: Open Court, 1974

1934, 'Über das Fundament der Erkenntnis', *Erkenntnis* 4, pp. 79–99, trans. by P. Heath as 'The Foundation of Knowledge', in M. Schlick, *Philosophical Papers Vol.* II *(1925–1936)*, ed. by Henk L. Mulder and Barbara van de Velde-Schlick, Dordrecht: Reidel, 1979, pp. 370–87 (previously trans. by D. Rynin in Ayer 1959, pp. 209–27)

Schmoller, Gustav, 1883, 'Zur Methodologie der Staats- und Sozialwissenschaft', *Jahrbuch für Gesetzgebung, Verwaltung und Volkswirtschaft im Deutschen Reich* 7, repr. in Schmoller, *Zur Literaturgeschichte der Staats- und Sozialwissenschaft*, Berlin, 1888

Schmolze, Gerhard, ed., 1969, *Revolution und Räterepublik in München 1918/19 in Augenzeugenberichten*, Düsseldorf: Rauch

Schnädelbach, Herbert, 1979, 'Über historistische Aufklärung', *Allgemeine Zeitschrift für Philosophie*, repr. in Schnädelbach, *Vernunft und Geschichte*, Frankfurt, a. M.: Suhrkamp, 1987, pp. 23–46

Schön, Manfred, 1987, 'Gustav Schmoller und Max Weber', in Mommsen and Osterhammel 1987, pp. 59–70

Schütte-Lihotzky, Margarete, 1982, 'Mein Freund Otto Neurath', in Stadler 1982a, pp. 40–2

Schullern von Schrattenhofen, Hermann R. von, 1902, 'Wilhelm Neurath', *Jahrbücher für Nationalökonomie* 79, pp. 161–6

Schumann, Wolfgang, 1973, 'Memories of Otto Neurath', in Neurath 1973, pp. 15–18

Simmel, Georg, 1890, *Über sociale Differenzierung. Sociologische und psychologische Untersuchungen*, Leipzig: Duncker and Humblot, repr. in Simmel 1989, pp. 109–296

1892a, *Einleitung in die Moralwissenschaft. Eine Kritik der ethischen Grundbegriffe*, 2 vols., Stuttgart: Cotta, repr. Frankfurt: Suhrkamp, 1989

1892b, *Probleme der Geschichtsphilosophie. Eine erkenntnistheoretische Studie*, Leipzig: Duncker and Humblot, 3rd rev. edn. 1907, 1st edn. repr. in Simmel 1989, pp. 297–421

1895, 'Über eine Beziehung der Selektionslehre zur Erkenntnistheorie', *Archiv für systematische Philosophie* I pp. 34–45, repr. in Simmel 1992a, pp. 62–74, trans. by I. Jerison as 'On a Relationship Between the Theory of Selection and Epistemology', in H.C. Plotkin, ed., *Learning, Development and Culture*, 1982, New York: Wiley and Sons

1898, 'Die Selbsterhaltung der socialen Gruppe. Sociologische Studie', *Jahrbuch für Gesetzgebung, Verwaltung und Rechtspflege des Deutschen Reiches* 22, pp. 589–640, repr. in Simmel 1992, pp. 311–72

1908, *Soziologie. Untersuchungen über die Formen der Vergesellschaftung*, Leipzig: Duncker and Humblot, repr. Frankfurt: Suhrkamp, 1992

1989, *Aufsätze 1887–1890. Über sociale Differenzierung. Die Probleme der Geschichtsphilosophie*, ed. by H.J. Dahme, Frankfurt: Suhrkamp

1992a, *Aufsätze und Abhandlungen, 1894–1900*, ed. by H.-J. Dahme and D.P. Frisby, Frankfurt: Suhrkamp,

Simon, Joseph T., 1979, *Augenzeuge. Erinnerungen eines österreichischen Sozialisten*, Vienna: Wiener Volksbuchhandlung

Soulez, Antonia, 1988, 'La construction des utopies comme tache de l'Ingenieur Social, selon O. Neurath en 1919', in P. Soulez, ed., *Les Philosophes et la guerre de 14*, Vincennes: Presses Universitaires, pp. 237–50

Stadler, Friedrich, 1979, 'Aspekte des gesellschaftlichen Hintergrunds und Standorts des Wiener Kreises am Beispiel der Universität Wien', in Berghel, Hüber and Köhler, 1979, pp. 41–59, trans. by T.E. Uebel as 'Aspects of the Social Background and Position of the Vienna Circle at the University of Vienna', in Uebel 1991a, pp. 51–79

1982a, ed., *Arbeiterbildung in der Zwischenkriegszeit. Ausstellungs katalog mit Forschungsteil*, Vienna: Österreichisches Gesellschafts- und Wirtschaftsmuseum

1982b, *Vom Positivismus zur wissenschaftlichen Weltauffassung*, Vienna: Löcker Verlag

1982c, 'Otto Neurath (1882–1945) – Zu Leben und Werk in seiner Zeit', in Stadler 1982a, pp. 2–22

1984, '"Wiener Methode der Bildstatistik" und politische Graphik des Konstruktivismus (Wien-Moskau 1931–1934)', in Historikersektion der österreichisch-sowjetischen Gesellschaft, ed., *Österreich und die Sowjetunion 1918–1955*, Vienna, pp. 220–49

1989, 'Otto Neurath (1882–1945) – Enzyklopedist, Schulreformer und Volksbildner', in *Erwachsenenbildung in Österreich* 40, pp. 33–6, trans. by T.E. Uebel as 'Otto Neurath – Encyclopedist, School Reformer and People's Educator', in Uebel 1991a, pp. 255–64

1993, ed., *Scientific Philosophy: Origins and Developments*, Vienna Circle Institute Yearbook 1, Dordrecht: Kluwer

Stölting, Erhard, 1986, *Akademische Soziologie in der Weimarer Republik*, Berlin: Duncker and Humblot

Tenbruck, Friedrich H., 1959, 'Formal Sociology', in Wolff 1959, pp. 61–99

1987, 'Max Weber and Eduard Meyer', in Mommsen and Osterhammel 1987, pp. 234–67

Thiele, Joachim, ed., 1979, *Wissenschaftliche Kommunikation. Die Korrespondenz Ernst Machs*, Kastellaun: A. Hehn

Toller, Ernst, 1934, *Eine Jugend in Deutschland*, trans. as *I was a German*, London: Bodley Head, 1934

Toulmin, Stephen, 1970, 'Der Metaphysiker Wittgenstein', *Neues Forum* 17, pp. 699–703

Tönnies, Ferdinand, 1887, *Gemeinschaft und Gesellschaft*, Leipzig: Fues's Verlag, 8th edn. 1935, trans. by C.P. Loomis as *Fundamental Concepts of Sociology*, New York: American Book Company, 1940

1899–1900, 'Philosophical Terminology', *Mind* 8, pp. 289–332, 467–91, 9, pp. 46–61, German orig. with appendices: *Philosophische Terminologie in psychologisch-soziologischer Ansicht*, Leipzig, 1906

1900, 'Zur Theorie der Geschichte (Exkurs)', *Archiv für systematische Philosophie* 8, pp. 1–38

1904, 'Ammons Gesellschaftstheorie', *Archiv für Sozialwissenschaft und Sozialpolitik* 19, pp. 88–111

1905–9, 'Zur naturwissenschaftlichen Gesellschaftslehre', *Jahrbuch für Gesetzgebung, Verwaltung und Volkswirtschaft im Deutschen Reich* 29, pp. 27–101, 1283–322, 30, pp. 121–45, 31, pp. 487–552, 33, pp. 879–94, 35, pp. 375–96, repr. in Tönnies 1925, pp. 133–329

1907, *Das Wesen der Soziologie*, Dresden: Zahn and Jaensch, repr. in Tönnies 1925, pp. 350ff

1911, 'Wege und Ziele der Soziologie', *Verhandlungen des Ersten Deutschen Soziologentages*, Tübingen: Mohr, repr. in *Soziologische Studien und Kritiken. Zweite Sammlung*, Jena: Fischer, 1926, pp. 125–43

1925, *Soziologische Studien und Kritiken. Erste Sammlung*, Jena: Fischer

Turner, Stephen and Regis Factor, 1984, *Max Weber and the Dispute over Reason and Value*, London: Routledge, Kegan and Paul

Twyman, Michael, 1975, 'The Significance of ISOTYPE', in *Graphic Communication through ISOTYPE*, Exhibition Catalogue, University of Reading, pp. 7–17

Uebel, Thomas E., ed., 1991a, *Rediscovering the Forgotten Vienna Circle. Austrian Studies on Otto Neurath and the Vienna Circle*, Dordrecht: Kluwer

1991b, 'Neurath's Programme for Naturalistic Epistemology', *Studies in the History and Philosophy of Science* 22, pp. 623–46

1992a, *Overcoming Logical Positivism From Within. The Emergence of Neurath's Naturalism in the Vienna Circle's Protocol Sentence Debate*, Amsterdam and Atlanta, GA: Rodopi

1992b, 'Rational Reconstruction as Elucidation? Carnap in the Early Protocol Sentence Debate', *Synthese* 93, pp. 107–40

1992c, 'Neurath vs. Carnap: Naturalism vs. Rationalism Before Quine', *History of Philosophy Quarterly* 9, pp. 445–70

1993a, 'Neurath's Protocol Statements: A Naturalistic Theory of Data and Pragmatic Theory of Theory Acceptance', *Philosophy of Science* 60, pp. 587–607

1993b, 'Wilhelm Neurath's Opposition to "Materialist" Darwinism', in Stadler 1993, pp. 209–28

1995a, 'Physicalism in Wittgenstein and the Vienna Circle', in *Physics, Philosophy and the Scientific Community. Essays in Honor of Robert S. Cohen*, ed. by K. Gavroglu, J. Stachel and M.W. Wartofsky, Dordrecht: Kluwer, pp. 328–56

1995b, 'Otto Neurath's Idealistic Inheritance: The Social and Economic Thought of Wilhelm Neurath', *Synthese* 103, pp. 87–121

Untermann, Ernst, 1907, *Dialektisches: Volkstümliche Vorträge aus dem Gebiete des proletarischen Monismus* Stuttgart: Dietz

Viesel, Hansjörg, ed., 1976, *Literaten an der Wand*, Ausstellung der Akademie der Künste, Berlin
Wachtel, Henry I., ed., 1955, *Security for All and Free Enterprise. A Summary of the Social Philosophy of Josef Popper-Lynkeus*. With an Introduction by Albert Einstein, New York: Philosophical Library
Waismann, Friedrich, 1930, 'Thesen', trans. by J. Schulte as 'Theses' in B. McGuinness, ed., *Wittgenstein and the Vienna Circle*, Oxford: Blackwell, 1967, pp. 233–61
  1945, 'Verifiability', *Proceedings of the Aristotelian Society*, suppl. vol. XIX, pp. 119–50, repr. in A Flew, ed., *Logic and Language* (First Series), Oxford: Blackwell, 1952, pp. 117–44
Weber, Max, 1903–6, 'Roscher und Knies und die logischen Probleme der historischen Nationaloekonomie', monograph published in 3 parts in *Jahrbuch für Gesetzgebung, Verwaltung und Volkswirtschaft* 27, 1903; 29, 1905; 30, 1906; trans. as *Roscher and Knies: The Logical Problems of Historical Economics*, London: The Free Press (Collier Macmillan Publishers), 1975
  1904, 'Die "Objektivität" sozialwissenschaftlicher und sozialpolitischer Erkenntnis', *Archiv für Sozialwissenschaft und Sozialpolitik* 19, repr. in Weber 1922 [1968], pp. 146–214, trans. as '"Objectivity" in Social Science and Social Policy', in Weber 1949, pp. 49–112
  1906, 'Kritische Studien auf dem Gebiet der kulturwissenschaftlichen Logik', *Archiv für Sozialwissenschaft und Sozialpolitik* 22, trans. as 'Critical Studies in the Logic of the Cultural Sciences', in Weber 1949, pp. 113–88
  1918, 'Der Sinn der "Wertfreiheit" der soziologischen und ökonomischen Wissenschaften', *Logos* 7, repr. in Weber 1922, pp. 489–540, trans. as 'The Meaning of "Ethical Neutrality"' in Weber 1949, pp. 1–47
  1921, *Gesammelte politische Schriften*, Drei Masken: München
  1922, *Gesammelte Aufsätze zur Wissenschaftslehre*, ed. by J. Winckelmann, Tübingen: Mohr, 3rd ed. 1968
  1949, *The Methodology of the Social Sciences*, ed. by Edward A. Shils, trans. by E.A. Shils and H.A. Finch, New York: The Free Press
  1988, *Zur Neuordnung Deutschlands. Schriften und Reden 1918–1920*, ed. by W.J. Mommsen and W. Schwentker, Max Weber Gesamtausgabe 16, Tübingen: Mohr
Wedin, M.V., 1992, 'Trouble in Paradise: On the Alleged Incoherence of the Tractatus', *Grazer Philosophische Studien* 42
Weiler, Gershon, 1970, *Mauthner's Critique of Language*, Cambridge: Cambridge University Press
Weissel, Erwin, 1976, *Die Ohnmacht des Sieges. Arbeiterschaft und Sozialisierung nach dem ersten Weltkrieg in Österreich*, Vienna: Europaverlag
Wellmer, Albrecht, 1985, *Zur Dialektik von Moderne und Postmoderne. Vernunftkritik nach Adorno*, Frankfurt: Suhrkamp
Wernitz, Axel, 1966, *Sozialdemokratische und Kommunistische Sozialisierungskonzeptionen. Eine Untersuchung zur deutschen Sozialgeschichte des 19. und 20. Jahrhunderts*, Dissertation, Universität Erlangen

Whimster, Sam, 1987, 'Karl Lamprecht and Max Weber: Historical Sociology within the Confines of a Historians' Controversy', in Mommsen and Osterhammel 1987, pp. 268–83

Wittgenstein, Ludwig, 1921, 'Logisch-Philosophische Abhandlung', *Annalen der Nat. u. K. Philosophie* 14, pp. 185–262, trans. by C. Ogden as *Tractatus Logico-Philosophicus*, London, 1922, rev. edn. 1933, repr. London: Routledge, Kegan, Paul, 1983

1953, *Philosophische Untersuchungen/Philosophical Investigations*, ed. by G.E.M. Anscombe, G.H. von Wright, and R. Rhees, Oxford: Blackwell

1969, *Über Gewißheit/On Certainty*, ed. by G.E.M. Anscombe, G.H. von Wright, trans. by D. Paul and G.E.M. Anscombe, New York: Harper Torchbooks

Wolff, Kurt H., 1959, ed., *Georg Simmel 1858–1918*, Columbus: Ohio State University Press

Zeisel, Hans, 1985, 'The Austromarxists in Red Vienna: Reflections and Recollections', in *The Austrian Socialist Experiment: Social Democracy and Austromarxism, 1918–1934*, ed. by A. Rabinbach, Boulder, Co.: Westview, pp. 119–33

Zilsel, Edgar, 1929, 'Philosophische Bemerkungen', *Der Kampf* 22, pp. 178–86

1931a, 'Materialismus und marxistische Geschichtsauffassung', *Der Kampf* 24 pp. 66–75

1931b, 'Partei, Marxismus, Materialismus, NeoKantianismus', *Der Kampf* 24 pp. 213–20

Zolo, Danilo, 1986, *Scienza e Politica in Otto Neurath*, Milan: Feltrinelli, enlarged edn. trans. D. McKie, *Reflexive Epistemology. The Philosophical Legacy of Otto Neurath*, Dordrecht: Kluwer, 1989

# Index

Adler, Friedrich, 145, 146, 242
Adler, Max, 57, 58, 59, 144, 145, 146, 235, 236, 237n, 238, 242
Adler, Victor, 145, 242
Adorno, T., 254n
Anduschuss, Martin, x
anti-foundationalism, 3, 93, 120, 129, 131, 149, 188, 204
  descriptive, 115–21
  normative, 111–5, 118
  methodological or metatheoretical, 125, 128
anti-reductionism, 2, 95–6, 100, 101, 107
anti-realism, *see* realism, instrumentalism
Arco-Valley, Graf, 45
Ariew, Roger, 116n, 118n
Aristotle, 112, 210
Arntz, Gerd, 65, 69, 85
Atlanticus (pseudonym), *see* Ballod, Karl
atomic statements, 196–7, 203n
Austrian school of economics, 112
Avenarius, Richard, 91, 147
Ay, E., 91n

Babeuf, Gracchus, 42
Ballod, Karl, 44, 47, 49n, 54, 175n, 221
*Ballungen*, 81, 153, 155, 156–8, 163, 171, 186, 187, 190–3, 195, 200, 203, 204, 205, 206, 208–36, 239–40, 243
  congestion of events, 210, 213–24, 227, 240
  density of concepts, 210, 224–9, 241
Bauer, Otto, 23–6, 28, 41, 54–5, 57, 58, 59n, 144n, 145, 146, 147, 230, 231, 233, 235, 238, 242
Bavarian revolution, 2, 4, 6, 21, 43–56, 173, 176–7, 230, 245–6
Beck, Hermann, 20n
Belke, Elisabeth, 96n
Bergmann, Gustav, 77
Bergmann, Hugo, 96n
Bergström, Lars, 91n
Bernath, Erwin, 65

Bernstein, Eduard, 27, 154, 236, 237
Beyer, Hans, 43n, 46n, 50n, 144n, 231n, 233n
Beyer, W.R., 91n
Bickel, Cornelius, 98n, 118n
Bieber, Karl A., 62
Blackmore, John, 101n
Block, Joseph, 237
Blum, M., 58n, 59n
Blumenberg, Hans, 90n
Böhm-Bawerk, Eugen, 94
Boltzmann, Ludwig, 105, 127
Bool, Flip, 72n
Born, Max, 207
Bottomore, Tom, 58n, 147n
Braithwaite, Richard, 249
Breisig, K., 99n
Breitner, Hugo, 63
Brentano, Lujo, 232, 233
Broos, Kees, 72n
Buek, Otto, 96n

Cahnman, Werner J., 118n
Capek, Milic, 100n
Carnap, Rudolf, x, 2, 5, 6n, 13, 72, 75n, 77, 78n, 79, 84, 85, 87, 91, 95, 96, 97, 137n, 142, 143, 144n, 146n, 148, 149, 150–9, 162, 163, 164n, 166, 168, 177n, 178n, 179, 181, 182, 183, 184n, 186, 187, 191–3, 194, 197, 199–200, 201, 202, 203, 228, 229, 234, 235n, 249, 250, 255
  formal and material modes, 151, 154, 180, 228
  *Principle of Tolerance*, 157
  'rational reconstruction', 149–52, 156, 182–3, 192, 199
causality, 217, 219, 229
Chaloupek, Günther K., 91n
Chang, Hasok, xii, 144n, 163n, 249n
Cherniak, Christopher, 90n
Childers, Timothy, x, xii
Cicero, 11

283

Coffa, Alberto, 150n, 196n
Cohen, Robert S., 91n, 101n, 145n, 148n
coherence theory
 of acceptance, 80, 159, 161, 219
 of truth *see also* truth, correspondence theory of
Cole, G.D.H., 86, 250
Collins, Harry, 247n
Communism, 9, 70
Comte, August, 75, 172
concepts
 historical conditioning of, 124, 126, 130, 142, 148, 156, 171, 188, 197
 abstract, 210–29, 239, 254
 cluster, *see Ballungen*
confirmation theory, 95, 191, 201, 213
Conventionalism, 99, 100, 116, 132–6
co-operation, 249–51
Couturat, Louis, 105
Creath, Richard, 150n
Crusoe, R., 149, 155

D'Acconti, Alessandra, 91n
Dahme, Heinz-Jürgen, 98n, 119n
Dahms, Hans-Joachim, 55n, 91n, 164n
D'Alembert, J., 84, 187, 188
Descartes, René, 129, 130, 131, 132, 138, 182
Deutsch, Julius, 63
Dewey, John, 91, 164
Diderot, 84
Dollfuss, Engelbert, 82, 181
Dorst, Tankred, 43n
Du Bois-Reymond, Emil, 129
Duhem, Pierre, 75, 93, 105, 108, 110, 115–18, 119–21, 124, 127, 131, 139, 145, 172, 173, 174, 194, 201, 203, 208–13, 239, 242
Durkheim, Emil, 242
Dvorak, Johann, 110n, 145n

Ebert, Friedrich, 43
economics
 administrative economy, 18, 21, 32
 ancient, 2, 233
 calculation in kind, 18, 32, 112–13
 capitalist economy, 28, 29, 40
 credit, 9, 10, 16, 17, 44
 economy in kind, 15, 17, 31, 35, 37, 128, 194, 221
 market economy, 29, 37, 227
 monetary economy, 9, 10, 15, 17, 32
 planned economy, 32, 46, 54
 political economy, 209, 213, 223
 profit, 9, 15, 27, 29, 35
 socialist economy, 20, 22, 24, 29, 32
 war, 2, 135, 174

economy of thought, 102, 121, 171
Einstein, Albert, 75, 105, 181, 198, 207n
Eisner, Kurt, 43, 45, 46, 232, 233
Elkana, Yehuda, 91n
Ellenbogen, Wilhelm, 242
Elsenhans, Th., 94
encyclopedism, 163, 179, 181, 187–8, 212, 255
Engels, Friedrich, 22n, 31, 37, 41, 44, 59, 143, 146, 147, 148, 163, 236, 237, 238, 248
Enlightenment, 92, 102, 106, 129–30, 131–2, 163, 165, 254
epistemology, 6, 93, 146, 252
Eschbach, Achim, 140n
evolutionism, 127
experts, role of, 37, 244, 246
Explanation, D-N model of, 217–18

Fabian, Reinhard, xii
Factor, Regis, 98n, 168, 169
falsificationism, 95, 191, 209, 213
Farrell, James T., 164n
Feigl, Herbert, 13, 77, 84
Feldbauer, Peter, 60n
Feuerbach, Ludwig, 163
Fine, Arthur, 94
Fischer, Heinz, 24n
Fischer, Irma, 107n
Fleck, Christian, 98n
Fledderus, Mary, 83
food,
 production of, 9, 14, 15, 26, 29, 35, 41, 43, 44
 distribution of, 14, 35
Frank, Josef, 10, 11, 12n, 20n, 61, 68
Frank, Philipp, 10, 72, 76, 78, 79, 83, 84, 102, 104n, 105, 106, 127n, 163, 179, 181
Frank, Wilhelm, 236, 238
Freud, Sigmund, 119n
Freudenthal, Gideon, 91n, 144n
Freund, Julian, 118n
Friday, 155
Friedman, Michael, 150n
Frisby, David, 98n, 119n
Fuhrmann, A., xii

Galison, Peter, xii, 2n, 165n
Galton, Francis, 128
Geddes, D.W.D., 90n
George, Henry, 15n
Gerstl, Max, 246
Giedymin, Jerzy, 116n
Gillies, Donald, 169n
Glaser, Ernst, 91n, 144n
Glöckel, Otto, 57
Gödel, Kurt, 77

## Index 285

Goethe, Johann Wolfgang von, 87, 88
Goode, P., 58n
Gorges, Imela, 98n
Gothein, Eberhart, 21
Greenaway, Frank, 167n
Grünberg, Karl, 147
Grunberger, Richard, 43n
Gulick, Charles, 58n
Gumbel, Ernst, 55n

Habermas, Jürgen, 254
Hacking, Rachel, x
Halfbrodt, Dick, 56
Hahn, Hans, 10, 13, 72, 75n, 76, 77, 79, 84, 99, 118, 146n, 153, 177n, 178n, 194, 203n
Hahn, Ludwig, 13, 77n
Hahn, Olga, 10, 13, 21, 72, 77, 83, 85
Haller, Rudolph, x, 72, 75n, 76, 77n, 90n, 91n, 101n, 119, 152n, 165n, 200n, 204
Hansen, Reginald, 98n
Harding, Sarah G., 116n, 118n
Harnisch, Dr, 54
Hebbel, F., 13
Hegel, G.W.F., 214, 217, 226, 238, 240, 241
Hegselmann, Rainer, 91n, 144n, 248n
Heidelberger, Michael, 90n
Hempel, C.G., x, 90n, 217
Hennis, Wilhelm, 98n
Herz, Rudolf, 56n
Heverley, Gerald, xii
Hilbert, David, 105, 194
Hilferding, Rudolph, 47
Hillmayr, Heinrich, 45n, 245n
historical materialism, 6, 41, 147, 209, 237–44
historical school, 209, 213–24, 241
Hitler, Adolf, 72, 165, 180
Hösl, Wolfgang, 60n
Hoffmann, Johannes, 46, 50–3, 61n, 62, 230, 246, 248
Hofmann-Grünenberg, Frank, 90n, 91n
holism, 92–3, 110, 116–18, 139, 172, 174, 225
see also underdetermination, indeterminacy
Hollitscher, Walter, 79
Holton, G., 101n
Hookway, Christopher, 90n
Horkheimer, Max, 164, 254n
Hung, Tscha, 87

ideal types, 125, 223
idealism, 199, 214, 217, 236, 241
Iggers, Georg C., 98n
indeterminacy
  of practical facts, 81, 117, 206
  of theoretical facts, 81, 93, 117, 120, 121, 139, 206, 210, 211

instrumentalism, 92, 106, 173–4, 176, 223
intersubjectivity, 107, 141, 150, 151, 153, 154, 155
  see also language, private
ISOTYPE, 85, 86
  see also picture statistics
Itelson, Gregorius, 75, 94n, 96

Jacoby, E.G., 107n, 118n
Jaffé, Edgar, 43, 46
Jahoda-Lazarsfeld, Marie, 79
James, William, 94, 147
Jankowsky, Heinz, xii
Jerusalem, Wilhelm, 94, 119n
Jevons, W.S., 194n
Jodl, Friedrich, 94
Jodlbauer, Josef, 63
Johnston, William M., 7
Jørgeson, Jørgen, 84

Kaempfert, Gertrud, 10
Käsler, Dirk, 98n
Kant, Immanuel, 9, 105, 129, 130, 131, 144, 146, 198, 214, 226, 248
Kauffmann, Felix, 77, 98n
Kautsky, Karl, 22, 145, 237, 238
Keuzenkamp, Hugo, 210n
Keynes, John Maynard, 210n
Kinross, Robin, 62n, 67n, 69n, 85n
Kitching, Gavin, 164n
Klose, Olaf, 106n
Knies, Karl, 214, 216–17
Köhler, Eckehart, 91n, 114n
Kolakowski, L., 239
Koopman, Tjalling, 210n
Koppelberg, Dirk, 90n, 91n, 157n
Korsch, Karl, 23–4, 26–9, 41
Kraft, Viktor, 77
Kranold, Hermann, 45, 47, 49, 54, 244, 248
Kritzer, Peter, 248n
Krüger, Dieter, 98n
Kulka, Heinrich, 62
Kun, Bela, 46

Labriola, Antonio, 209, 210, 235–44, 241n
Lakatos, Imre, 249
Lamprecht, Karl, 220
Landauer, Gustav, 49, 50, 232
language
  ideal, 139, 152, 186
  imprecise, 81, 195
  linguistic turn, 142
  natural, 123, 131, 139, 153, 186, 195
  physical, 151–3

language (cont.)
  private (or phenomenological), 148, 153, 154, 156
  precise, 195, 210
Laplace, Pierre Simon, 172, 174, 184, 193
Laplacean ideal, 129, 174, 184–5, 197, 202, 243
Laplanche, J., 119n
Lassalle, Ferdinand, 57
Lazarus, M., 119n
Leibniz, Gottfried Wilhelm, 171
Leichter, Otto, 144n
Lenin, Vladimir, 146, 146
Lepenies, Wolf, 98n
Levien, Max, 52
Leviné, Eugen, 50, 52, 55, 180, 231, 232
Leviné-Meyer, Rosa, 232
Lichtenberg, G.C., 164n
Liebersohn, Harry, 98n, 118n, 165n
Lindenlaub, Dieter, 98n
Lindler, Dr, 246n
Llull, Raimon, 171
Loos, Adolf, 61
Lukács, Georg, 54n
Ludwig III of Bavaria, 21, 43
Luxemburg, Rosa, 41n

Mach, Ernst, 75, 77, 94, 99, 100, 101, 102–11, 119, 127, 128, 133, 135, 136, 138–9, 145, 146, 165, 196, 237
Mach, Ernst Verein, 77, 145, 181
Mandel, Ernst, 16n
Mannheim, Karl, 97
Marlow (pseudonym) see Wolfram, Ludwig Hermann
Martinez-Alier, Juan, 37n, 175n
Marut, Red, 49
Marx, Karl, 7, 13, 22, 31, 37, 38, 41, 44, 59, 101, 143, 144, 145, 146, 148, 163, 164n, 165, 237, 238, 248, 256
Marxism, 6, 9, 22, 31, 55, 57, 78, 145, 146, 147, 163, 177, 178, 181, 182, 233–4, 235, 237–44
  see also Neurath's Marxism
Mauthner, Fritz, 96n, 140n
Meja, Volker, 97n
Menger, Carl, 98, 125, 126, 127n, 167, 194n, 213, 218, 226
Mergner, Gottfried, 38n
*Methodenstreit*, 98, 125–6, 213–16, 229, 243
methodological solipsism, 142, 149, 151–2, 154, 156, 158
Meyer, Eduard, 11, 94, 98, 214, 220
Michelson, A.A., 198
Mill, John Stuart, 209, 215, 218, 225, 228, 239
Miller, D.C., 198, 206, 207
Mitchell, Allan, 43n, 46n

Mohn, Erich, 90n, 144n
Moebius, 87
Morentz, Ludwig, 43n
Morgan, Mary, 210n
Morley, E.W., 198
Mormann, Thomas, 163n
Morris, Charles, 84, 163, 182, 250
Müller, K.H., 91n, 144n, 163n, 164n
Müller, Richard, 16n
Müller-Meiningen, Ernst, 4n
von Müller, K.A., 245
Mühsam, Erich, 43n, 50, 55, 56, 231, 232, 251
Munz, Erwin, 43n
museums, 2, 20, 63, 92, 249
Musil, Robert, 106
Mussolini, Benito, 180

Nagel, Ernst, 113, 164n
Natkin, Marcel, 77
naturalism, 90, 136, 141
negotiation, 3, 42, 93, 243, 248
Neider, Henrich, 11n, 72n, 77, 152, 165n
Nemeth, Elisabeth, 3n, 91n, 144n, 163n, 175n
neo-Kantianism, 125, 144, 145, 146, 214, 235, 244n
Neumann, G., 19n
Neurath, Gertrud, xii
Neurath, Marie, xii, 10n, 21n, 65n, 70n, 83n, 85, 159n
  see also Reidemeister, Marie
Neurath, Otto
  anti-philosophy, 89, 91, 143–4, 164, 255
  attack on epistemology, 93, 96
  attack on metaphysics, 5, 75, 76, 78, 143, 177–8, 227, 235
  attack on method, 200, 202–8, 222, 224, 253
  auxiliary motive, 134, 168, 174, 247, 249
  Marxism, 143, 144, 208, 229, 256
  naturalised epistemology, 90–1, 156, 159ff
  Neurath's Boat, 89, 91–4, 130–1, 138–9, 155, 163, 165
  non-cognitivism, 97
  Neurath's General Principle, 169, 202–8
  Neurath's Special Principle, 120, 202–8
  on value, 111–13
  social Epicureanism, 30, 112, 114–15
Neurath, Paul (Otto's son), 10n, 13, 91n, 97n, 98n, 144n
Neurath, Wilhelm (Otto's father), 8–10, 94, 97n, 99, 100n
Neurath, Wilhelm (Otto's brother), 11
Niekisch, Ernst, 7, 7n, 43n, 45, 46, 47n, 50, 53, 55
Nielsen, C. R., xii
Nyiri, J.C., 91n

## Index

Oaks, Guy, 118n
Oberdan, Thomas, 150n
observation statements, 80, 123, 206n, 210
  see also protocol statements

Paulsen, Friedrich, 107n
Peirce, Charles Sanders, 90
persuasion, 248
phenomenalism, 183–4
  see also methodological solipsism
von Philippovich, Eugen, 94
physicalism, 81, 143, 148, 151, 154, 168, 178, 199, 229, 235–6, 243
Picard, Emile, 167
Pick, Käthe, 144n
picture statistics, Vienna method of, 20, 34, 65, 67, 85, 249
  see also ISOTYPE
Planck, Max, 77
Plato, 13, 89, 90
Plekhanov, G.W., 41, 209, 210, 235–44
pluralism, 255
Plutarch, 89, 90
Poincaré, Henri, 75, 90, 104, 105, 115, 116, 127, 147
Pontalis, J.B., 119n
Popper, Karl, 103, 161, 169n, 185, 187n, 193, 195n, 200, 201, 202–8, 202n, 203n, 205n, 222, 243, 246, 250
Popper-Lynkeus, Josef, 42, 44, 94, 96, 99, 100n, 175n
protocol sentence debate, 5, 80, 96, 142, 148, 154, 200, 235
protocol statements, Neurath's proposal for, 80, 159–62, 197–8
  entitlement requirement, 200–2
  intersubjectivity requirement, 200–2
Proust, Joelle, 150n
pseudo-rationalism, 91, 129, 137, 184, 190, 205, 208
pseudo-science, 91
psychologism, Popper against, 202, 220
Putnam, Hilary, 255

Quine, W.V.O., 90, 93, 118n, 152, 161

Rabinbach, A., 57
Rammstedt, Otthein, 98n, 119n
Rand, Rose, 77, 79
Rauscher, Franz, 83n
realism, 106, 169, 173, 209, 214–15, 217, 225, 253
reductionism, 95
  phenomenological reductionism, 184
  physicalist reductionism, 183–4
  see also anti-reductionism

Reichenbach, Hans, 1, 6, 24, 77, 144, 248n
Reidemeister, Kurt, 63
Reidemeister, Marie, 63, 83, 85–8
Reisch, Georg, 91n, 173n
relativism, 137, 139–40, 146, 225–7
Rey, Abel, 118n
Richardson, Alan, 150n
Rickert, Heinrich, 125, 215, 220, 222, 224, 225, 226, 227, 241
Ringer, Fritz K., 97n, 98n, 110n
Roscher, Wilhelm, 214, 217
Rotha, Paul, 86
Rothschild, Kurt W., 26n
Rossi-Landi, Ferrucio, 148n
Rougier, Louis, 84
Rühle, Otto, 49n
Rutte, Heiner, 77n, 78, 84n, 90n, 152n, 165n
Russell, Bertrand, 76, 77, 196
Ryckman, Thomas, 150n

Schäfer, Lothar, 116n
Schapire, Anna, 10, 12, 13, 128n
Schapire-Neurath, Anna see Schapire, Anna
Scheerer, Josef, 65, 85
Scheideman, Friedrich, 43
Schleier, Hans, 98n
Schlick, Moritz, 2, 5, 77, 78, 80, 84, 87, 144, 149, 150, 157, 158, 185, 194, 196, 197n, 200, 243
Schlittenbauer, Sebastian, 47, 251
Schmidt, Georg, 47n
Schmoller, Gustav, 11, 12, 94, 98, 125, 167, 209, 213, 221
Schmolze, Gerhard, 43n, 45n
Schnädelbach, Herbert, 254
Schneppenhorst, Ernst, 50
Schön, Manfred, 98n, 214n
Schoenlank, Bruno, 22n
Schröder, Ernst, 13, 105
Schütte-Lihotzky, Grete, 55n, 62, 63n, 71n, 72n
Schumann, Wolfgang, 13, 19n, 20–1, 29n, 31n, 40n, 45, 46, 47, 49, 244, 248, 248n, 251
scientific world conception, 24, 77, 78, 94, 178, 253
Seeger, R.J., 101n
Segitz, J.M., 248
Seitz, Karl, 63
Simmel, Georg, 94, 98, 100, 111, 112n, 118, 119, 125, 127, 165, 224
Simon, Josef, 46, 47, 50, 245, 246, 251
socialisation, 8, 18, 21, 22, 23, 246
  debate on, 23
  models for, 45; Korsch, 26, 27, 34, 35, 40; Neurath, 41, 43, 44, 51, 57, 171, 173–5, 230
  full socialisation, 46, 47, 145, 171, 173–5, 177, 231–2

288  *Index*

socialisation (*cont.*)
  nationalisation, 23, 25, 26, 27, 40, 43, 47, 233
  revolutionary conception of, 23, 27,31
  reformist conception of, 23, 41
  social planning, 87, 173, 246
socialism, 22, 24, 37, 40, 53, 177, 179, 247
Soulez, Antonia, 91n
Spanger, Eduard, 140n
Spengler, Oswald, 2, 76, 91, 136–7, 139, 141, 226–7, 240
Sraffa, Piero,148n
Stadler, Friedrich, 55n, 72n, 77n, 91n, 101n, 110n, 145n, 147n, 249n
Stalin, Joseph, 72, 165
standards of living, 14, 15, 25, 30, 31, 35, 37, 112
  education, 14, 29, 31, 42, 44, 56, 57, 58, 249
  entertainment, 14, 29, 31, 44
  happiness, 21, 29, 30, 31, 62, 86
  housing, 14, 29, 31, 42, 44, 60, 62, 86
  life mood, 30
  nutrition, 14, 29 31, 42, 44
statistics, 2, 5, 18, 32, 32, 44, 249,
Starkenburg, Heinz, 237
Stehr, Nico, 97n
Steinthal, H., 119n
Stöckinger, Frau, xii
Stölting, Erhard, 98n
Strand, Oskar, 62
Suárez, M., x, xii
Suess, Eduard, 167n
synthetic *a priori*, 100, 105, 125, 131, 148

Tenbruck, Friedrich, 220n
Thiele, Joachim, 101n
Tinbergen, Jan, 210n
Tönnies, Ferdinand, 10, 21n, 94, 98, 100n, 107n, 111, 112n, 118, 119, 121–3, 125, 131, 141, 156, 165, 224, 228
Toller, Ernst, 43n, 50, 55, 56, 231, 232
totalitarianism
  Fascism, 180–1
  Nazism, 2, 165, 181
  Stalinism, 165, 172
Toulmin, Stephen, 140n
Traven, B. (pseudonym) *see* Marut, Red
truth, correspondence theory of, 219, 253
Turner, Stephen, 98n, 167, 169

Umrath, Heinz, 47
underdetermination, 93, 108, 117, 188, 203, 207, 209
unity of science, 75, 76, 78, 81, 102–3, 107, 127, 164, 175
  Carnap's conception of, 150, 169–70, 179, 182
  Mach's conception of, 102, 104, 127, 146
  Neurath's conception of, 81, 108–9, 169, 175, 184, 209, 223, 254, 256
Unity of Science movement, 2, 3, 167, 179
Untermann, Ernst, 242
Urchs, Max, xii
utilitarianism, 30, 114–15
utopianism, 22, 41, 42, 54

Vaihinger, Hans, 94
visual education, 20, 63, 84, 86, 182
voluntarism, 94, 132, 139, 225, 237

Waismann, Friedrich, 77, 194, 196, 203n, 204n
Walras, Leon, 194n
Watkins, John, 4n
Watt, Susan, xii
Weber, Alfred, 221
Weber, Max, 54, 94, 98, 111, 112n, 118, 119, 125, 165, 167, 168, 169, 171, 209–10, 213–24, 225, 227, 229, 231, 239, 241, 243
  on value, 223
  on concepts, 213–28
  on reality, 169, 187
Webster, Douglas, 198
Wedin, M.V., 196n
Weiler, Gershon, 140n
Weiser, Rosl, 63
Weissel, Erwin, 26, 42
Wellmer, Albrecht, 254
Wernitz, Axel, 22n
*Werturteilsstreit*, 96, 98, 99, 101, 111–14
Whimster, Sam, 98n
Wiedemann, Conrad, 242n
Wilhelm II, Emperor, 21, 43
Williams, A.V., 86
Wittgenstein, Ludwig, 5, 140, 143, 144, 148n, 150, 164, 196
Wlach, Oskar, 62
Wohtinsky, Wladimir, 249
Wolff, Kurt H.,119n
Wolfram, Ludwig Hermann, 11, 128
Wundt, Wilhelm, 75, 76, 107, 131, 172

Zeisel, Hans, 57n
Zilsel, Edgar, 77, 146n, 235, 236, 238
Zolo, Danilo, 91n, 107n, 157n, 163n, 196n, 202n
Zouros, George, x
Zuckermann, Bruno, 65

*Ideas in Context*

Edited by Quentin Skinner (general editor), Lorraine Daston,
Wolf Lepenies, Richard Rorty and J.B. Schneewind

1 RICHARD RORTY, J.B. SCHNEEWIND and QUENTIN SKINNER
(eds.)
Philosophy in History
*Essays in the Historiography of Philosophy**

2 J.G.A. POCOCK
Virtue, Commerce and History
*Essays on Political Thought and History, Chiefly in the Eighteenth Century**

3 M.M. GOLDSMITH
Private Vices, Public Benefits
*Bernard Mandeville's Social and Political Thought*

4 ANTHONY PAGDEN (ed.)
The Languages of Political Theory in Early Modern Europe*

5 DAVID SUMMERS
The Judgment of Sense
*Renaissance Nationalism and the Rise of Aesthetics**

6 LAURENCE DICKEY
Hegel: Religion, Economics and the Politics of Spirit, 1770–1807*

7 MARGO TODD
Christian Humanism and the Puritan Social Order

8 LYNN SUMIDA JOY
Gassendi the Atomist
*Advocate of History in an Age of Science*

9 EDMUND LEITES (ed.)
Conscience and Casuistry in Early Modern Europe

10 WOLF LEPENIES
Between Literature and Science: The Rise of Sociology*

11 TERENCE BALL, JAMES FARR and RUSSELL L. HANSON (eds.)
Political Innovation and Conceptual Change*

12 GERD GIGERENZER *et al.*
The Empire of Chance
*How Probability Changed Science and Everyday Life*\*

13 PETER NOVICK
That Noble Dream
*The 'Objectivity Question' and the American Historical Profession*\*

14 DAVID LIEBERMAN
The Province of Legislation Determined
*Legal Theory in Eighteenth-Century Britain*

15 DANIEL PICK
Faces of Degeneration
*A European Disorder, c.1848–c.1918*

16 KEITH BAKER
Approaching the French Revolution
*Essays on French Political Culture in the Eighteenth Century*\*

17 IAN HACKING
The Taming of Chance*

18 GISELA BOCK, QUENTIN SKINNER and MAURIZIO VIROLI (eds.)
Machiavelli and Republicanism*

19 DOROTHY ROSS
The Origins of American Social Science*

20 KLAUS CHRISTIAN KOHNKE
The Rise of Neo-Kantianism
*German Academic Philosophy between Idealism and Positivism*

21 IAN MACLEAN
Interpretation and Meaning in the Renaissance
*The Case of Law*

22 MAURIZIO VIROLI
From Politics to Reason of State
*The Acquisition and Transformation of the Language of Politics 1250–1600*

23 MARTIN VAN GELDEREN
The Political Thought of the Dutch Revolt 1555–1590

24 NICHOLAS PHILIPSON and QUENTIN SKINNER (eds.)
Political Discourse in Early Modern Britain

25 JAMES TULLY
An Approach to Political Philosophy: Locke in Contexts*

26 RICHARD TUCK
Philosophy and Government 1572–1651*

27 RICHARD R. YEO
Defining Science
*William Whewell, Natural Knowledge and Public Debate in Early Victorian Britain*

28 MARTIN WARNKE
The Court Artist
*The Ancestry of the Modern Artist*

29 PETER N. MILLER
Defining the Common Good
*Empire, Religion and Philosophy in Eighteenth-Century Britain*

30 CHRISTOPHER J. BERRY
The Idea of Luxury
*A Conceptual and Historical Investigation*\*

31 E.J. HUNDERT
The Enlightenment's 'Fable'
*Bernard Mandeville and the Discovery of Society*

32 JULIA STAPLETON
Englishness and the Study of Politics
*The Social and Political Thought of Ernest Barker*

33 KEITH TRIBE
German Economic Thought from the Enlightenment to the Social Market

34 SACHIKO KUSUKAWA
The Transformation of Natural Philosophy
*The Case of Philip Melancthon*

35 DAVID ARMITAGE, ARMAND HIMY and QUENTIN SKINNER (eds.)
Milton and Republicanism

36 MARKKU PELTONEN
Classical Humanism and Republicanism in English Political Thought 1570–1640

37 PHILIP IRONSIDE
The Social and Political Thought of Bertrand Russell
*The Development of an Aristocratic Liberalism*

38 NANCY CARTWRIGHT, JORDI CAT, LOLA FLECK and THOMAS E. UEBEL
Otto Neurath: Philosophy between Science and Politics

Titles marked with an asterisk are also available in paperback

Printed in the United States
26372LVS00001B/345